PROTON EXCHANGE MEMBRANE FUEL CELLS

ELECTROCHEMICAL ENERGY STORAGE AND CONVERSION

Series Editor: Jiujun Zhang
National Research Council Institute for Fuel Cell Innovation
Vancouver, British Columbia, Canada

Published Titles

Electrochemical Supercapacitors for Energy Storage and Delivery: Fundamentals and Applications
Aiping Yu, Victor Chabot, and Jiujun Zhang

Proton Exchange Membrane Fuel Cells
Zhigang Qi

Forthcoming Titles

Electrochemical Polymer Electrolyte Membranes
Yan-Jie Wang, David P. Wilkinson, and Jiujun Zhang

Lithium-Ion Batteries: Fundamentals and Applications
Yuping Wu

Solid Oxide Fuel Cells: From Fundamental Principles to Complete Systems
Radenka Maric

ELECTROCHEMICAL ENERGY STORAGE AND CONVERSION

PROTON EXCHANGE MEMBRANE FUEL CELLS

Zhigang Qi

Wuhan Intepower Fuel Cells Co., Ltd
Hubei, China

CRC Press
Taylor & Francis Group
Boca Raton London New York

CRC Press is an imprint of the
Taylor & Francis Group, an **informa** business

CRC Press
Taylor & Francis Group
6000 Broken Sound Parkway NW, Suite 300
Boca Raton, FL 33487-2742

© 2014 by Taylor & Francis Group, LLC
CRC Press is an imprint of Taylor & Francis Group, an Informa business

No claim to original U.S. Government works

ISBN-13: 978-1-4665-1370-9 (hbk)
ISBN-13: 978-1-138-07511-5 (pbk)

Library of Congress Cataloging-in-Publication Data

Qi, Zhigang.
   Proton exchange membrane fuel cells / Zhigang Qi.
      pages cm. -- (Electrochemical energy storage and conversion)
   Includes bibliographical references and index.
   ISBN 978-1-4665-1370-9 (hardback)
   1. Proton exchange membrane fuel cells. I. Title.

TK2933.P76Q33 2013
621.31'2429--dc23   2013027624

Visit the Taylor & Francis Web site at
http://www.taylorandfrancis.com

and the CRC Press Web site at
http://www.crcpress.com

Contents

Series Preface

Electrochemical Energy Storage and Conversion

Clean energy technologies, which include energy storage and conversion, will play the most important role in overcoming fossil fuel exhaustion and global pollution for the sustainable development of human society. Among clean energy technologies, electrochemical energy storage and conversion are considered the most feasible, environmentally friendly, and sustainable. Electrochemical energy technologies such as batteries, fuel cells, supercapacitors, hydrogen generation and storage, as well as solar energy conversion have been or will be used in important application areas including transportation and stationary and portable/micro power. To meet the increasing demand in both the energy and power densities of these electrochemical energy devices in various new application areas, further research and development are essential to overcome major obstacles, such as cost and durability, which are considered to be hindering their application and commercialization. To facilitate this new exploration, we believe that a book series that covers and gives an overall picture of all the important areas of electrochemical energy storage and conversion technologies is highly desired.

The book series will provide a comprehensive description of electrochemical energy conversion and storage in terms of fundamentals, technologies, applications, and the latest developments including secondary (or rechargeable) batteries, fuel cells, supercapacitors, CO_2 electroreduction to produce low-carbon fuels, electrolysis for hydrogen generation/storage, and photoelectrochemical for water splitting to produce hydrogen. Each book in this series will be self-contained and written by scientists and engineers with excellent academic records and strong industrial expertise, who are at the top of their fields on the cutting edge of technology. With a broader view of various electrochemical energy conversion and storage devices, this book series will be unique and an essential read for university students including undergraduates and graduates, and scientists and engineers working in related fields. We believe that this book series will enable readers to easily locate the latest information concerning electrochemical technology, fundamentals, and applications.

Jiujun Zhang, PhD

Preface

When I first received the invitation to write this book from the Series Editor, Dr. J. J. Zhang of the National Research Council of Canada, I initially turned it down because there had already been quite a few books and numerous review articles published covering nearly all aspects of fuel cell technology.

I was particularly influenced by the wonderful book, *Fuel Cell Systems Explained* by James Larminie and Andrew Dicks. Seemingly unsurpassable, I thought about what impact yet another book on this topic would have on the subject. At Dr. Zhang's insistence and encouragement, I finally agreed to write this book because he believed that my expertise brought a fresh approach to the current understanding of fuel cell technology.

This book focuses on proton exchange membrane fuel cells (PEMFCs) in the areas of fundamentals, practices, applications, and developmental status. It does so, taking into consideration the skill level of many kinds of readers, from the novice to the skilled professional. Scientists, engineers, high-level management personnel, college students, and even high school students can benefit from this book.

Modeling that relies on complicated and expensive software was avoided. Instead, simple approaches were taken. Using simple approaches may lead to some minor inaccuracies. However, you should be able to quickly see the essence of the physics behind the analyses without getting lost in complicated mathematics. All the calculations in the book require only a rudimentary knowledge of high school math and can be done using either Excel or a calculator.

Many statements and descriptions are plain and brief, but they are the crystallization of many years of my experience and research. You'll explore fundamental questions: Will the catalyst particles be able to participate in the electrochemical reaction if they are fully covered by a thin ionomer film? What can the limiting current density be based on the mass transport of air? How severe will the voltage loss be if a thin layer of liquid water forms? How can you quickly assess catalytic activity difference based on voltage difference in V-I curves? How can a direct methanol fuel cell work using neat methanol? How high can H_2 and O_2 gases be pressurized within a PEM electrolyzer?

You should be able to read this handbook from any chapter. If you are a lay reader, I suggest that you read the entire book to gain a fundamental knowledge of PEMFCs. If you are an academic or industrial researcher, focusing on the engineering aspect of this book may be particularly helpful. This book will increase your fundamental understanding of fuel cells and engineering problems. Your ability to troubleshoot problems will be greatly improved. Ultimately, you will gain a solid, workable knowledge of fuel cells.

As a developing technology, fuel cells are not fully understood. Contradictory thinking exists among researchers, conflicting results appear in publications, and many companies do not make solutions and proprietary information available to the public. The key intention of writing this book is to help the reader to not only understand the fundamental concepts and challenges, but to construct and evaluate various components, modules, subsystems and the entire fuel cell system based on the current state of this technology, as well as to think critically about this technology as a whole.

Although I have been working in the fuel cell arena in both academia and industry since 1996, keeping communication with researchers worldwide, and periodically reading scientific publications, there are still many things yet to be understood about this emerging field. Writing this book has been a journey of discovery. I was the first person who reaped the benefits of the information contained in this book, and sincerely hope that you will do the same.

Zhigang Qi, PhD
zhigangqi2005@yahoo.com

Acknowledgments

I am very grateful to many friends and colleagues: Andy Marsh, Everett Anderson, Peng Lim, Reid Hislop, Ricarda Gauder, Richard Okine, Randal Perry, Alfred Wong, Sandra Saathoff, Lijun Bai, Mick Hicks, Hexiang Zhong, Lifang Zhang, Shijun Liao, Pingwen Ming, Shumao Wang, Lijun Jiang, Ruoguo Zhang, Liqing Hu, Weiyu Shi, Jianxin Ma, Cunman Zhang, Bojin Qi, Yunliang Miao, Xinjian Zhu, Jiannong Wang, Xianggen Gao, Haolin Tang, Wu Pan, Zhigang Shao, and all my coworkers at Wuhan Intepower, who generously provided much of the data and figures for this book.

I sincerely appreciate the assistance of Acquiring Editor Allison Shatkin, Project Coordinator Jill Jurgensen, Project Editor Robin Lloyd-Starkes, Assistant Editor Florence Kizza, and many others at Taylor & Francis Group, for their tireless work coordinating and editing this book. Without the assistance of these wonderful people, this book would not be possible.

I am especially thankful to my beautiful wife, Min, and my lovely kids, Brian, Albert, and Lydia, for their encouragement, constant support, and genuine love.

About the Author

Dr. Zhigang Qi, currently the chief executive officer of Wuhan Intepower Fuel Cells Co., Ltd. (China), is a seasoned electro/analytical chemist and materials scientist, and has received diverse technical and managerial training. He earned his B.Sc. in chemistry and an M.Sc. in analytical chemistry from the University of Science and Technology Beijing (China) in 1985 and 1988, respectively. He earned another M.Sc. in electrochemistry from Memorial University of Newfoundland (Canada) in 1993, and a PhD in polymer and materials chemistry from McGill University (Canada) in 1996.

After doing post-doctoral research for a year and a half, Dr. Qi joined H Power Corp. (U.S.) as a senior staff scientist and the supervisor of the electrochemistry lab. In 2003, he moved to Plug Power (U.S.) as the manager of the analytical and fundamentals group, and was subsequently appointed as a corporate fellow in late 2004. He joined MTI Micro Fuel Cells (U.S.) as a corporate fellow in November 2007, managing DMFC technology advancement and intellectual properties, until he moved to Wuhan Intepower Fuel Cells Co., Ltd. (China) in 2009 as a vice president overseeing research and development (R&D) and domestic and international collaboration in the area of renewable energy. He was appointed as the chief executive officer of Wuhan Intepower Fuel Cells Co., Ltd. in 2013.

Dr. Qi has broad experience in the areas of proton exchange membrane fuel cells (PEMFCs), direct methanol fuel cells (DMFCs), and phosphoric acid fuel cells (PAFCs). He has published 56 articles in peer-reviewed journals, 6 book chapters, has 9 US patents, and 7 Chinese patents. He has voluntarily reviewed a large number of manuscripts in the fuel cell area for technical journals. He is a technical advisor of the Knowledge Foundation (U.S.), a member of the China Fuel Cell and Flow Cell Standardization Committee, a technical committee member of the Fuel Cell Key Laboratory in Hubei Province, the chief scientist and manager of two key projects of the 863 Plan (China) with a total funding of $14 million, and was an editorial board member of the *Journal of Power Sources* (EU).

List of Acronyms

a	activity; or rate of acceleration
A	area
AC	alternating current
A_F	frontal area of a vehicle
AFC	alkaline fuel cell
ANSI	American National Standards Institute
ASM	air supply module
avg	average
BET	Brunauer–Emmett–Teller
BOL	beginning of life
BOP	balance of plant
BPS	bulk pressure sensor
C	capacitance
c_{bulk}	bulk concentration
CCB	catalyst coated backing
CCM	catalyst coated membrane
C_D	aerodynamic drag coefficient
CD	current density
CDMA	code division multiple access
CE	Conformité Européenne (i.e., European Conformity)
CEN	European Committee for Standardization
CHP	combined heat and power
CL	catalyst layer
CM	control module
COM	communications module
C_p	heat capacity
CPU	central processing unit
C_R	rolling resistance between tires and road
c_s	surface concentration
CSA	Canadian Standards Association
CT	coolant tank
CV	check valve; or cyclic voltammetry
CVM	cell voltage monitoring module
d	density; or diameter
D	diffusion coefficient
dB	decibel
DB	D-subminiature
dc/dx	concentration gradient
DC	direct current

DE	diesel engine
D_{eff}	effective diffusion coefficient
DFAFC	direct formic acid fuel cell
D.I.	de-ionized
DLFC	direct liquid fuel cell
DMFC	direct methanol fuel cell
D^{o}	bond strength
DOE	Department of Energy
DS	desulfurization
DSC	differential scanning calorimetry
E	voltage or potential
$E_{1/2}$	half-wave voltage or potential
ECD	electrolysis current density
ECSA	electrochemically active surface area
E_k	kinetic energy
EMM	exhaust management module
EN	European Standards
env.	environment
E^{o}	thermodynamic voltage or potential
EOL	end of life
ePTFE	expanded polytetrafluoroethylene
Eq.	equation
EW	equivalent weight
f	revolutions per minute in RDE test
F	Faraday constant (= 96485 C)
FC	fuel cell
FCHV	fuel cell hybrid vehicle
FCHV-adv	fuel cell hybrid vehicle advanced
FCV	fuel cell vehicle
FM	flow meter
FSM	fuel supply module
FT	filter
FWHM	full width at half peak height
g	gas; or acceleration of gravity (= 9.8 m s^{-2})
GDE	gas diffusion electrode
GDM	gas diffusion medium
gge	gasoline gallon equivalent
GHG	green-house gas
GSM	global system for mobile communications
HDS	hydrodesulfurization
HEDP	high energy demand period
HEX	heat exchanger
HF	humidifier
HHV	higher heating value

HMM	heat management module
HOR	H_2 oxidation reaction
HPR	high pressure regulator
HPS	high pressure sensor
HTS	high temperature shift
HTSV	high temperature solenoid valve
Hyflon®	trade name of the perfluorosulfonic acid developed by Solvay Solexis
I	current
ICE	internal combustion engine
$i_{crossover}$	crossover current
i_d	diffusion limiting current
IE	ion exchanger
IPM	internal power supply module
iR	voltage loss when current i passes through resistance R
J	flux of species
$j_{ap}^{\,o}$	apparent exchange current density
j_{lim}	limiting current density
j^o	exchange current density
K	reaction equilibrium constant
k_f	kinetic reaction rate
l	liquid
L	thickness of GDM, PEM, MPL, CL, or diffusion layer
LCD	liquid-crystal display
LEDP	low energy demand period
LHV	lower heating value
LHV_{gas}	lower heating value when reactant methanol in the gas form
LHV_{liq}	lower heating value when reactant methanol in the liquid form
L_P	thickness at pressure P
LPR	low pressure regulator
LPS	low pressure sensor
LTS	low temperature shift
LTSV	low temperature solenoid valve
m	mass
M	metal in general
MCFC	molten carbonate fuel cell
MD	machine direction
MDF	methanol diffusion film
MEA	membrane electrode assembly
MeOH	methanol
MFT	mesh filter
MOR	methanol oxidation reaction
MPL	microporous layer
n	amount of a species; or number of electrons per molecule reacted
Nafion®	trade name of the perfluorosulfonic acid developed by DuPont

NEBS	network equipment-building system
n_r	number of electrons involved in the rate determining step
NSTF	nanostructured thin film
NTP	normal temperature and pressure
OCV	open circuit voltage
ORR	O_2 reduction reaction
P	pressure
PAFC	phosphoric acid fuel cell
P_{Aux}	auxiliary power
PBI	polybenzimidazole
PCB	printed circuit board
PCM	power conditioning module
PEM	proton exchange membrane
PEMFC	proton exchange membrane fuel cell
PFSA	perfluorosulfonic acid
P_G	gas pressure in porosimetry experiment
P_L	pressure applied to mercury in porosimetry experiment
PMW	pulse-width modulation
P^o	standard pressure (=1 bar)
ProE	Pro/ENGINEER
Prox	preferential oxidizer
PS	pressure sensor
PTFE	polytetrafluoroethylene
PVD	physical vapor deposition
Q	charge
q	heat flow
R&D	research and development
R.	reaction
r	radius
R	resistance; or gas constant (= 8.31 J K^{-1} mol^{-1}, 0.0831 L bar K^{-1} mol^{-1})
RDE	rotating disc electrode
RH	relative humidity
R_P	resistance at pressure P
RWGS	reversed water gas shift
S/C	steam-to-carbon ratio
S	surface area
SEM	scanning electron microscopy
SNMP	simple network management protocol
SOFC	solid oxide fuel cell
SPEEK	sulfonated poly(ether ether ketone)
STP	standard temperature and pressure
SUV	sport utility vehicle
SV	solenoid valve
T	temperature

TD	transverse direction
TEM	transmission electron spectroscopy
T_g	glass transition temperature
TS	temperature sensor
UL	Underwriters Laboratories
UPS	uninterruptible power supply
US or USA	United States of the America
USB	universal serial buss
UTC	United Technologies Corporation
v	velocity; or volume of a particle
V	volume
V-I	voltage versus current or current density
V_{LHV}	thermodynamic voltage based on lower heating value
VRLA	valve regulated lead-acid (battery)
WGS	water-gas shift
W-I	power versus current or current density
WMM	water management module
XPS	X-ray photoelectron spectroscopy
XRD	X-ray diffraction
YSZ	yttria-stabilized zirconia
Z	H_2 compressibility factor
Z'	real impedance
ΔE_A	activation overpotential
ΔE_M	mass transport overpotential
ΔE_R	resistance overpotential
$\Delta_f G$	Gibbs free energy of formation
$\Delta_f G^\circ$	Gibbs free energy of formation under standard condition (i.e., 298 K and 1 bar)
$\Delta_f H$	enthalpy of formation
$\Delta_f H^\circ$	enthalpy of formation under standard condition
ΔG	Gibbs free energy change of a reaction
ΔG°	Gibbs free energy change of a reaction under standard condition
ΔH	enthalpy change of a reaction
ΔH°	enthalpy change of a reaction under standard condition
ΔS	entropy change of a reaction
ε	porosity
η	efficiency; or viscosity
η_{DC}	efficiency of either DC–DC converter or DC–AC inverter
η_F	fuel utilization
η_P	parasitic power loss
η_S	stack electrical efficiency
η_{Sys}	system electrical efficiency
θ	incidence angle of X-ray beam; or contact angle; or slope angle
λ	wavelength of X-ray in XRD test; or number of water molecules held by each $-SO_3H$ group in a PEM

ρ	density of air
σ	conductivity; or surface tension
σ_P	conductivity at pressure P
ω	angular velocity ($= 2\pi f$)
v	flow rate; or kinematic viscosity
α	charge transfer coefficient

List of Units

$	US dollar	mmol	millimole
%	percentage	mol	mole
A	ampere	MPa	megapascal
Ah	ampere-hour	mV	millivolt
C	coulomb	mW	milliwatt
°C	degree Celsius	mΩ	milliohm
cm	centimeter	nm	nanometer
eV	electronvolt	Pa	pascal
g	gram	ppm	parts per million
h	hour	psi	pounds per square inch
Hz	hertz	psia	psi absolute
J	joule	psig	psi gauge
K	kelvin	s	second
kg	kilogram	S	siemens
kJ	kilojoule	slpm	standard liter per minute
kPa	kilopascal	V	volt
kW	kilowatt	V_{ac}	volt of alternating current
kW_e	kilowatt electrical power	V_{dc}	volt of direct current
kWh	kilowatt-hour	W	watt
L or l	liter	Wh	watt-hour
m	meter	y	year
mA	milliampere	Ω	ohm
meq	milliequivalent	μA	microampere
min	minute	μm	micrometer
mL or ml	milliliter	μV	microvolt
mm	millimeter		

List of Chemicals

•OH	hydroxide radical
Al	aluminum
Al_2O_3	aluminium oxide
AlH_3	aluminium hydride
AlH_4^-	alanate
Ar	argon
Au	gold
B	boron
Be	beryllium
C	carbon
C_2H_5OH	ethanol
C_2H_5SH	ethanethiol
C_2H_6	ethane
C_3H_8	propane
C_8H_{18}	octane
$Ca(AlH_4)_2$	calcium aluminum hydride
$Ca(BH_4)_2$	calcium borohydride
Ca	calcium
CaO	calcium oxide
Cd	cadmium
$-CF_3$	tri-fluorocarbon group
CH_3OH	methanol
CH_4	methane
CHOOH	formic acid
CO	carbon monoxide
CO_2	carbon dioxide
CO_3^{2-}	carbonate
$-COOH$	carboxylic acid group
CrN	chromium nitride
e^-	electron
F^-	fluoride
Fe^{2+}	ferrous
Fe_2O_3	ferric oxide
Fe^{3+}	ferric
H	elemental hydrogen
H^+	proton
H_2	hydrogen
H_2O	water
H_2O_2	hydrogen peroxide

H_2S	hydrogen sulfide
H_2SO_4	sulfuric acid
HF	hydrogen fluoride; or hydrofluoric acid
Ir	iridium
IrO_2	iridium oxide
K	potassium
K_2CO_3	potassium carbonate
KOH	potassium hydroxide
La	lanthanum
$LaNi_5$	alloy of lanthanum with nickel with atomic ratio 1:5
Li	lithium
Li_2CO_3	lithium carbonate
$LiAlH_4$, Li_3AlH_6	lithium aluminum hydride
$LiAlO_2$	lithium aluminate
$LiBH_4$	lithium borohydride
$LiMg(AlH_4)_3$	lithium magnesium aluminum hydride
$Mg(AlH_4)_2$	magnesium aluminum hydride
$Mg(BH_4)_2$	magnesium borohydride
Mg	magnesium
Mg_2Ni	alloy of magnesium with nickel with atomic ratio 2:1
MH_x	metal hydride
Mn	manganese
N	elemental nitrogen
N_2	nitrogen
N_2H_4	hydrazine
Na	sodium
Na_2CO_3	sodiium carbonate
$NaBH_4$	sodium borohydride
$NaBO_2$	sodium borate
NaF	sodium fluoride
NaOH	sodium hydroxide
NH_3	ammonia
NH_4^+	ammonium
Ni	nickel
$NiMH_x$	nickel metal hydride
NiMo	alloy of nickel with molybdenum
NO	nitrogen monoxide
NO_2	nitrogen dioxide
NO_x	nitrogen oxide
O	elemental oxygen
O^{2-}	oxide ion
O_2	oxygen
OH^-	hydroxide
Pb	lead

Pt/C	Pt supported on carbon
Pt/Mo	platinum supported on molybdenum
Pt/W	platinum supported on tungsten
Pt	platinum
Pt^0	Pt atom with zero oxidation state
PtCo	alloy of platinum with cobalt
Pt-CO	platinum with CO molecule adsorption
$Pt-H_{ad}$	Pt with hydrogen atom adsorption
Pt^{II}	Pt atom with +2 oxidation state
PtIr	alloy of platinum with iridium
Pt^{IV}	Pt atom with +4 oxidation state
PtM_x	alloy of platinum with metal M with atomic ratio 1:x
PtM_xN_y	alloy of platinum with metals M and N with atomic ratio 1:x:y
PtNi	alloy of platinum with nickel
PtO	platinum oxide
PtO_2	platinum dioxide
PtO_x	surface oxidized platinum
PtPd	alloy of platinum with palladium
PtRu	alloy of platinum with ruthenium
$Pt-S_{ad}$	Pt with sulfur coverage
Ru	ruthenium
RuIr	alloy of ruthenium with iridium
Ru-OH	ruthenium with OH coverage
S	sulfur
SiC	silicon carbide
SiO_2	silicon dioxide
Sn	tin
SO_2	sulfur dioxide
$-SO_3H$	sulfonic acid group
Ti	titanium
TiFe	alloy of titanium with iron
$TiMn_2$	alloy of titanium with manganese with atomic ratio 1:2
TiO_2	titanium dioxide
WO_3	tungstun oxide
Zn	zinc
ZnS	zinc sulfide
Zr	zirconium
$Zr(HPO_4)$	zirconium hydrogen phosphate
ZrO_2	zirconium dioxide

1

Proton Exchange Membrane Fuel Cells

1.1 Fuel Cells

A fuel cell is an electrochemical device that converts chemical energy directly into electrical energy with the reactants being supplied externally. It possesses the following characteristics:

1. It is an energy converting device.
2. The energy conversion is via an electrochemical process.
3. It converts the energy from chemical to electrical in one step.
4. It does not store the reactants within the reactor called stack.

A fuel cell bears some similarity to a battery because they both convert chemical energy directly into electrical energy. Since the conversion is a one-step process, the conversion efficiency is high and it does not suffer from the Carnot limit. But they are distinctly different with respect to where the reactants are stored. A battery stores the reactants (i.e., the active materials) within it. When the reactants are used up in a primary battery, it is discarded; when the reactants are used up in a secondary battery, it needs to be recharged by electricity. How much electrical energy a battery can provide depends on the amount of reactants that are stored within the battery and the discharging rate (a higher discharging rate leads to lower releasable electrical energy). The recharging process for a secondary battery normally takes several hours to complete. The recharging is a reverse of the battery discharging process during which the products are changed back to the original reactants by externally applied electrical energy. Therefore, a battery is an energy storing and converting device. For a secondary battery, it is more like a device that converts electrical energy to storable chemical energy and converts the stored chemical energy to electrical energy when the battery is in use (i.e., electrical to chemical to electrical energy). In contrast, a fuel cell does not store any reactants within the reactor called stack. The reactants are stored outside the fuel cell stack, and only when the fuel cell is required to produce electrical energy will the reactants be supplied to the fuel cell stack.

When the requirement for producing electrical energy is lifted, the reactant supply to the fuel cell stack is stopped. How much electrical energy a fuel cell can provide depends on the amount of reactants that are stored in the storage tanks. When the reactants are used up, the tanks that store the reactants need to be refueled, and this process takes only a few minutes or sometimes even less. A fuel cell stack cannot go through a recharging process.

A fuel cell also bears some similarity to a heat engine, such as an internal combustion engine (ICE) or a diesel engine (DE), because they both store the reactants outside the reactors. How much electrical energy they can provide depends on the amount of reactants (e.g., fuels) that are stored in the storage tanks. When the reactants are used up, the fuel tanks can be refueled quickly. But, a heat engine is much less efficient than a fuel cell in converting the chemical energy into electrical energy because it suffers from the Carnot limit, and it normally involves several steps such as chemical energy to heat energy and heat energy to mechanical energy.

Therefore, a fuel cell possesses the high conversion efficiency of a battery and the fast refueling characteristics of a heat engine. A fuel cell can replace either a battery or a heat engine. A fuel cell can also provide the functionalities of the combination of a battery and a heat engine. Table 1.1 compares fuel cells with batteries and heat engines.

Since a fuel cell converts the chemical energy to electrical energy through an electrochemical process, it is an electrochemical reactor. An electrochemical reactor must have two electrodes, one for the oxidation of the fuel and the other for the reduction of the oxidant. The electrode where the fuel oxidation occurs is called the *anode*, and the electrode where the oxidant reduction occurs is called the *cathode*. The anode and the cathode are separated by an electrolyte to prevent them from getting into direct contact and it also allows ionic species to transport between the two electrodes. The oxidant loses electrons at the surface of the anode, and the resulting electrons pass through an external circuit to reach the cathode, where the electrons are taken by the oxidants. In the process of the electron moving from the anode to the cathode,

TABLE 1.1

Comparisons of Fuel Cells, Batteries, and Heat Engines

Parameter	Battery	Heat Engine	Fuel Cell
Reaction type	Electrochemical	Combustion	Electrochemical
Carnot limit	No	Yes	No
Efficiency	High	Low	High
Fuel location	Inside	Outside	Outside
Refueling	Electrical charging	Add fuel	Add fuel
Refueling time	Long (hours)	Short (minutes)	Short (minutes)
Running time per refueling	Short	Long	Long
Reaction noise	No	High	Low

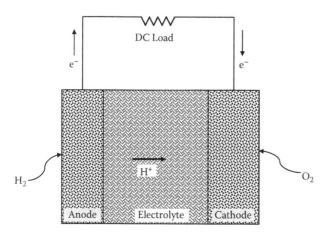

FIGURE 1.1
Schematic representation of an H_2/O_2 fuel cell.

electrical energy is provided to the apparatus that requires electrical energy. The apparatus is part of the external circuit. The apparatus can vary greatly, and it is generally referred to as a *load*, or more precisely an *electrical load*.

Figure 1.1 schematically shows the electrochemical reactions using hydrogen (H_2) as the fuel, oxygen (O_2) as the oxidant, and a proton conductor as the electrolyte. At the anode, H_2 dissociates into protons (H^+) and electrons (e^-) according to Reaction 1.1. Protons transport through the electrolyte to the cathode, and electrons move along the external circuit to the cathode too, where they combine with oxygen to form water (H_2O) according to Reaction 1.2. When the electrons pass through the electric load along the external circuit, it powers the load. Reaction 1.3 shows the overall reaction.

$$\text{Anode:} \quad H_2 = 2H^+ + 2e^- \quad \text{(R. 1.1)}$$

$$\text{Cathode:} \quad 0.5O_2 + 2H^+ + 2e^- = H_2O \quad \text{(R. 1.2)}$$

$$\text{Overall:} \quad H_2 + 0.5O_2 = H_2O \quad \text{(R. 1.3)}$$

It can be quickly recognized that R. 1.3 is the same as hydrogen chemically reacting with oxygen in a combustion process. But in a combustion process, hydrogen and oxygen will be in direct contact, and the hydrogen molecules lose electrons directly to the oxygen molecules. In such a chemical combustion process, heat instead of electrical energy is produced. In other words, a chemical reaction cannot produce electrical current directly. In contrast, the reactions of hydrogen and oxygen in Figure 1.1 occur on two different electrodes, which are separated by the electrolyte, and hydrogen and oxygen do not come in direct contact with each other. Since the electrolyte only allows the protons to transport through, electrons have to go through the external circuit; therefore, a flow of electrons is produced and the electric load is powered on. The anode, the external circuit including the load, the cathode, and the electrolyte form a complete circuit. In the absence of any of these components, there will be no flow of electrons (and protons). Meanwhile, the flows of the electrons and the protons occur simultaneously with equal amounts of charge.

Theoretically, many reactants can be used for a fuel cell. The fuel can be any substance as long as it is oxidizable. Potential fuels include carbon, hydrogen, hydrazine, alcohols, hydrocarbons, metal hydrides, and so on. The oxidant can be any substance as long as it is reducible. Potential fuels include oxygen, ozone, chlorine, bromine, iodine, sulfur, and so on. However, since many fuels and oxidants have difficulty reacting on an electrode surface or are very corrosive or toxic, the best fuel is H_2, followed by methanol and formic acid; the best and the cheapest oxidant is the oxygen from air.

Hydrogen gas does not exist naturally on earth, and it has to be produced through either electrolysis or thermal splitting of water, or extraction from other substances that contain elemental hydrogen. The cheapest source would be the H_2 by-products from some chemical processes such as the chloralkali industry and the iron and steel industry. If H_2 is produced from water electrolysis, it will go through an electrical energy (i.e., electricity) to chemical energy (i.e., H_2) process. When the fuel cell is in use, it will convert the chemical energy to electrical energy. Obviously, such an electrical-to-chemical and then to electrical double-step process will waste plenty of electrical energy; therefore, it seems to be an awkward choice. However, producing H_2 through water electrolysis or thermal splitting is likely to be the ultimate and best choice in the future for several reasons.

First, fossil fuels are predicted to be depleted within a few hundred years (around 40, 100, and 250 years for oil, natural gas, and coal, respectively). It may not seem to be a problem to produce H_2 from hydrocarbon materials such as methane (the primary component of natural gas), but when natural gas depletes in the future, it will become impossible to extract H_2 from these kinds of fuels. People can then use coal to produce hydrogen, but coal will soon deplete as well. People can also rely on biomasses to produce H_2, but it may be questionable that the Earth can produce enough biomass yearly for producing H_2 to meet the energy demand of the growing population; growing biomasses for producing H_2 will also reduce the land available for producing food to feed human beings. But renewable energies, such as solar energy, wind energy, hydro energy, tidal wave energy, and geothermal energy, are more or less unlimited. When these energies are used to produce H_2 through water electrolysis, they can supply endless H_2 to meet human demands.

Second, since we do not use energy consistently—there is always some excess energy that is either wasted or not utilized during a day. Highest energy demand occurs between 6:00 a.m. to 11:00 p.m. (let us call it the *high energy demand period*, or HEDP), especially during the morning and the evening. But from 11:00 p.m. to 6:00 a.m., the energy demand is quite low (let us call it the *low energy demand period*, or LEDP). In order to manage this uneven energy demand during a day, power plants have to either scale down their energy generation capability or somehow store the excess energy during the LEDP. Currently, there is no easy and economical way to store the excess energy. Scaling energy generation capability up and down requires careful operation, and it may shorten the life of power plants. If we use the excess energy during the LEDP to produce H_2, the H_2 can be regarded as free; at the same time, the power plants do not have to go through frequent scaling up and down cycles, e.g., the peak-and-valley modulation cycles.

Third, not all electrical energy is equal in importance and price. Generally speaking, the grid power is at the lower end, the motive power (using gasoline or diesel) is in the middle, and the portable power is at the higher end. In the United States, the cost of grid energy for residential houses is about 10 cents per kWh, and reaches about 20 cents per kWh when the supply fees are included. Gasoline is about 90 cents per liter, which contains about 8.7 kWh of energy (based on the lower heating value of octane, C_8H_{18}), translating to 10.3 cents per kWh, similar to the cost of the grid energy. However, since only a portion of the energy can be converted to electrical power in the energy conversion process, the cost of electrical energy obtained from gasoline will be a few times higher than that of the grid energy. For portable power, a triple-A primary battery that provides 0.4–1.8 Wh of energy costs about $1, corresponding to $556–2500 per kWh, which is several thousand to several tens of thousand times higher than the grid energy. So, if the grid energy is converted to chemical energy, such as H_2, and the H_2 is converted to electrical energy for motive and portable applications, even though these two conversion steps waste some energy, the already established higher sale price for the end product can well justify the energy losses.

Backup power becomes more and more important because it can provide the crucial energy needed when the grid power is suddenly off due to either natural events or human operation errors. Lead acid batteries are the primary backup power sources today due to their higher safety and lower cost compared with other types of batteries. The purchase price of lead (Pb)-acid batteries is about $100 per kWh, which is much higher than that of the grid energy, gasoline, and diesel, but the important roles they play outweigh their price. Fuel cells using H_2 are believed to be the best backup power. They can be cheaper than Pb-acid batteries at a large enough manufacturing volume, and at a 10-year lifetime, the cost associated with H_2 generation through water electrolysis becomes a trivial concern (Pb-acid batteries need to be replaced every few years and the environmental temperature needs to be maintained at around 25°C, which consumes a large amount of energy annually).

1.2 Types of Fuel Cells

Based on the electrolyte used, there are five types of fuel cells. Their key features are listed in Table 1.2.

A proton exchange membrane fuel cell (PEMFC) uses a solid membrane that transports protons. It can operate from about 0°C to 80°C with the output power ranging from a few watts to several hundred kilowatts. H_2 is the best fuel for a PEMFC, and the anode and cathode reactions are shown in Reactions 1.1 and 1.2, respectively.

When methanol is oxidized directly at the anode as the fuel, the fuel cell is called a *direct methanol fuel cell* (DMFC). The anode, the cathode, and the overall reactions are shown in Reactions 1.4, 1.5, and 1.6, respectively. It is important to note that methanol oxidation needs the presence of water. In other words, if there is no water at the anode, methanol will not be oxidized. So, supplying enough water to the anode is a necessity for a DMFC.

Anode: $CH_3OH + H_2O = CO_2 + 6H^+ + 6e^-$ (R. 1.4)

Cathode: $1.5O_2 + 6H^+ + 6e^- = 3H_2O$ (R. 1.5)

Overall: $CH_3OH + 1.5O_2 = CO_2 + 2H_2O$ (R. 1.6)

When formic acid is oxidized directly at the anode as the fuel, the fuel cell is called a *direct formic acid fuel cell* (DFAFC). The anode, the cathode, and the overall reactions are expressed by Reactions 1.7, 1.8, and 1.9, respectively. Since the formic acid oxidation does not require water, highly concentrated formic acid can be used.

Anode: $CHOOH = CO_2 + 2H^+ + 2e^-$ (R. 1.7)

Cathode: $0.5O_2 + 2H^+ + 2e^- = H_2O$ (R. 1.8)

Overall: $CHOOH + 0.5O_2 = CO_2 + H_2O$ (R. 1.9)

When a DMFC and a DFAFC use a PEM as the electrolyte, as presented in Reactions 1.4 to 1.9, they can be regarded as special cases of PEMFCs.

A *phosphoric acid fuel cell* (PAFC) uses liquid phosphoric acid as the electrolyte with protons as the charge transport species. The acid is normally imbedded in a solid matrix, such as porous silicon carbide (SiC) or a polybenzimidazole (PBI) membrane. The fuel cell operates best between 160°C and 210°C. The output power ranges from a few kilowatts to a few hundred kilowatts. The anode, the cathode, and the overall reactions are the same as Reactions 1.1, 1.2, and 1.3, respectively.

Please note that Reactions 1.2, 1.5, and 1.8 are the same. For the PEMFCs, DMFCs, DFAFCs, and PAFCs presented here, protons are the charge

TABLE 1.2

Types of Fuel Cell Systems and Typical Parameters

Parameter	PEMFC	AFC	PAFC	MCFC	SOFC
Electrolyte	H^+-exchange ionomer	KOH	H_3PO_4	$Li_2CO_3+K_2CO_3$ $Li_2CO_3+Na_2CO_3$	YSZ
Ionic conductor	H^+	OH^-	H^+	CO_3^{2-}	O^{2-}
Typical operation temp. (°C)	0–80	60–250	120–210	650	750–1000
Typical power (kW)	0.001–100	0.05–10	1–200	50–1000	0.01–1000
Direct fuel	H_2, MeOH, formic acid	H_2	H_2	H_2	H_2
CO poisoning tolerance	<10 ppm	Yes	<0.5%	Yes	Yes
Typical applications	Portable, backup, stationary, transportation	Outer space[a]	Stationary, power plant, transportation	Power plant	Portable, power plant
Activation losses	High	Higher	Higher	Lower	Lowest
Thermal cycling impact	Low	Medium	Medium	Highest	High
Electrical efficiency (%)	30–50	20–35	20–35	40–60	40–60
Reactant humidification	Yes	No	No	No	No
Location of water formation	Cathode	Anode	Cathode	Anode	Anode
State of water	Vapor + Liquid	Vapor + Liquid (<100°C) Vapor (>100°C)	Vapor	Vapor	Vapor
Internal reforming	Impossible	Impossible	Impossible	Possible	Possible
Number of phases	3	3	3	3	2
Water management	Critical	No issue	No issue	No issue	No issue

[a] Terrestrial applications may become possible when durable hydroxide exchange membranes are developed and used.

transport species within the electrolyte. Also, protons are all produced at the anode and transport to the cathode.

An *alkaline fuel cell* (AFC) typically uses a KOH aqueous solution as the electrolyte with hydroxide (OH^-) as the charge transport species. Its operating temperature ranges from about 60°C to 250°C with an output power up to kilowatts. Since the carbon dioxide (CO_2) in the earth's atmosphere can react with KOH to form K_2CO_3, resulting in a reduction in KOH concentration and the blockage of a porous electrode, this type of fuel cell is not suitable for terrestrial applications, but it is perfectly good for applications in outer space where there is no CO_2. Research was recently carried out to develop hydroxide exchange solid membranes to minimize the impact of CO_2, but the durability of the membrane is not good enough yet, and how well such a fuel cell can handle CO_2 needs to be thoroughly evaluated. The anode, the cathode, and the overall reactions are expressed in Reactions 1.10, 1.11, and 1.12, respectively. Please note that OH⁻ is produced at the cathode and transports to the anode.

$$\text{Anode:} \quad H_2 + 2OH^- = 2H_2O + 2e^- \tag{R. 1.10}$$

$$\text{Cathode:} \quad 0.5O_2 + 2e^- + H_2O = 2OH^- \tag{R. 1.11}$$

$$\text{Overall:} \quad H_2 + 0.5O_2 = H_2O \tag{R. 1.12}$$

A *molten carbonate fuel cell* (MCFC) uses a molten salt, such as a mixture of K_2CO_3 and Li_2CO_3 (or Na_2CO_3 and Li_2CO_3), as the electrolyte. Obviously, an MCFC can only operate above the melting point of the carbonate salt mixture (~500°C). An MCFC normally operates at around 650°C in order to achieve higher ionic conductance. The molten electrolyte is typically embedded in a high-temperature porous $LiAlO_2$ matrix. The fuel cell generates power of a few hundred kilowatts or higher. The anode, the cathode, and the overall reactions are expressed by Reactions 1.13, 1.14, and 1.15, respectively. Please note that CO_3^{2-} is produced at the cathode and transports to the anode. The cathode reaction involves CO_2, and it is typically obtained from the anode through circulation. In Reaction 1.15, "ca" and "an" are short for *cathode* and *anode*, respectively, and they should be shown in the reaction because the CO_2 concentrations at the anode and the cathode are often different.

$$\text{Anode:} \quad H_2 + CO_3^{2-} = H_2O + CO_2 + 2e^- \tag{R. 1.13}$$

$$\text{Cathode:} \quad 0.5O_2 + CO_2 + 2e^- = CO_3^{2-} \tag{R. 1.14}$$

$$\text{Overall:} \quad H_2 + 0.5O_2 + CO_2 \text{ (ca)} = H_2O + CO_2 \text{ (an)} \tag{R. 1.15}$$

A *solid oxide fuel cell* (SOFC) uses a solid oxide (O^{2-}) conductor, such as yttria-stabilized zirconia (YSZ), as the electrolyte. Since the oxide conductivity is strongly affected by the temperature, this type of fuel cell typically operates at 750–1000°C. The output power ranges from a few watts to megawatts. The anode, the cathode, and the overall reactions are expressed by Reactions 1.16, 1.17, and 1.18, respectively. Please note that O^{2-} is produced at the cathode and transports to the anode.

Anode: $H_2 + O^{2-} = H_2O + 2e^-$ (R. 1.16)

Cathode: $0.5O_2 + 2e^- = O^{2-}$ (R. 1.17)

Overall: $H_2 + 0.5O_2 = H_2O$ (R. 1.18)

It is worth emphasizing that although the overall reactions are the same for the PEMFCs/PAFCs, AFCs, MCFCs (except for the presence of CO_2 in the reaction equation), and SOFCs, their anode and cathode reactions are dramatically different.

1.3 Advantages of Fuel Cells

It is believed that fuel cells will become one of the most important technologies in the twenty-first century. The following are the major advantages of fuel cells.

A fuel cell can convert chemical energy into electrical energy with high efficiency, because the energy conversion is complete in one step, and it does not suffer from the Carnot limit. Also, the efficiency is independent of the size of the fuel cell. First, the efficiency of a physically small fuel cell does not have to have a lower efficiency than a physically large fuel cell, and vice versa. Second, a low-power-output fuel cell can be as efficient as a high-power-output fuel cell. Third, when a fuel cell operates at a power output level lower than its designed nominal power output, it does not necessarily suffer from efficiency loss; actually, the fuel cell stack will have a higher electrical efficiency when it operates at a lower power output level than its nominal power output level, as discussed in Chapter 2. In addition, if the heat generated by a fuel cell is collected and used, the total efficiency of the combined heat and electrical power can go beyond 80%. In Europe and Japan, combined heat and power (CHP) is quite popular. The heat can be used for hot water preparation and space heating.

Compared to ICEs, DEs, and turbines, a fuel cell is much quieter. Since the energy conversion is an electrochemical process, it does not generate any sound, just like the behavior of a battery. This will not only reduce environmental noise pollution, but also enable a fuel cell to be used in situations where a low noise level is critical. However, since a fuel cell system normally needs devices to supply reactants for the reactions, or circulate coolant for heat management, or ventilate air to prevent the electronic or electrical components from overheating, these devices, such as compressors, blowers, fans, and pumps, will generate some noise. As a compressor can be very noisy, and it normally consumes a significant portion of the electrical energy produced by the fuel cell, it is best to avoid using such a device whenever possible. The acceptable noise level of a fuel cell system is determined by its applications. For example, for stationary applications, the noise level of a fuel cell system should be less than 65 dB at a distance of 1 meter.

A fuel cell is reliable because the electrochemical reactions do not involve any moving parts. This is similar to a battery, but distinctly different from an ICE, where continuous piston movement is involved when it generates power. However, at present, fuel cell reliability still needs improvement because it may fail prematurely and its lifetime is not as good as some traditional technologies. Further work in the following areas is needed to improve the reliability of a fuel cell system: improvement of the durability and reduction of cost of some key materials, such as the catalyst and the membrane; better understanding of material decay mechanisms; enhancing the reliability of some balance-of-plant (BOP) components; and optimization of the control algorithm and the system operation.

A fuel cell stack is relatively easy to expand by increasing either the active area of the unit cell or the number of unit cells in a stack. For example, when the active area of a unit cell doubles or when the number of unit cells doubles, the power output of the resulting stack will be nearly doubled under the same operating conditions. Larger areas or more cells are more likely to cause less uniform distribution of the reactants within a unit cell or among unit cells, leading to slightly lower performance per unit electrode area or reduced reliability of the stack. Normally, the active area of the unit cell and the number of unit cells are considered together when designing a stack in order to produce the required power. Also, when the active area is changed, the flow-field channels for the flow of fluids may need to be redesigned.

A fuel cell will provide power as long as the reactants are supplied. When multiple high-pressure H_2 cylinders are used to supply the H_2, the empty cylinders can be hot-swapped while the other cylinders still supply H_2 to the fuel cell stack, and thus the cylinder replacement process does not interrupt the operation of the fuel cell system. The replacement process may only take a few minutes. In contrast, when a secondary battery becomes empty, the recharging process will take hours. When the H_2 is supplied by pipelines in the future, the use of H_2 will become extremely convenient.

A fuel cell can effectively use fuels such as H_2 produced by using renewable energy such as solar and wind power. As fossil fuels proceed toward depletion, H_2 will become more and more important as an energy carrier. The utilization of H_2 will enable solar and wind power to be indirectly used in the transportation sector. Although H_2 can also produce power through combustion in devices such as an ICE, only a fuel cell can use it most effectively.

A fuel cell is safe. Since the anode and the cathode reactions occur separately without direct contact between the fuel and the oxidant, and most of the fuel entering the fuel cell stack is consumed, a fuel cell is very safe. A fuel cell system also has various installed sensors, such as H_2 concentration sensors, that will enable the system to cut off the H_2 supply when there is H_2 leakage or when the fuel cell stack is in an unsafe condition. Fuel cell combustion or explosion has not been reported in the past few decades.

1.4 Proton Exchange Membrane Fuel Cells

This book is about PEMFCs. For information on other types of fuel cells, please refer to relevant books and publications (Larminie and Dicks 2003).

Figure 1.2 schematically shows the major components of a PEM single cell. A single cell is a unit that contains only one membrane electrode assembly (MEA), i.e., one anode, one cathode, and one membrane placed between the anode and the cathode.

1.4.1 Membrane

A PEMFC uses a solid polymer that conducts protons as the electrolyte. The first PEM used in a PEMFC was polystyrene sulfonic acid for the Gemini project, and it was reported that such a membrane could only last for about 500 hours (Larminie and Dicks 2003). Later, DuPont produced Nafion, a polymer with a polytetrafluoroethylene main chain and a perfluorinated side chain ending with a sulfonic acid group $-SO_3H$. The mass of the polymer repeating units per mole of the sulfonic acid groups is called the *equivalent weight* (EW). The proton conductivity arises from the sulfonic acid groups. Because a C–F bond is stronger than a C–H bond (their bond strength, D^o_{298K}, in diatomic molecules is 552 and 338 kJ mol^{-1}, respectively) (Kerr 1994) and the C–C bond is well shielded by the F atoms, a perfluorinated polymer possesses higher chemical and electrochemical stabilities than the nonfluorinated counterparts. In addition, since F atoms have much higher electron affinities than H atoms (their atomic electron affinities are 3.40 and 0.75 eV, respectively) (Miller 1994), they make the sulfonic acid group more acidic (i.e., higher proton conductivity). The chemical structures of several perfluorosulfonic

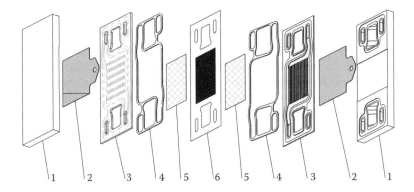

FIGURE 1.2
Major components of a PEMFC single cell (1 = end plates; 2 = current collectors; 3 = flow-field plates; 4 = gaskets; 5 = gas diffusion media; 6 = membrane electrode assembly).

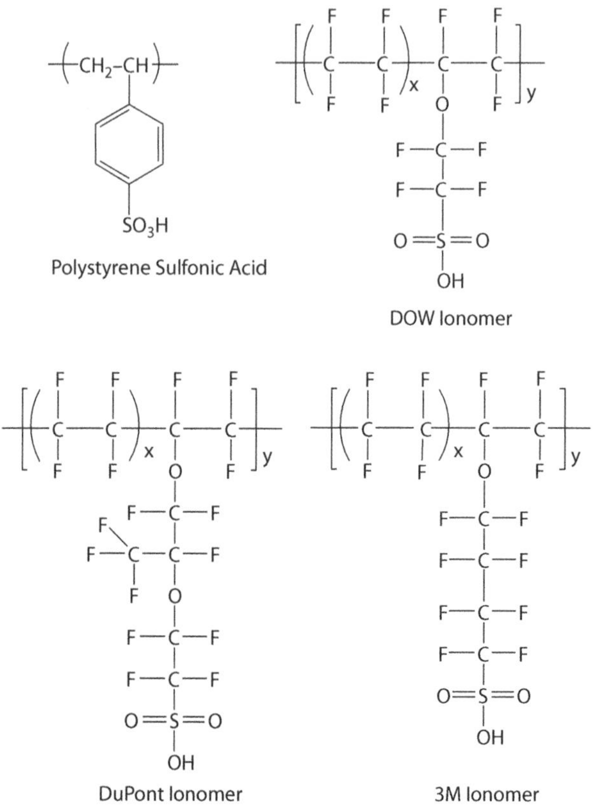

FIGURE 1.3
Chemical structures of some ionomers.

acids (PFSAs) along with polystyrene sulfonic acid are shown in Figure 1.3. Table 1.3 lists the major specifications of DuPont's Nafion 115 membrane.

A perfluorosulfonic acid like Nafion needs water to become proton conductive. A hydrated Nafion membrane consists of hydrophobic domains and hydrophilic domains. The hydrophobic domains are formed by the main chain (i.e., the backbone) of the PFSA, while the hydrophilic domains are formed by the sulfonic acid groups that are associated with water molecules. Proton transportation is via the hydrophilic domains. Normally, the higher the water content, the higher the proton conductivity because the hydrophilic domains become larger and the connections between these domains get better. For Nafion with an EW of 1100, a sulfonic acid group can hold for about 22 water molecules when fully hydrated in boiling water. Studies have shown that for Nafion with EW of 1100, the proton conductivity of the membrane is negligible before each sulfonic acid group possesses about 3 water molecules, then the conductivity increases quickly up to about 13 water molecules

TABLE 1.3

Properties of DuPont Nafion 115 Membrane

Property		Unit	Value
Thickness		μm	127
Tensile modulus	23°C, 50%RH	MPa	249
	Water soaked, 23°C		114
	Water soaked, 100°C		64
Max. tensile strength	23°C, 50%RH	MPa	43 (MD), 32 (TD)
	Water soaked, 23°C		34 (MD), 26 (TD)
	Water soaked, 100°C		25 (MD), 24 (TD)
Elongation at break	23°C, 50%RH	%	225 (MD), 310 (TD)
	Water soaked, 23°C		220 (MD), 275 (TD)
	Water soaked, 100°C		180 (MD), 240 (TD)
Initial tear resistance	23°C, 50%RH	$g\ mm^{-1}$	6000 (MD, TD)
	Water soaked, 23°C		3500 (MD, TD)
	Water soaked, 100°C		3000 (MD, TD)
Specific gravity		$g\ cm^{-3}$	1.98
Min. conductivity	25°C in D.I. water	$S\ cm^{-1}$	0.10
Acid capacity	Total	$meq\ g^{-1}$	0.95–1.01
	Min. available		0.90
Water content	23°C, 50%RH	%	5
	Water soaked, 100°C		38
Thickness change	Water soaked, 23°C	%	10
	Water soaked, 100°C		14
Linear expansion	Water soaked, 23°C	%	10 (MD, TD)
	Water soaked, 100°C		15 (MD, TD)

Note: Membrane initially conditioned at 23°C, 50%RH; MD = machine direction, TD = transverse direction.

Source: Adapted from the DuPont website, http://www2.dupont.com/FuelCells/en_US/assets/downloads/dfc101.pdf.

per sulfonic acid group. Afterward, the increase in conductivity slows down with further increase in water molecules.

In order to increase the proton conductivity, a lower EW ionomer can be made. However, when the EW is too low (e.g., less than 700), the mechanical strength of the membrane becomes unacceptable, especially after the membrane is fully hydrated, which will in turn shorten the membrane lifetime. The proton conductivity and the mechanical strength of a membrane also depend on the side chain length and structure. The PFSAs made by Dow and 3M have shorter side chains than DuPont's Nafion, as shown in Figure 1.3.

The structure of the PFSA developed by Solvay Solexis has the same chemical structure as the Dow ionomer, and is called Hyflon® Ion for both the membrane and the ionomer solutions. PFSAs with shorter side chains

FIGURE 1.4
Fumatech membrane in rolls.

possess higher crystallinity than those with longer side chains, and thus they can be made with lower EW (e.g., 850 g equiv^{-1}) and higher proton conductivity while still having good enough mechanical properties.

Membranes are typically made by two methods, casting and extrusion. The *casting method* applies a solution of PFSA on a flat surface and lets the solvent evaporate gradually. The solvent evaporation rate needs to be controlled in order to make pinhole-free membranes. This method can make membranes as thin as 15 microns. The *extrusion method* extrudes a melt precursor polymer through a preset gap, followed by a few steps of posttreatment, one of which is converting the precursor polymer to the final acidic form. This method can make membranes as thin as 50 microns. Thinner membranes are difficult to produce using this method because the membrane may suffer from some mechanical damage during the several posttreatment steps. Figure 1.4 shows the PEM in rolls made by Fumatech.

The dimensions of PFSA membranes change significantly between the dry and the hydrated states, which can cause detachment of the gas diffusion medium, such as carbon paper, from the membrane, and accelerate the deterioration of the membrane at places that are near the edges of the carbon paper. One way to reduce the dimensional change in the membrane planar directions is to incorporate in the membrane a thin porous sheet, such as expended PTFE (ePTFE), that does not change dimensions with hydration levels. Such a membrane is called a *reinforced membrane*. Since ePTFE does not conduct protons, it is typically made at 25 microns or thinner with about 95% porosity in order to have the resulting composite PEM possess good enough proton conductance. An ePTFE-ionomer composite membrane manufacturing process is illustrated in Figure 1.5. It mainly involves the unwinding of

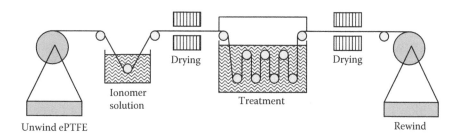

FIGURE 1.5
Illustration of ePTFE-ionomer composite membrane manufacturing process.

an ePTFE web, soaking ePTFE in an ionomer solution, drying, treatment, and rewinding of the resulting e-PTFE-ionomer composite membrane. The treatment step is either for the hydration of the membrane if it is already in an H⁺ form, or for hydrolysis of the membrane into an H⁺ form if it is originally not in an H⁺ form. If one-step ePTFE soaking in the ionomer solution is not adequate, additional steps can be used. Another way to reduce the dimensional change of a PEM is to incorporate in the membrane some fibrous filler whose dimensions do not change with hydration levels.

Due to fluorination, PFSA materials are relatively more costly than non-fluorinated hydrocarbon membranes. The latter such as poly (ether ether ketone) and poly (arylene ether sulfone) have also been widely investigated for fuel cell applications. In addition to lower cost, some of them also have better dimensional stability with regard to hydration levels. The problem is their lower chemical stability in the fuel cell environment.

When a dry Nafion membrane is immersed in liquid water, its water uptake increases with temperature. For example, the water uptake is about twice the amount at 100 than at 20°C (Alberti 2008). It is logical that the hydrophilic domains and the channels connecting them are better opened at higher temperatures, and thus the membrane can host more water. Hydration of the membrane by immersing it in liquid water is much more effective than exposing it to water vapor. For example, Nafion membrane soaked in boiling water can take in 38% water (corresponding to 22 water molecules per $-SO_3H$ group), while that exposed to water vapor at 100% RH and 100°C can only take in 23% water (corresponding to 13 water molecules per $-SO_3H$ group). Therefore, the conformational structure of Nafion is likely to be different under different hydration conditions. During the hot bonding in the MEA making process, the membrane loses a significant amount of water, and its conformation may also change; then the membrane can hardly regain the 38% water during the operation of a fuel cell. The same is true for the ionomer within the catalyst layer.

At lower hydration level water adsorbs onto the pore walls of the hydrophilic domains, and thus proton transport is confined to this thin water layer, offering a lower conductivity. When more water gets into the membrane, the pores will be gradually filled with liquid water, and some protons associate with water molecules to form protonated water clusters (e.g., H_3O^+, $H_5O_2^+$, $H_9O_4^+$), which allow protons to move along, and therefore, higher proton conductivity is achieved. The turning point or the percolation was found to be around 37% RH that is used to hydrate the membrane and the membrane contains about 3~4 water molecules per $-SO_3H$ group (Costamagna 2008). In addition, it is expected that at higher water contents the hydrophilic domains and the channels linking them are wider and better connected, which also favors the proton conductance.

The proton conductivity decreases dramatically at subzero temperatures. Hou et al. reported that the proton conductivity of Nafion 212 membrane dropped from 86.5 to 10 and 5.8 mS cm^{-1} when the temperature decreased from 25 to –10 and –20°C, respectively (Hou 2008). Although the conductivity drop is not a good thing for fuel cell performance due to larger iR losses, it does offer a benefit during the startup at subzero temperatures because the larger iR loss can help the fuel cell to achieve to above the freezing temperature faster.

After a PEM is cooled at very low temperatures (e.g., less than –50°C) differential scanning calorimetry (DSC) shows an endothermic peak at around 0°C as the temperature scans up. This seems to indicate that the PEM contains water that freezes at the water freezing temperature. From the peak area the amount of such "freezable" water can be determined, and its difference from the total amount of water within the membrane that is pre-determined by weighing is used to represent the amount of water that is not freezable (called non-freezable water). For example, Hou et al. found that fully hydrated Nafion 212 membrane in liquid water at 25°C contains 5.3 and 15.2 freezable and non-freezable water molecules per $-SO_3H$ group, respectively; while the same membrane hydrated in water vapor at 75% RH and 25°C contains 0 and 6.2 freezable and non-freezable water per $-SO_3H$ group, respectively (Hou 2008).

The presence of an endothermic peak at around 0°C during the DSC test led to one belief that there are two types of water in the PEM, "non-freezable" and "freezable". The non-freezable water is strongly associated with $-SO_3H$ and its dissociated species, $-SO_3^-$ and H^+. Due to the presence of those ionic species and the nano-sized space confining the liquid, water will have a much lower freezing temperature than 0°C, and thus those water are termed non-freezable. A question is raised as to where the freezable water stays within the PEM? If it really freezes at around 0°C, it means that it is electrolyte-free. The regions that are electrolyte-free are the hydrophobic domains, but highly polarized water should not be able to get into those regions. Water likes to enter and stay in the hydrophilic domains, but then how can some water (the "freezable" portion) is not associated with H^+ that dissociates from the $-SO_3H$ group and thus is able to freeze at 0°C? Even an intermediate region or domain is proposed, water still should behave differently from free water or bulk water, and thus it should not be able to freeze at 0°C.

Another belief that the Author believes more correct is that all water within the PEM is non-freezable up to certain ultra low temperatures. Some water migrates to the surface of the PEM during the cooling process, and it is such water that freezes and shows the endothermic peak at around 0°C during the DSC test. Water remaining inside the membrane does not form solid ice even at −100°C (Pineri 2007). The study by Pineri et al. also showed that the water content per −SO_3H group decreased from 16 to less than 12 in about 3300 s for Nafion 117 that was quenched at 203 K and annealed to 223 K, which indicated that the migration of water to the surface of the PEM was able to occur at lower than −100°C. Therefore, it appears to be plausible to conclude that the migration of water toward the membrane surface is a continuous process during cooling; also, the surface water may not start to freeze at 0°C but rather at a much lower temperature (e.g., −20°C) because it is still more or less associated with the membrane. Pineri et al. also indicated that if the PEM contains fewer than ca. 5 water molecule per −SO_3H group, there was no water migrating to the surface of the PEM, and thus no endothermic peak at around 0°C was expected during the DSC test. This correlates quite well with the results by Hou that membrane hydrated in water vapor with 75% RH at 25°C contains 6.2 non-freezable water per −SO_3H group without any freezable water detected (Hou 2008).

Since PFSA membranes have glass transition temperatures, T_g, around 130°C (lower for PFSA with longer side chains) and they cannot hold water at above the boiling temperature of water, they can only be used at temperatures lower than 100°C, typically less than 80°C (T_g drops significantly when the membrane is hydrated). This makes the Pt-based catalysts used in a PEMFC easily poisoned by ppm levels of CO, and a bulky cooling system is needed in order to eject heat out of the fuel cell system. For example, the radiator could be reduced for ca. 60% if the fuel cell temperature is increased to 120°C from 80°C. Adding some hydroscopic particles such as SiO_2, TiO_2, ZrO_2, Al_2O_3, WO_3, $Zr(HPO_4)$ and heteropolyacids (e.g., phosphotungstic acid) helps the PEM to retain some water at above 100°C, but the proton conductivity is still too low, and more importantly, the softening of the PFSAs at such temperatures will cause disintegrate of the membranes quickly.

Although the T_g is low for PFSAs, they are chemically stable up to about 280°C (for Nafion), above which the sulfonic acid groups start to be lost as SO_2.

A PEM may become yellowish during storage due to the adsorption of volatile organic compounds and salts. If this happens the PEM should be cleaned in boiling solutions such as 3% H_2O_2 (to remove organic contaminants), 0.1 M H_2SO_4 (to remove metallic ions), and finally deionized water (to remove H_2SO_4). This procedure is typically carried out in lab tests, but it increases the complexity and cost for larger scale manufacturing process. Therefore, it is best to use the PEM once or shortly after it is manufactured.

1.4.2 Catalysts

The H_2 oxidation reaction (HOR) and the O_2 reduction reaction (ORR) require catalysts to achieve useful reaction rates. The best catalysts are platinum (Pt) and its alloys with certain transition metals. The catalysts are normally made into nanoparticles with diameters between 1.5 and 5 nm. H_2 oxidation proceeds faster than O_2 reduction, and thus it requires less of a catalyst. With pure H_2 as the reactant, a Pt loading of 0.05 mg cm^{-2} is enough, while the catalyst loading for O_2 reduction is typically several times higher.

The widely used catalyst is Pt supported on carbon (Pt/C), such as Pt/Vulcan XC-72. Pt nanoparticles are often deposited onto the support during the synthesis process. The support has several functions. It helps the Pt to form nanoparticles during the formation process. It acts as an anchor for the Pt nanoparticles through some chemical–physical interactions, so that the Pt nanoparticles are less likely to grow in size during use. It also allows electrons to transport among Pt nanoparticles and throughout the catalyst layer. It may also alter the activity of the supported catalysts. A TEM picture of Pt/C is shown in Figure 1.6.

The support needs to meet certain criteria. It should be a good electron conductor to allow electrons to move within the catalyst layer. It would be ideal if the support possesses both good electron and proton conductivities, because such a support can enhance catalyst utilization. It should be stable in the fuel cell environment; otherwise, the Pt nanoparticles can detach from the support and potentially become useless. For example, the fuel cell

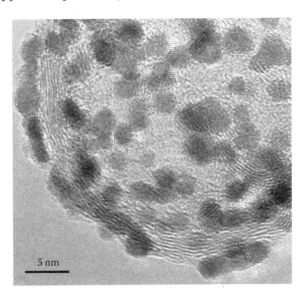

FIGURE 1.6
TEM of Pt/C. (Courtesy of East China University of Science and Technology.)

cathode is highly oxidative in the presence of oxygen, water vapor, and high potential (~1 V at open circuit voltage, and can be as high as 1.6 V during startup and shutdown). Common metals cannot survive such an environment, and thus they are not suitable as catalyst supports. Due to the presence of highly active Pt nanoparticles, the oxidation of the support can be catalyzed and accelerated. Carbon is the best and most popular catalyst support currently, but it is far from ideal. Thermodynamically, carbon will be oxidized at potentials as low as 0.21 V, but the kinetics are slow, even in the presence of Pt. But at the open-circuit voltage (OCV) of around 1 V, the corrosion of carbon accelerates. This is a major reason that the fuel cell cathode decays faster at the OCV. If a hydrogen–air boundary is formed in the anode during the startup or shutdown process of the fuel cell, a portion of the cathode can experience a potential as high as 1.6 V. The carbon will be oxidized quickly, causing the thickness of the catalyst layer to reduce significantly and eventually collapse (Tang et al. 2006). Graphitized carbon has higher anti-oxidation capability than amorphous carbon, favoring better durability, but it normally does not favor Pt nanoparticle deposition because of reduced surface area and defects, resulting in larger Pt nanoparticles and lower Pt content.

The presence of Pt is likely to catalyze the corrosion process of the carbon-type supports, although there are contradictory reports on this point. Supports can also alter the activity and durability of the catalyst particles. For example, doping carbon-type supports with boron (B) or nitrogen (N) was found to be able to enhance the durability (and activity) of the catalysts. Doping means some carbon atoms are replaced with other foreign atoms such as B or N.

Accompanying the oxidation of carbon support, the double layer capacitance of the catalyst layer increases gradually, and a new redox process may appear at ca. 0.6 V from quinone/hydroquinone couple.

At voltage near the OCV, Pt can also be oxidized according to Reactions 1.19 and 1.20:

$$Pt + H_2O = PtO + 2H^+ + 2e^- \qquad E^\circ = 0.98 \text{ V} \qquad \text{(R. 1.19)}$$

$$Pt = Pt^{2+} + 2e^- \qquad E^\circ = 1.19 \text{ V} \qquad \text{(R. 1.20)}$$

Some Pt^{2+} will be converted back to Pt at lower potentials during the operation of a fuel cell, but it may not deposit to the original particles. Some Pt^{2+} may migrate into the PEM, where it is reduced by H_2 diffusing into the PEM from the anode to form Pt according to Reaction 1.21. With time a band of Pt may appear in the PEM when observed under SEM. The location of the Pt band is determined by where Pt^{2+} and H_2 meet in the PEM according to their transport rates.

$$Pt^{2+} + H_2 = Pt + 2H^+ \qquad \text{(R. 1.21)}$$

Since R. 1.19 occurs at a lower potential than R. 1.20, the formation of PtO layer can prevent or minimize the underneath Pt to go through R. 1.20. However, Pt^{2+} can also form from PtO via chemical reaction 1.22:

$$PtO + 2H^+ = Pt^{2+} + H_2O \qquad\qquad (R.\ 1.22)$$

Particle size effects on the ORR, the peak positions of the Pt/PtO_x redox process, and the CO oxidation reaction are nicely reviewed by Hayden and Maillard (Hayden 2009; Maillard 2009).

In order to maximize the usage of Pt atoms within Pt particles, Pt particles should be made as small as possible so that the percentage of Pt atoms on the outer surface of the particle will increase. However, when Pt particles are made smaller than about 3 nm, their specific (i.e., mass) catalytic ability decreases, and their tendency to coalesce into larger particles becomes stronger. The size of most Pt particles in widely used Pt/C is between 2 and 4 nm. The size of the Pt particles often depends on the Pt content on carbon, and a higher Pt content typically leads to larger Pt nanoparticles.

Since only the Pt atoms on the Pt particle surface can participate in the catalysis process, one way to maximize Pt atom utilization is to deposit an extremely thin layer of Pt on a non-Pt particle to form a core–shell structure with the Pt layer as the shell and the non-Pt particle as the core. The electron conductivity of the core particles is not important because they are covered by a highly electrical conducting Pt shell. In addition, when a submonolayer to several monolayers of Pt are made on certain core metal particles, the catalytic ability of Pt improves due to electronic interactions between the shell Pt atoms and the interior core metal atoms that increase the Pt 5d orbital vacancies and thus increases the π electron donation from O_2 to Pt atoms (electronic effect), and due to the decrease in the Pt-Pt atomic distance (geometric effect). This mechanism is similar to the improved catalytic ability of PtM_x and PtM_xN_y alloys, where M and N represent different metals and x and y their atomic contents in the alloy. For the Pt/core, PtM_x or PtM_xN_y, if there is some leaching out of the non-Pt metals, their corresponding cations can replace the protons of the PFSA either in the catalyst layer or in the membrane to reduce its proton conductivity as well as the catalyst–PFSA–reactant three-phase boundaries, and thus decrease the fuel cell performance. The shape and crystalline facet of the Pt nanoparticles can also affect the catalytic activity.

Carbon monoxide (CO), even at trace amounts such as a few ppm levels, can poison Pt catalysts because it can strongly adsorb on the Pt surface, leaving a very small percentage of the Pt surface (e.g., less than 5% at 80°C in the presence of 10 ppm CO) for the HOR. Pt alloys with ruthenium (Ru) and tin (Sn) possess higher CO tolerance, and are thus popular (especially PtRu) as the anode catalysts when H_2 is not CO-free. The mechanisms are mainly the accelerated oxidation of CO on PtRu and the reduction of CO adsorption strength on PtSn, respectively. Reaction 1.23 shows how Ru accelerates the

CO oxidation through the formation of Ru-OH. The formation of Ru-OH is ca. 0.2 V lower than the formation of Pt-OH. The formation of Mo-OH occurs at an even lower potential, and thus PtMo is also a good CO-tolerant catalyst.

$$Pt\text{-}CO + Ru\text{-}OH = Pt + Ru + CO_2 + H^+ \qquad (R.\ 1.23)$$

CO poisoning is fully recoverable when CO-free H_2 is subsequently used. A more severe poisoning species is S-containing compounds such as H_2S, which can poison the anode at ppb levels. The poisoning effect is cumulative due to the formation of an S layer on the surface of the catalysts, and it is not recoverable when switching to S-free H_2 (Shi et al. reported that recovery is achievable by applying a voltage of 1.5 V (Shi 2007)). Therefore, any S-containing compounds must be completely removed from the fuel stream.

Common poisons for the cathode include NO_x, SO_2, and volatile organic compounds (e.g., benzene). Therefore, the air filter should not only remove particulates but also those poisoning species through chemical absorptions or physical adsorptions. The filter should be cleaned, regenerated, or replaced when saturated with any of those species.

1.4.3 Catalyst Layer

Catalysts are typically made into thin 3-dimensional layers in order to increase the total catalytic surface area. The layer should be porous to allow the reactant and the reaction product to transport in and out, and it should conduct both electrons and protons because the reaction involves both electrons and protons. The thickness of the catalyst layer (CL) typically ranges from a few microns to around 20 microns. Thicker catalyst layers have higher total catalytic surface area per geometric area, which favors higher performance, but they also have higher resistance to the transport of electrons, protons, reactants, and the product, which negatively affects fuel cell performance. The porosity of the CL is ca. 30% with primary pores less than 10 nm and secondary pores less than 100 nm. Tests should be performed to determine the optimal catalyst layer thickness based on its components, structure, and fuel cell operating condition. A good catalyst layer should be one with a higher total catalytic surface area and a reduced thickness.

Some binding materials need to mix with the Pt/C catalyst particles to hold the latter in three dimensions. One binding material is PTFE. Pt/C mixes with PTFE aqueous suspension to form a homogeneous mixture, which is applied to a gas diffusion medium (GDM) and is subsequently dried at suitable temperatures. One of the temperatures should be around 335°C, slightly above the melting point of PTFE (327°C), to aid the flow of PTFE within the resulting catalyst layer. In addition to binding Pt/C particles, PTFE also makes the resulting catalyst layer more hydrophobic, facilitating the mass transport of the gaseous reactant. Since PTFE is an insulating material, it will reduce the electronic conductance of the resulting catalyst layer.

Earlier electrodes were made with PTFE as the binding agent at a Pt load-ing larger than 4 mg cm^{-2}. Most of the catalysts could not participate in the electrochemical reaction due to the lack of proton conductance throughout the entire catalyst layer. The situation was improved when some ionomer solution was impregnated into the catalyst layer. Another binding material is an ionomer such as Nafion. Pt/C mixes with a Nafion solution in the pres-ence of a suitable solvent or solvent mixture, such as water and isopropanol, to form a homogeneous mixture, which is then applied to either a GDM or a decal and subsequently dried at suitable temperatures. The drying tem-perature should be lower than about 120°C to avoid damaging the Nafion. A temperature slightly higher than 100°C is sufficient to remove the solvents. If vacuum drying is used, a drying temperature lower than 100°C is sufficient. It is important to mix Pt/C with water first, and then add the organic solvent such as isopropanol. If Pt/C is directly mixed with isopropanol, the catalyst may ignite at room temperatures. Besides binding Pt/C particles, Nafion also provides the resulting catalyst layer with proton conductance, facilitating the transport of protons. Since 1997 most of the catalyst layers are made using ionomer as the binding agent, and the Pt loading is lowered to less than 0.4 mg cm^{-2}.

Not all catalyst nanoparticles can participate in the fuel cell reaction due to the requirement of three-phase regions (or boundaries). The reported Pt surface utilization ranges from as low as 35% to as high as 100%. Three-phase regions are those that are accessible to the reactant, electrons, and pro-tons, as shown in Figure 1.7. For a catalyst layer consisting of Pt/C and PTFE, there are no three-phase regions due to the lack of proton conductance; therefore, only the region that is in contact with the membrane can meet the three-phase region criteria, and the remainder of the catalyst layer is basi-cally wasted. For a catalyst layer consisting of Pt/C and an ionomer such as Nafion, the three-phase regions distribute throughout the entire catalyst layer, allowing most of the catalyst particles to participate in the electrochemical

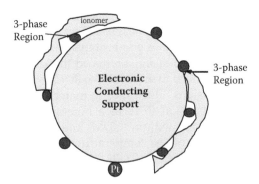

FIGURE 1.7
Illustration of 3-phase regions.

reaction. The inactive Pt nanoparticles are mainly those that are either far apart from the ionomer or those that do not form continuous proton transport paths to the membrane. It is expected that Pt particles will be able to participate in the electrochemical reactions if they are not too far away from the ionomer even though they are not in direct contact with the ionomer, because protons may be able to "spillover" the surface of the catalyst support for certain distance and water may significantly increase its proton conductivity in a charged space with dimensions of nanometers. Wang et al. found that the ionic conductivity of pure water in a charged matrix could reach to around 10^{-3} S cm^{-2} (Wang 2012). The distance might be much larger than what we could imagine (e.g., up to 1 micrometer?), but it has not been experimentally determined yet. It is worth the efforts to design experiments to determine this distance. Pt particles that have no continuous electron path to the GDM cannot participate in the fuel cell reaction either, but the number of such Pt particles is not expected to be high. A good catalyst layer should maximize the Pt nanoparticle surface area that can participate in the fuel cell reaction.

The total catalyst surface area that participates in the fuel cell reaction cannot be accurately obtained. The closest estimate is the so-called electrochemically active surface area (ECSA), which is obtained through cyclic voltammetry. It can be carried out in two environments. One uses the hydrogen adsorption–desorption peak areas in the hydrogen underpotential deposition region. It is also important to switch the cathodic potential scan to the anodic potential scan before the hydrogen evolution starts; however, sometimes this potential reversal point may not be determined very accurately, and it varies slightly with different catalyst layers. Very often, several prescans are performed to pinpoint this reversal potential. The other method uses the CO oxidative stripping area by allowing CO to form a monolayer on the catalyst surface though adsorption, followed by a subsequent oxidation of CO to CO_2, according to Reaction 1.24:

$$CO + H_2O = CO_2 + 2H^+ + 2e^- \qquad \text{(R. 1.24)}$$

The CO oxidative stripping peak is very sharp, and only the Pt surface that meets the three-phase regions can contribute.

There are several techniques (e.g., spraying, screen printing, inkjet printing) to apply catalyst ink onto a substrate (either GDM or decal). Spraying typically uses an airbrush with the aid of a pressurized gas such as nitrogen to disperse the catalyst ink onto the substrate. The application may repeat in order to get the desired Pt loading. In order to maximize the dispersion of the catalyst ink electro-spray by applying a high voltage (e.g., several thousands of volt) to charge the ink droplets is also used. The catalyst ink should be made with suitable viscosity to suit for the application techniques.

Catalyst layer can also be made by other methods such as electrochemical deposition (not practical for large scale production) and physical vapor deposition (PVD). By using the PVD method 3M has constructed novel nanostructured thin film (NSTF) electrodes with the Pt loading as low as 0.05 mg cm^{-2} and achieved excellent performance and durability (Debe 2006).

The electrode can also be made with multiple catalyst layers. Each layer has different composition and structure (e.g., catalyst type and loading, support type and loading, ionomer content, PTFE content, porosity, and thickness) in order to achieve higher performance and durability and lower total precious metal loading. Of course, the manufacturing process will be slower and more costly.

1.4.4 Gas Diffusion Medium

As mentioned previously, the catalyst layer is applied to either a GDM or to a decal (applying to the PEM directly is also explored). The GDM allows the reactant to transport from the flow-field channels on the plates to the catalyst layer, and allows the product to transport from the catalyst layer to the flow-field channels on the plates; so it must be porous, typically with porosity about 80% and with pore size in 10's μm. The GDM also provides mechanical support for the catalyst layer or the catalyst-coated membrane (CCM). The GDM transports heat and electrons between the plates and the catalyst layers as well, so it must conduct electrons and heat.

The most popular GDM is carbon paper, as shown in Figure 1.8. It is made of carbon fibers (typically with diameters less than 10 μm) bound together by certain resins. Before the carbon fibers are graphitized, the GDM can be soft enough so that it can be made roll by roll, lowering the production cost and shortening the production time. After the carbon fibers are graphitized at high temperatures, the resulting carbon paper becomes quite stiff, making

FIGURE 1.8
SEM of carbon paper. Courtesy of Jiangsu Tianniao High Tech. Company.

it less likely to intrude into the flow-field channels on the plates. The porosity and the density of a carbon paper are typically around 75% and 0.4 g cm^{-3}, respectively.

Another popular GDM is carbon cloth, made of carbon threads woven together. It is quite soft and thus could intrude into the flow-field channels on the plates due to deformation. It appears that carbon cloth type GDM started to fade out since around 2000.

1.4.5 Microporous Layer

Often a thin layer of a mixture of carbon powder and PTFE is applied to the GDM. Such a thin layer is called a *microporous layer* (MPL). It is generally applied to the GDM surface that faces the catalyst layer. A catalyst layer–coated GDM is called a gas diffusion electrode (GDE) or a catalyst coated backing (CCB). The hydrophobicity of an MPL can be adjusted by the percentage of PTFE, with a higher PTFE content leading to a more hydrophobic MPL. The major function of an MPL is to aid the water management of the CL and even the entire MEA. An MPL is more useful when the CL uses Nafion as the binder (an ionomer-type catalyst layer) because such a catalyst layer is more likely to be flooded. An MPL with proper hydrophobicity can help such a CL achieve a water balance. Since water can transport through the membrane, an MPL can help water transport from the side with a more hydrophobic MPL to the other side with a less hydrophobic MPL or no MPL. For example, if the cathode has a hydrophobic MPL while the anode has no MPL, more of the water produced at the cathode can transport through the membrane to the anode side, making operation possible without externally humidifying the anode reactant. Another function of an MPL is to enhance the adhesion between the CL and the GDM. A third function of an MPL is to prevent deep penetration of the catalyst into the GDM, because catalysts at such locations are less effective. Studies also found that the presence of MPL helped the subzero startup of the fuel cell (Tabe 2012).

On the other hand, the presence of a highly hydrophobic MPL may cause flooding of the CL because it hinders liquid water transport out from the CL to the GDM.

Experiments are often necessary to determine the electrode that an MPL is useful, the PTFE percentage within the MPL, the thickness of the MPL, and the type of carbon that is used in the MPL.

In order to reduce the CO poisoning impact on the anode an additional layer containing catalysts such as Au and Au-Fe$_2$O$_3$ that are highly active toward CO oxidation can be made on top of the regular MPL; or those catalysts are incorporated into the MPL. With such a structure CO is fully oxidized in conjunction with the *air bleeding* technique when CO-containing H$_2$ passes through, and thus the amount of CO that will adsorb on the surface of anode catalyst is minimized.

1.4.6 Membrane Electrode Assembly

The anode GDM (with or without MPL), the anode CL bound to either the GDM or the PEM, the PEM, the cathode CL bound to either the GDM or the PEM, and the cathode GDM (with or without MPL) are often hot-bonded together to form a multilayered structure. Such a structure is called a *membrane electrode assembly* (MEA). When two catalyst layers are separately applied to either side of a PEM, the resulting PEM is a CCM. Hot-bonding is carried out under a suitable pressure and a suitable temperature for a few minutes. The temperature is chosen such that the membrane softens. For Nafion, it is often around 130°C. The pressure is often around 100 bars.

The bonding procedure is typically as follows, as illustrated by Figure 1.9:

- A bottom flat metal plate of around 1–2 mm thick is laid on a table.
- A bottom thin flexible plastic or metal sheet is laid on top of the metal plate.
- A bottom gasket with an opening the size of the GDM is laid on top of the plastic or metal sheet.
- A GDE (or GDM with or without MPL) is laid within the opening of the bottom gasket.
- A PEM (or CCM) is laid on top of the GDE.

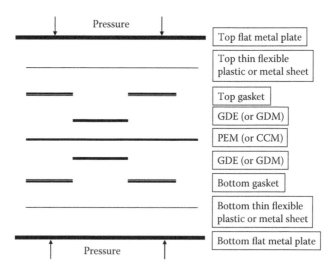

FIGURE 1.9
Illustration of MEA bonding process.

- A top gasket with an opening the size of the GDM is laid on top of the PEM (or CCM) with the openings of the bottom and the top gaskets being aligned exactly with each other.
- Another GDE (or GDM with or without MPL) is laid within the opening of the top gasket.
- A top thin flexible plastic or metal sheet is laid on top of the GDE.
- A top flat metal plate of around 1–2 mm thick is laid on top of the thin flexible plastic or metal sheet.
- The entire assembly is then placed on the bottom plate of a hot press whose top and bottom plates have been set at the bonding temperature.
- The gap between the two plates of the hot press is closed and a small amount of pressure is applied on the assembly.
- The temperature of the hot press will decrease slightly when the two plates are in contact with the assembly; once the temperature returns to the set value, the pressure is increased to the bonding pressure and is kept there for a few minutes.
- The press is opened and the assembly is removed.
- The resulting assembly is either laid on a table or transferred to a cold press to let its temperature drop to around room temperature.
- The assembly is opened by removing the top flat metal plate and the top flexible plastic or metal sheet, and the MEA is removed from the gaskets.

The use of the thin flexible plastic or metal sheets prevents the GDE or GDM from sticking to the flat metal plate. If sticking happens, the GDE or GDM may loosen or detach from the PEM when the flat metal plate is removed. Under the hot pressing condition, the flexible plastic or metal sheet should be stable and should not stick to either the flat metal plates or the GDM. If it sticks to the GDM, it should be peeled off carefully by pressing the GDM down slightly with one hand to prevent detachment of the GDM from the membrane.

Some researchers do not bond the GDE (or GDM) to the PEM (or CCM), and the contact between the two is achieved within the cell or the stack under its assembling compression. This author suggests that it is better to bind the GDE (or GDM) and the PEM (or CCM) together to achieve better performance and longer lifetime.

The use of the gaskets is to make the two GDEs or GDM align exactly with each other, and to prevent overcompression of the GDM during the hot bonding process. For example, if the thickness of the GDM is supposed to be compressed no more than 30% of its original thickness, the thickness of the gasket should not be 30% thinner than the GDM. Overcompression of the GDM may cause permanent damage to its structure and therefore affect its functionality.

1.4.7 Plate

A plate is for distributing reactants and products, transferring electrons between unit cells, conducting heat generated by the unit cells, separating unit cells, and offering mechanical support for the entire stack. Therefore, the plate must be able to conduct both electrons and heat (electrical and thermal conductivity should be higher than 10 S cm^{-1} and 10 W m^{-1} K^{-1}, respectively), possesses good mechanical strength (tensile and flexural strength should be larger than 40 and 60 MPa, respectively), impermeable to H_2 and O_2 (permeability less than 2×10^{-6} cm^3 cm^{-2} s^{-1}), and have flow channels on the surface. Since the fuel cell environment is corrosive, the plate also needs to be of high corrosion resistance (corrosion current density lower than 16 μA cm^{-2}). Potential materials for plates include metals and graphite. Except for the corrosion resistance, a metal meets all the other requirements. Also, a thin metal plate can be manufactured by stamping with all the needed features made in one step, increasing the manufacturing rate and lowing the cost. Widely investigated metal plates include stainless steel, aluminum, nickel, and titanium. However, in order to prevent a metal plate from corrosion, it has to be coated with a thin protective layer that has high corrosion resistance. Before coating, the metallic plate should be thoroughly cleaned, degreased, and polished. The protective layer can be made of noble metals (e.g., gold), metal nitrides (e.g., CrN), and graphite, which increases the cost and manufacturing time. A single coating layer may contain pinholes and thus multiple coating layers with alternating and different materials are probably needed. In addition, the protective coating may locally get loose or detached from the underneath metal during use to gradually lose its protective function. A US company called TreadStone coats metallic plate by first growing evenly distributed conductive dots (e.g., gold) on its surface, followed by coating the entire plate surface with cheap corrosion-resistant material (e.g., polymers). Since the entire coating layer has many conductive dots evenly distributed within the polymeric material, the electrical conduction between the plate and the GDM is achieved. Such a coating methodology significantly reduces the cost of coated metallic plates.

Currently, graphite is commonly used for making plates due to its good corrosion resistance. Molded plates cost much less than machined plates, and both thermoplastic resins such as poly(vinylidene fluoride) and thermosetting resins such as epoxy, phenolics, and vinyl esters can be used as the binding agent for plate molding. A curing agent such as diaminodiphenylsulphone may also be added. The molding process is much faster when a thermosetting resin is used because the resulting plate can be removed from the mold without waiting for cooling. But a graphite plate has lower electrical and thermal conductivities, lower mechanical strength, and higher porosity

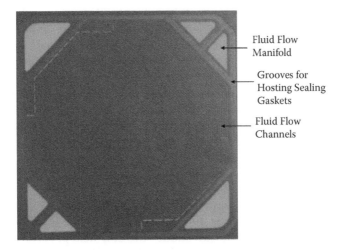

Fluid Flow
Manifold

Grooves for
Hosting Sealing
Gaskets

Fluid Flow
Channels

FIGURE 1.10
Picture of a molded graphite plate. Courtesy of Zhejiang Shentong Hydrogen Fuel Cell
Technology Company.

and specific thermal capacity than a metal plate. Due to the lower mechanical
strength and higher porosity, a graphite plate has to be made much thicker
than a metal plate, which makes the plate account for about 80% of the total
volume and mass of the entire stack. Because of the higher specific thermal capacity and mass, it is difficult to start a stack using graphite plates at
subzero temperatures. For vehicle applications that may experience ambient
temperatures as low as −40°C, such a stack will not be able to start unless it
is heated. The flow fields on the graphite plates were exclusively machined in
the early days and machining is still used by many manufacturers. Due to the
long time needed to machine a plate, machining significantly increases the
cost of the plate. Machined plates are the most costly components in a stack,
even though graphite material itself is much cheaper than Pt and PFSAs.
Molded graphite plates are much cheaper than machined plates, and thus
the molding technology is under vigorous development. Figure 1.10 shows a
molded graphite plate. The channels in the middle are for the flow and distribution of a fluid and the openings on the corners are for leading the fluid
to or away from the plate, and they form the manifolds within a stack. For
the three pairs of openings shown in Figure 1.10, one pair is for the entrance
and exit of H_2, another pair is for the entrance and exit of liquid coolant, and
the third pair is for the entrance and exit of air. The wider grooves along the
peripherals of the plate are for hosting sealing gaskets.

Typically a plate should be made as impermeable as possible to prevent
crossover of fluids. However, UTC Power creatively uses porous water trans-
portable plate achieving superior performance, especially in the high current
density region, due to its unique ability to manage liquid water. The plate
has hydrophilic pores with a diameter of ca. 1 μm and its bulk porosity is

about 30%. Excess liquid water will be taken by those pores so that flooding due to liquid water accumulation is avoided. At the same time, water in those pores humidifies and saturates the inlet dry gaseous reactants all over the active area of the electrodes. The liquid water in the pores forms a wet seal so that the gaseous reactants will not be able to get into the liquid coolant stream though the plates. Besides careful selection of the size of the hydrophilic pores, the pressure of the gaseous reactants should be slightly higher (e.g., 0.1 bar) than that of the liquid coolant to assure that liquid water flows from the GDM to the plate from where excess liquid water is removed.

1.4.8 Single Cell

A cell fixture that contains only one MEA is called a *single cell*. It sequentially consists of an anode-side end plate, an anode-side insulating gasket, an anode-side current collector, an anode-side flow-field plate, an anode-side sealing gasket, an MEA, a cathode-side sealing gasket, a cathode-side flow-field plate, a cathode-side current collector, a cathode-side gasket, and a cathode-side end plate. Sometimes the current collector and the plate are made into one piece. The insulating gaskets are used to insulate the end plates from the current collectors. The sealing gaskets are used to prevent the reactants from leaking out (or ambient air leaking in) between the flow-field plates and the GDM from the cell peripherals. The sealing gaskets should have good compressibility, strength, elasticity, and chemical and physical stabilities. They will be compressed to seal the gap between the flow-field plates and the GDM when the single cell is tightened, normally using tie-rods and screws. One or several springs or nonflat washers should be placed under the screws because the cell may expand during operation and shrink when cooled. The thickness of the sealing gaskets is chosen such that when compressed, the flow-field plates and the GDM are in good electrical contact, and the GDM is not overcompressed. Overcompression will cause damage (or even crashing) to the GDM, while undercompression will result in higher contact resistance between the flow-field plates and the GDM. The best compression must be determined through experiments. A rough guide is to reduce the GDM thickness by about 25%. The sealing gasket should maintain its compressibility, strength, and elasticity during the operation of the single cell; otherwise, if it relaxes by losing its strength or elasticity, the reactants will leak out (or ambient air will leak in) from between the flow-field plates and the GDM. The inner peripherals of the sealing gaskets are likely to come in contact with the reactants, so they should be chemically inert against the reactants and physically stable against the hot and humid inner cell environment. The sealing gasket can be an O-ring or a flat band. The former needs to be placed in grooves made on the surface of the flow-field plate in the area surrounding the flow channels, while the latter is normally placed on the flat surface of the flow-field plate in the area surrounding the flow channels.

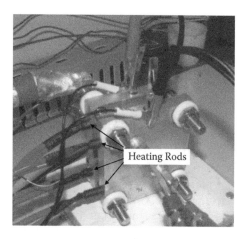

FIGURE 1.11
Picture of a single cell test fixture. Courtesy of Dalian Institute of Chemical Physics, Chinese Academy of Sciences.

In order to control the single cell temperature, heating and/or cooling means are needed. One simple heating method is to place heating rods within the flow-field plates. The plates should be thick enough to allow a hole drilled to host the heating rod. The hole is parallel to the plate surface and is normally along the center position of the plate. If two holes are drilled on one plate, they should be kept at a suitable distance in order to achieve even heating as much as possible. Figure 1.11 shows a single cell test fixture with two heating rods in each plate.

Another simple heating method is to place a heating pad on the outer surface of each of the flow-field plates. Because the pad can cover a larger area than a rod, it can provide a more uniform temperature distribution across the MEA.

A more complex heating method is to allow a liquid coolant to flow within the flow-field plates. This requires making paths within the plates for the coolant to flow and using a pump to circulate the coolant. This method can achieve the most uniform temperature distribution across the entire MEA. This method also allows quick cooling of the cell by circulating colder coolant.

No matter what heating method is used, a thermal couple needs to be placed within the flow-field plates (not the end plates); it is better to place the thermal couple as close as possible to the GDM in order to estimate the temperature of the MEA more accurately. If the purpose is to evaluate the temperature distribution across the entire MEA, multiple thermal couples that are placed at different positions will be needed. The thermal couples can be placed in the anode flow-field plate, the cathode flow-field plate, or both.

The active area of a single cell can range from a few square centimeters to several hundred square centimeters, depending on the purpose. If the purpose is to evaluate the performance of an MEA, the active area is generally 5,

10, 25, or 50 cm². If the purpose is to evaluate the flow-field design, the active area should be the same as that in actual use.

1.4.9 Stack

In order to meet the voltage and the power requirements of end users, many MEAs are typically stacked together along with other necessary components, such as the plates and the sealing gaskets, to form a structure called a *stack*. Apparently, the structure of a stack is similar to a single cell, but it contains more than one MEA. Each MEA along its neighboring anode flow-field plate and cathode flow-field plate (and possibly a coolant plate) can be called a *unit cell*, to distinguish it from a single cell. Figure 1.12 illustrates a two-cell stack.

It takes more care to assemble a stack than a single cell, and it is important to avoid misalignment of all the components such as the MEAs, the plates, and the sealing gaskets. Assembly normally starts by placing an end plate (anode side or cathode side) on a flat surface, followed by placing several vertical guiding pins (or rods) onto the end plate. Holes corresponding to those guiding pins are made on the MEAs, the flow-field plates, the coolant plates, and possibly the sealing gaskets when they are manufactured, and they are stacked together in the right order using the guiding pins. After the other end plate is finally put in position, the stack will be tightened at a pressure of around 15 bars. Then the guiding pins can be removed.

A stack can contain up to several hundred MEAs, generating a lot of heat. If the average unit cell voltage is 0.63 V, the heat generation rate of a stack will be close to its electrical power. The heat must be effectively removed to

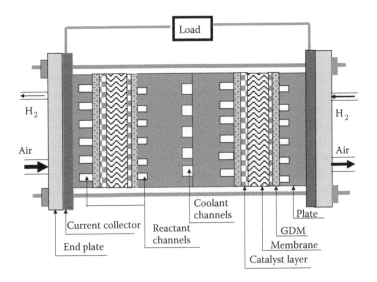

FIGURE 1.12
Illustration of a two-cell stack.

prevent the MEAs from overheating. Overheating will shorten the lifetime of the PEM because a regular PEM should not be used at a temperature higher than about 80°C, and possibly cause the MEA to dry up, which will affect both its performance and lifetime. Either liquid or air is used to remove heat, and correspondingly, the stack is called a *liquid-cooled* or *air-cooled* stack. Since the heat capacity of liquid is many times higher than that of air, liquid cooling is more effective, but the corresponding stack and the fuel cell system become more complicated. The specific heat capacity of water and air are 4.2 and 1.0 kJ kg^{-1} K^{-1}, respectively, and the air volume is about 1244 times of that of liquid water for the same mass. So, liquid water can dissipate over 5200 times of heat than air at the same volumetric flow rate.

The liquid is typically water based, either pure water or a mixture of water with an antifreeze such as ethylene glycol and propylene glycol. A mixed coolant will not freeze at a temperature higher than about −40°C, which is suitable for a fuel cell being used in a cold environment. Within the stack, plates with channels for the coolant to flow are needed, and those plates are called *coolant plates*. It is best for each MEA to share a coolant plate. Sometimes, two or more MEAs share one coolant plate to minimize the volume of the entire stack, but the MEAs that are not in direct contact with the coolant plate will have a higher temperature than the ones that are in direct contact with the coolant plate, resulting in uneven temperature distribution among MEAs, which may make water management more difficult.

Since all the unit cells within a stack are connected via the liquid coolant, it is important that the coolant is ion-free; otherwise, although weakly, the unit cells are short-circuited by the coolant, resulting in internal power loss. Unavoidably, ionic species from the plates or metallic piping will gradually accumulate within the initially ion-free coolant; thus, it needs to be completely replaced when its ionic conductivity reaches an unacceptably high value. Alternatively, an ion-exchange device can be placed along the coolant flow path, and this device is replaced or regenerated when its ion-exchange capability becomes inadequate.

Air cooling passes air through the stack. This can be done in two ways. One way combines the cooling air with the reaction air, and passes the combined air through the air channels that are open to the environment. The other way is to separate the reaction air from the cooling air. The reaction air passes through the open channels as described above, but the cooling air passes along the flow paths made on the cooling plates, which is similar to a liquid-cooled stack.

For the configuration with combined reaction and cooling air, air will provide oxygen for the cathode reaction as well as for taking away some heat from the stack. When the air flow rate is high enough, typically more than 15 times the stoichiometric ratio (depending on the stack efficiency and the temperature difference between the stack and the environment), it can control the stack at a desired temperature range. In such a circumstance, the air flow rate is determined by the stack cooling need instead of the reaction need. A fan is used instead of an air blower or an air compressor to provide air to the stack. Since a fan consumes much less power and is quieter than either a blower or a compressor, the parasitic power loss and the noise level will be lower, favoring higher system efficiency and quieter operation. However, since a fan has a much lower air pressure boost capability, the air channels must be short and wide enough. Short (e.g., ~10 cm) and wide channels (e.g., ~2 mm height × 2 mm wide) also help heat removal. The fans can be mounted on the stack or separated from the stack. In the former configuration, it normally requires more than one fan to cover all the air flow channels of the stack. In the latter configuration, the fan(s) needs to be at a suitable distance from the stack in order for every air channel to receive enough air; also, any components placed between the fan(s) and the stack should not obstruct the air flow to the stack. In both cases, the fan(s) should withdraw air through the stack, not blow air into the stack, in order for air to pass through the channels of the stack effectively.

For the configuration with separate reaction and cooling air, the reaction air is supplied to the stack using fans that are similar to the above description. But since the air is for the cathode reaction instead of stack cooling, its flow rate is determined by the reaction needs, and around three times the stoichiometric ratio may be enough. The cooling air requires an air blower or an air compressor to be sent into the stack through the channels on the cooling plates. Since an air blower or an air compressor can send more air into the stack, this configuration should be able to handle a higher environmental temperature better than the configuration with combined reaction and cooling air, but the system becomes more complicated and the parasitic power loss will be higher.

With either air or liquid cooling, the temperature will be higher near the coolant outlet than near the coolant inlet on each plate because heat accumulates from the inlet to the outlet along the cooling channels. This helps the removal of water from the stack in vapor form. The coolant inlet-to-outlet temperature difference is an important factor to control for liquid-cooled stacks, and is mainly controlled by the coolant flow rate and the heat exchanger used. Higher coolant flow rates and lower heat dissipation by the heat exchanger make the temperature difference smaller. If the difference is not high enough, the stack will be flooded because not enough water is removed; but if the difference is too high, the stack will dry up because too much water is removed. The optimal temperature difference depends on the stack design and the stack operating conditions. For a single stack, the optimal temperature difference can be different at different stack output powers. As a rough guide, the optimal temperature difference is normally between 5°C and 10°C when the stack coolant outlet temperature is around 65°C, and should be determined through experiments.

Since a stack contains more than one unit cell, the temperature might be different among cells. Typically, the two outermost cells adjacent to the two end plates have a slightly lower temperature than the other cells, because the end plates may take away some heat from the two outermost cells due to their lower temperature. The stack design should take this into consideration to minimize the temperature difference between the end cells and the other cells. Methods that can be considered include well insulating the outermost plates from the end plates, making the coolant flow channel geometry of the two outermost cells different from that of the other cells, and covering the entire stack with insulating materials.

It is crucial for the fluid (both the reactants and the coolant) to pass through every cell evenly. For example, if less air passes through some cells (than the other cells), those cells may experience lower performance. Even worse, some cells are flooded because they receive less air and other cells are dried up because they receive too much air. With time, the situation can get worse. Cells that are flooded due to flooding of the flow-field channels offer higher resistance to the flow of the reactant, causing the cells to receive even less reactant, which in turn are more flooded. These two factors reinforce each other and can cause the cells to lose their function quickly. Similarly, drier cells may receive more and more reactant, which will make them even drier with time. If this happens, it is impossible to keep the entire stack in optimal operating condition, leading to shortened stack life. To minimize the likelihood of this situation, components such as the catalyst layers, the GDM, the MEAs, and the plates should be identical, and after the reacting or cooling fluid gets into the stack, it should be evenly distributed among all the cells.

In order to minimize the nonuniform distribution of a fluid among cells, the fluid is generally sent to a common chamber within the stack first, from which the fluid is distributed into each cell at the same time. When MEAs, plates, and gaskets are made, separate openings for hydrogen, air, and coolant are made on them, with the dimensions of the openings being in accordance with the flow rate of each fluid that enters the stack. For example, since the air flow rate is several times that of hydrogen, the opening for air will be much larger than that for hydrogen. After those components are assembled to form a stack, all the openings for hydrogen align together to form a chamber called a *hydrogen manifold*. The same is true for the openings for air and for the coolant; the resulting chambers are called the *air manifold* and the *coolant manifold*, respectively. For each fluid, there is an inlet manifold and an outlet manifold. Holes and conduits for the flow of hydrogen, air, and coolant are made on the two end plates, and these holes and conduits are connected with the inlet and the outlet manifolds for each fluid. A fluid enters the stack from the inlet conduit on one end plate, fills the corresponding inlet manifold, distributes among all the cells, gathers in the corresponding outlet manifold, and finally exits the stack through the outlet conduit on the same or the other end plate. Of course, for air-cooled stacks, there are no manifolds for the reaction air.

Before a stack is put in use to generate power, a leak test should be performed. A leak can be external, that is, a fluid leaks out of (or into) the stack to (or from) the environment. A leak can also be internal, that is, a fluid leaks into chambers for other fluids; for example, hydrogen leaks into either the air chamber or the coolant chamber. Helium, nitrogen, and air can be used as the testing gas, but helium is best used to test the hydrogen chamber because its molecule size is small and similar to that of hydrogen. The following procedure can be used to test the hydrogen flow path: Let helium into the stack at a suitable pressure from the hydrogen inlet and out of the hydrogen outlet for a few minutes to replace the air initially present in the hydrogen path. Close the hydrogen outlet and fill all the space within the stack that is related to hydrogen, such as the H_2 inlet and outlet manifolds and the H_2 flow-field channels on the plates. After waiting for a few minutes, close the hydrogen inlet valve and monitor the pressure drop of helium in the stack. If there is no pressure drop at all, it means that the hydrogen flow path is perfectly gas tight. Very often, the pressure will drop with time because there will always be some leakage. According to the amount of pressure drop and the corresponding time period, the leak rate can be calculated based on the ideal gas law, $P_1V = n_1RT$, $P_2V = n_2RT$, where P_1/n_1 and P_2/n_2 are the pressure/moles of helium at the beginning and the end of the test, respectively. The leak rate is $(n_1 - n_2)/t$, where t is the test duration. As long as the leak rate is in an acceptable range, the hydrogen flow path can be considered as leak-proof. Different users may have different acceptable leak rate targets. As a rough guide, if the leak rate is less than 10 μL cm^{-2} min^{-1} corresponding to a H_2 leaking current density of 1.4 mA cm^{-2} ($2 \times 96485 \times 10$ μL $\times$ 1000 mA A^{-1}/(22.4 L mol^{-1} $\times$ 60 s min^{-1} $\times$ 10^6 μL L^{-1})), it should be good enough. During the above test, the air and the coolant inlet and outlet valves are kept open. The leakage test for the air flow path and for the coolant flow path can be carried out in the same way using air or nitrogen.

If the leak is not acceptable, determining whether it is an external leak or internal leak is necessary. Using the same test mentioned above, after both the air and coolant inlet valves and the outlet valves are closed, if the leaking result is still unacceptable, then it is an external leak for the hydrogen flow path. If the leaking result becomes acceptable, then it is an internal leak (i.e., helium leaks into either the air chamber or the coolant chamber or both). For the latter case, keep the air (or coolant) inlet and outlet valves closed, but open the coolant (or air) inlet and outlet valves. If the leak result is still unacceptable, it indicates that hydrogen is leaking into the coolant (or air) chamber.

Leakage can be due to inadequate sealing or component defects. If there are holes on an MEA, internal leakage across the MEA will happen (i.e., hydrogen gets to the air side and air gets to the hydrogen side). If there are cracks on a plate, a fluid can get to the other side of the plate, also resulting in internal leakage. If the sealing gasket between a plate and an MEA does not seal adequately, the reactant can leak out of the stack (external leakage). If the insulating gasket on the end plate does not seal properly, a fluid can leak from one manifold to the other manifold.

Here we share a real case involving manifold leakage. During the test of a stack, the hydrogen sensor in the lab was triggered due to hydrogen leakage. Two engineers spent several hours figuring out the origin of the hydrogen leak. They checked all the connections related to hydrogen flow path, from the first point of hydrogen entering the lab to the last point of hydrogen leaving the lab, but no leak point was found. They even thought that the hydrogen supply system outside the lab leaked and hydrogen somehow miraculously got into the lab. But they did not find a leak in the hydrogen supply system outside the lab either. When they disconnected the stack from the hydrogen loop, the hydrogen leakage stopped, which proved that the hydrogen leakage was related to the stack. They then checked to see if there was hydrogen leaking out of the stack between plates, but no leakage was found. Unexpectedly, after they used a long tube to send the air exhaust from the stack to outside the lab, the hydrogen sensor in the lab was not triggered anymore. To this point, it appeared that hydrogen leaked from the anode side to the cathode side. If the leakage was through the MEAs, all the hydrogen leaking to the cathode side should be oxidized quickly in the cathode catalyst layer during the operation of the stack, then there should have been no hydrogen getting into the cathode exhaust. Finally, it was concluded that hydrogen leaked from its inlet manifold directly to the air outlet manifold, and then got into the lab with the cathode exhaust. If the leak test had been performed more thoroughly and the acceptable leakage rate had been set more stringently, this kind of time-consuming process would have been avoided.

1.4.10 System

A stack is only useful when it is part of a system. The stack is a critical module, but it needs other modules to form a workable fuel cell system. Figure 1.13

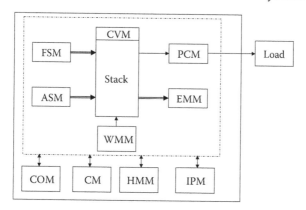

FIGURE 1.13
Functional block diagram of a fuel cell system.

shows a functional block diagram of a fuel cell system. The system consists of the following modules: cell voltage monitoring module (CVM), fuel supply module (FSM), air supply module (ASM), exhaust management module (EMM), heat management module (HMM), water management module (WMM), internal power supply module (IPM), power conditioning module (PCM), communications module (COM), and controls module (CM).

1.4.10.1 Cell Voltage Monitoring Module (CVM)

Since a stack contains more than one cell, and the performance uniformity among cells is not guaranteed, it is often necessary to monitor the voltage of every cell. The lowest performing cell is most troubling and receives the most attention. If the voltage of a cell drops below 0 V for whatever a reason, the cell may be damaged in seconds. Even if the average voltage of this cell is slightly above 0 V, the local voltage of some portions of this cell may have already dropped below 0 V, and those regions can be damaged quickly. Negative voltage is often due to an inadequate supply of hydrogen to the cell. When the hydrogen supply is lower than the stoichiometric ratio, some anode components will have to participate in the reaction as fuel. Since the catalyst layer is typically thinner than 20 microns, it will be oxidized away in a very short time. Further reaction will quickly damage the PEM due to high temperatures. Once a cell is damaged, the entire stack must be replaced. Continuing operation of a stack with damaged cells should be absolutely avoided. Since water can be oxidized to sustain quite high current densities at around 1.5 V (refer to Figure 3.4), the presence of water in the catalyst layer will reduce the extent of CL component oxidation. In other words, the damage to the carbon support and the catalyst will be less severe in the presence of water. Unfortunately, once water is consumed in a very short time (less than 1 minute), severe damage to the CL components will not be avoidable.

The CVM measures the voltage difference between a cathode flow-field plate and a neighboring anode flow-field plate separated by an MEA. Except for the two flow-field plates at the ends of the stack, all the other flow-field plates will give out the anode potential of one cell and the cathode potential of the neighboring cell. The challenge is making the measurement connections simple and sturdy. One way is to bury a wire in each flow-field plate during the plate manufacturing process. Another way is to inset a wire in a hole made on each flow-field plate and ensure that the wire cannot easily be removed or dislodged from the hole. A third way is to make the plates with an extra portion that allows multiple spring fingers to be inserted between every two neighboring plates. The spring force should be adequate enough to ensure that the fingers will not be easily detached from the plates, but not strong enough to cause cracking of the plates. A fourth way is to make each plate with an extra portion that allows a flat conducting bar or strip to be

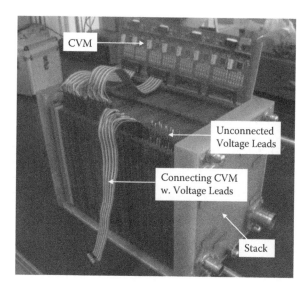

FIGURE 1.14
Picture of CVM. Courtesy of Shanghai Shenli High Tech. Company and Wuhan Intepower Fuel Cells Co., Ltd.

pressed on; neighboring bars or strips are equidistant from each other with the same distance as the neighboring plates. These bars or strips are insulated from each other. This method requires that the edges of the plates on this extra portion are on a straight surface; otherwise, some of the flat bars or strips may not be in good contact with the corresponding plates. Figure 1.14 shows a CVM that is partially connected to a stack, where metal wires are used to connect the CVM to the plates of the stack.

1.4.10.2 Fuel Supply Module (FSM)

A fuel cell is an energy conversion device, not an energy storage device. It is the stack that performs the function of converting chemical energy into electrical energy. As described earlier, the stack stores no fuel inside it. The fuel, such as H_2, is stored outside the stack and is fed to the stack when the fuel cell is required to generate power. All of the parts, from the fuel storage tank to the fuel inlet on the end plate of the stack, comprise a fuel supply module.

The FSM is simplest when pure H_2 is used, but it is still composed of several critical devices. Currently, the cheapest and most widely used H_2 storage tanks are industrial steel cylinders. These cylinders weigh about 54 kg and store 40 L of H_2 at a pressure slightly lower than 150 bars, typically around 125 bars. Each bottle stores about 0.37 kg of H_2. The opening–closing valve is located on top of the cylinder. Steel piping, along with several monitoring or controlling devices, is used between the outlet of the H_2 storage cylinder and the stack fuel inlet port.

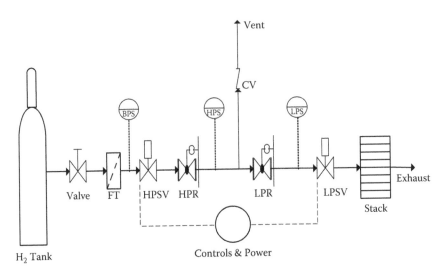

FIGURE 1.15
Scheme of H_2 supply module. (FT = filter, HPSV = high pressure solenoid valve, LPSV = low pressure solenoid valve, HPR = high pressure regulator, LPR = low pressure regulator, BPS = bulk pressure sensor, HPS = high pressure sensor, LPS = low pressure sensor, CV = check valve.)

When high-pressure lightweight cylinders are used, several additional devices are needed to safely handle the high pressure (350 or 700 bars). A scheme is shown in Figure 1.15. Sequentially, from the hydrogen cylinder to the stack, the devices are an on/off valve, a filter, a bulk H_2 pressure sensor, a high-pressure solenoid valve, a high-pressure regulator, a high-pressure sensor, a low-pressure regulator, a low-pressure sensor, and a low-pressure solenoid valve. There is a check valve between the high-pressure regulator and the low-pressure regulator for venting H_2 out when needed. The on/off valve is used to open or close the H_2 cylinder. The filter is used to remove any particulates potentially contained in the H_2 source. The bulk pressure sensor measures the H_2 pressure within the cylinder (initially at 350 or 700 bars) and sends the results to the control board so that the system knows how much H_2 is left within the cylinder and when to replace empty cylinders with full ones. When the bulk H_2 pressure drops to a preset lower value, the control board will instruct the high-pressure solenoid valve to remain closed. The H_2 in the cylinder should never be completely used up to prevent air from getting into it. If some air gets into the H_2 cylinder, the cylinder could explode when it is refilled with H_2. The high-pressure solenoid valve will open only when the fuel cell is instructed to generate power. The high-pressure regulator reduces the H_2 pressure from its bulk pressure to a preset much lower pressure (such as 40 bars), which is monitored by the high-pressure sensor. The low-pressure regulator further reduces the H_2 pressure to a preset low value (typically less than a few bars) that is suitable for the stack, and this pressure is monitored by the low-pressure sensor. If the low-pressure regulator fails to reduce the H_2 pressure to the preset value, the low pressure solenoid valve will not open in order to prevent the stack from being damaged by H_2 whose pressure is too high for the stack. There may be a venting valve between the high-pressure regulator and the low-pressure regulator. It will open to vent H_2 out in emergency situations, or if the H_2 pressure after the high-pressure regulator is somehow too high, or if the low-pressure regulator has problems. If the high-pressure regulator can reduce the cylinder pressure directly to a low pressure that is suitable for the stack, the low-pressure regulator can be eliminated. If H_2 needs to be humidified before getting into the stack, there will be a humidifier between the low-pressure solenoid valve and the stack.

After H_2 gets into the stack, it can be arranged to circulate within and through the stack in two ways. One way is called a *dead-ended operation*. A solenoid valve that is located at the H_2 exhaust port is closed for most of the time during the operation of the stack. But periodically, the solenoid valve is opened to vent the gas mixture out of the stack, and this process is called *purging*. The gas mixture is composed of H_2, water vapor, and nitrogen. Nitrogen comes from the cathode side by diffusing through the PEM during the operation of the stack. Water vapor and nitrogen gradually accumulate at the anode side and need to be vented out in order to avoid affecting the anode performance. Of course, some H_2 is also purged along with the water vapor and nitrogen, causing some waste of H_2. How often to purge (i.e., the purging frequency) depends on the stack quality and the operating condition, and should be determined through experiments. It is typically at a level of around 1 minute. The user can set up a voltage drop limit, and when the voltage of the stack drops to this limit, the fuel cell system will open the solenoid valve at the exhaust end to purge. The purging time period should be as short as possible to waste less H_2 and is normally between 0.1 and 0.5 s. During the no-purging time period, all the H_2 that gets into the stack is consumed, that is, the H_2 stoichiometric ratio is 1. The consumption of H_2 within the catalyst layer results in lower H_2 concentration and pressure than those within the GDM and the flow-field channels, and thus hydrogen diffuses into the catalyst layer due to the concentration or pressure gradient. For a dead-ended operation, there is no need to add a flow meter to monitor the H_2 flow rate. The H_2 flow rate during the no-purging time period will equal the H_2 consumption rate at the anode, assuming that there is no H_2 leakage and the loss of H_2 through the PEM is negligible. During the short purging time period, the inlet H_2 pressure will drop slightly (for a few psig), then quickly go back when the exhaust solenoid valve is closed. For a dead-ended operation, the H_2 is rarely humidified to avoid too frequent purging.

The other configuration allows H_2 to flow through the stack with the H_2 exhaust valve open for most of the time; this is called a *flow-through operation*. The H_2 flow rate is higher than the stoichiometric ratio and is typically in the range between 1.05 to 1.20, depending on the stack quality and the fuel cell operating conditions. Lower stoichiometric ratios may waste less H_2, but the fuel cell performance may be slightly lower, and it can run the risk that some cells may not get enough H_2. When a cell receives less than the stoichiometric amount of H_2, it is called *fuel starvation*, which will cause the cell to be damaged quickly. In order to maximize the use of H_2, the H_2 exhaust is often recirculated back to the H_2 inlet port. Since the pressure of the exhaust is much lower than that at the H_2 inlet, a pressure boost device such as a pump may be needed in the H_2 recirculation loop, making the FSM more complicated. An ejector is widely used for H_2 recirculation.

With a flow-through operation, water vapor and nitrogen also accumulate gradually within the anode, and periodic purging cannot be avoided. But the purging frequency is often much lower than that in the dead-ended operation.

If more than one H_2 cylinder is used, they can all be used together or in separate groups. If they are all used together, their outlet can be combined before the filter or the bulk H_2 pressure sensor, and the remaining portion is the same as described previously. In this configuration, all of the cylinders will be replaced altogether when they become empty. The disadvantage of this configuration is that the cylinders cannot be hot swapped. In other words, the fuel cell needs to be shut down during the change of the cylinders. The advantage of this configuration is that the fuel cell system can run for a longer time before cylinder replacement because all of the cylinders are used. When the cylinders are used in groups, such as in two groups, each group will need a bulk H_2 pressure sensor and a high-pressure solenoid valve, but the portion after the high-pressure solenoid valve will be the same as described previously. When the fuel cell is in use, one high-pressure solenoid valve will open while the other remains closed. When the bulk H_2 pressure sensor detects that this group of cylinders is empty, it will close the corresponding high-pressure solenoid valve, and at the same time, open the high-pressure solenoid valve for the other group of cylinders. The disadvantage of this configuration is that two bulk H_2 pressure sensors and two high-pressure solenoid valves will be needed, and the replacement of cylinders may be twice as often as the other configuration. The advantage of this configuration is that the empty group of H_2 cylinders can be replaced without interrupting the operation of the fuel cell system. Of course the replacement of any cylinder is possible for either configuration during any time as long as there is a switch between the cylinder and the main gas line.

If the fuel is not H_2 but a hydrocarbon-type fuel such as natural gas, propane, methanol, or ethanol, the FSM becomes very complicated. Briefly, the fuel needs to go through fuel processing in several steps, such as reforming, water–gas shift (WGS) reactions, and preferential oxidation to get a low CO content (<10 ppm) H_2-rich gas mixture called *reformate*, and the reformate is supplied to the stack. Some details are discussed in Chapter 3.

1.4.10.3 Air Supply Module (ASM)

The air supply module provides air for the fuel cell system, mainly for the cathode reaction. For air-cooled stacks with combined reaction and cooling air, the ASM is simply the fans. For air-cooled stacks with separate reaction and cooling air, the ASM consists of the fans for the reaction air and an air blower or an air compressor for the cooling air. For liquid-cooled stacks, the ASM provides the reaction air using an air blower or an air compressor. Since the ASM for liquid-cooled stacks is more complex than for air-cooled stacks, the following discussion will be about the ASM for liquid-cooled stacks.

The ASM typically consists of an air blower or an air compressor, a filter, a flow meter, a pressure sensor, a humidifier, and a temperature sensor, as shown in Figure 1.16. The blower or compressor sends ambient air with boosted pressure into the fuel cell system. A picture of an air blower is shown in Figure 1.17. The mesh filter (MFT) prevents large objects from

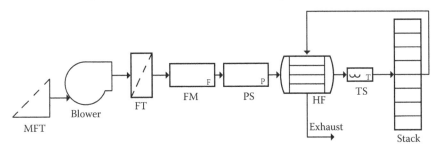

FIGURE 1.16
Schematic of air supply module. (MFT = mesh filter, FT = filter, FM = flow meter, PS = pressure sensor, HF = humidifier, TS = temperature sensor.)

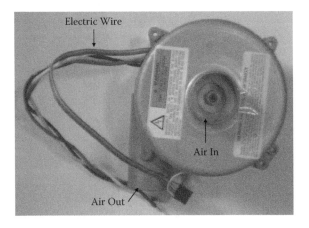

FIGURE 1.17
Picture of a blower.

From
Stack

To
Stack

Ambient
Air

Cathode
Exhaust

FIGURE 1.18
Picture of humidifiers.

getting into the blower. The filter (FT) after the blower removes particulates from air. By incorporating some suitable materials (e.g., KOH), the filter may also function to remove some gases such as SO_2, NO_x, and hydrocarbons that can cause poisoning of the cathode catalysts. These gases are likely to exist in the exhaust from vehicles using international combustion engines. Parameters such as capacity, effectiveness, and pressure drop should be considered in designing the filter. The flow meter measures the air flow rate, and the pressure sensor shows the air pressure. The humidifier is used to humidify the inlet dry air. The tubular humidifiers shown in Figure 1.18 are easy to use. They are composed of hundreds to thousands of thin tubules made of materials such as Nafion that allow water to transport through the tubule walls. The diameter of the tubules is around 1.5 mm and the wall thickness is ca. 0.2~0.3 mm. The transfer of water and heat is more effective with thinner tubule walls. Inlet dry air passes through the inside (or outside) of the tubules, while air exhaust from the stack passes through the outside (or inside) of the tubules. Since the air exhaust from the stack is hot and humid, some water and heat will transport through the tubule wall to the inlet dry air to make it achieve a humidification level of around 75%. The temperature sensor measures the inlet gas temperature before getting into the stack. Depending on the system design, some devices such as the flow meter, the pressure sensor, the humidifier, and the temperature sensor may not be needed. The best system is one that eliminates all the measuring and sensing devices.

The ASM may also provide air for other purposes in accordance with the fuel cell system design. For example, it will provide the combustion air for the reformer whose temperature is typically a few hundred degrees Celsius (natural gas reformer around 900°C and methanol reformer around 250°C). The ASM may also supply air for the preferential oxidizer to lower the CO concentration to a few ppm levels within the reformate through the following selective reaction:

$$2CO + O_2 = 2CO_2 \qquad \text{(R. 1.25)}$$

In addition, air may also be needed in order to use the air bleed technique to minimize CO poisoning to the anode. A few percentage points of air are added into the fuel stream to oxidize CO according to R. 1.25. Some H_2 is also be oxidized in this process (e.g., 400 H_2 molecules per CO molecule (Gottesfeld 1988)), and therefore air bleeding slightly reduces the fuel utilization (~2% for 50 ppm CO). An air blower has a lower power requirement, noise level, weight, and volume than an air compressor, so it should be chosen over an air compressor. However, an air blower offers a lower air pressure boost capacity, and therefore requires the stack to have a lower air pressure drop. Straighter, shorter, and wider channels will offer a lower pressure drop.

The air pressure boost ability of an air blower or an air compressor is related to the air flow rate: the higher the air flow rate, the smaller the pressure boost achievable from the same blower or at the same input power. When a stack operates at the highest designed power output, the air flow rate will reach the maximum. The pressure boosted by the blower or compressor must be higher than the stack pressure drop at this air flow rate; otherwise, the blower or compressor is inadequate. The user should carefully study the pressure versus flow rate characteristics offered by the manufacturer of the blowers or compressors before a selection is made. A blower or a compressor offering higher pressure boost and higher flow rate will consume more power.

It is suggested that a user not operate a blower or a compressor near its upper limit in order to prevent it from being burned. Long-time use near the upper limit should definitely be avoided. Many blowers and compressors have temperature- or current-protection mechanisms. When the temperature or the current exceeds the preset limit, the blower or compressor will automatically shut down.

1.4.10.4 Exhaust Management Module (EMM)

An exhaust management module is mainly for managing the anode exhaust. For both the dead-ended and the flow-through H_2 operations, some H_2 will be vented out during the purging time period. Even though the purging only lasts for 100–500 ms, the amount of H_2 can be significant, especially for high-power output stacks. If the fuel cell system is used in a closed environment, some H_2 may accumulate inside the room if it is not ducted out of the room, imposing some safety risks. Therefore, it is best to destroy all the H_2 within the anode exhaust before being vented out to the environment. This is the responsibility of the EMM.

Since H_2 is reactive, it can be consumed through chemical reactions. A reactor is the major component of an EMM. In order to aid the chemical reaction of H_2, some noble metal catalysts can be used. A good reactor should allow the H_2 to react quickly and thoroughly at a reasonably low temperature. If the reaction temperature is too high, the materials cost will be high, and the lifetime of the reactor can be shortened. The purging frequency can be as short as 10 s, which demands that the reactor should be able to offer a complete H_2 reaction in about 10 s.

Without air, H_2 will not react. So, air must be supplied to the reactor. The air can come from the cathode exhaust or the environment. The reaction rate also depends on the H_2–to-air ratio, and the optimal ratio should be determined beforehand in real situations. If only the cathode exhaust is sent to the reactor, the amount of air may be either too high or too low, leading to an ineffective reaction. If it is too high, only a portion of the air exhaust should be sent to the reactor; if it is too low, extra air should be sent to the reactor by using fans. If the air exhaust (and the purged H_2) has a high vapor concentration, it may make the chemical reaction between H_2 and air difficult. In such a circumstance, the vapor should be removed from the exhaust(s) before being sent to the reactor. Also, H_2 should mix well enough with air for a quick and thorough reaction.

The reaction rate will be in proportion to the total surface area of the catalysts in the reactor. The total surface area is very likely to become smaller during the use of the reactor because catalyst particles tend to aggregate, especially at high temperatures. It may be better to use porous ceramic with the catalyst deposited onto the surface of the pores as a thin layer with a thickness of a few atoms.

The reactor should be thoroughly tested under various fuel cell operation conditions to make sure that it will function properly. Figure 1.19 shows a catalytic reactor developed by Beijing Jonton Hydrogen Tech. Company. It consists of two sections: The bottom section weighs 3.6 kg and is for hosting a check valve (to prevent the anode exhaust from getting back into the stack), a gas–water separator (to remove vapor from the anode exhaust), a power source (to power a fan mounted on the top section), and controls. The top section weighs 3.7 kg and is for hosting the reactor, a fan for withdrawing air into the reactor to achieve a proper H_2-to-O_2 ratio, and an exhaust port. Multiple top sections can be used when the amount of H_2 being purged out is high by laying two or more top sections on top of each other. The reactor shown in Figure 1.19 contains two top sections. The catalytic reactor uses noble metal–coated metal mesh to catalyze the chemical reaction between H_2 and O_2. The reaction becomes fast enough when the temperature rises to a few hundred degrees Celsius. Tests show that when the amount of H_2 purged out was 40 L min^{-1}, the concentration of H_2 emitted to the environment was 40 ppm when one top section was used and became undetectable when two top sections were used. Tests also showed that the reactor imposed little resistance to the flow of the anode exhaust.

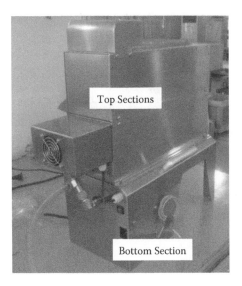

FIGURE 1.19
Picture of a catalytic reactor for anode exhaust management. Courtesy of Beijing Jonton Hydrogen Tech. Company.

1.4.10.5 Heat Management Module (HMM)

A heat management module is used to dissipate heat from the fuel cell stack and some other components. The heat generated by the stack is quite signifi-cant. Normally, the average cell voltage of a PEMFC is lower than 0.70 V, which means that at least 44% (0.55 V/1.25 V) of the chemical energy within H_2 and air is converted to heat. A heat exchanger is typically used to radiate the heat out of the fuel cell system. Figure 1.20 is a picture of a heat exchanger. Of the 56% electrical energy, about 10% of it is turned into heat by the DC–DC converter, and another ~10% consumed for the system parasitic power needs also turns into heat. Then, the total heat generated from a fuel cell system is slightly more than 55% of the total chemical energy (44% + 56% × 10% + 56% × 10% = 55.2%).

Controlling the coolant temperatures at the points of entering and leaving the stack is the major function of the HMM. As described earlier, if the tem-peratures are too high, the fuel cell lifetime will be shortened. If the tempera-tures are higher than about 90°C, the stack may not be able to work because the PEM will be too dry at such temperatures. For air-cooled stacks with combined reaction and cooling air, the heat is removed by fans, which is the simplest HMM. For air-cooled stacks with separate reaction and cooling air, the flow rate of the cooling air is driven by a blower or a compressor.

For liquid-cooled stacks, the coolant flow rate and the heat exchanger together control the coolant temperatures at the stack inlet and outlet. Figure 1.21 illustrates a coolant loop. A pump is the driving force that circulates the liq-uid coolant in the loop. It sends coolant from the coolant tank into the stack

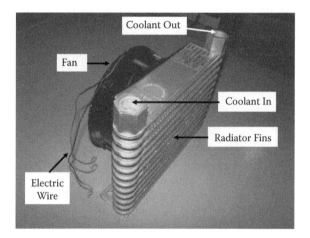

FIGURE 1.20
Picture of a heat exchanger.

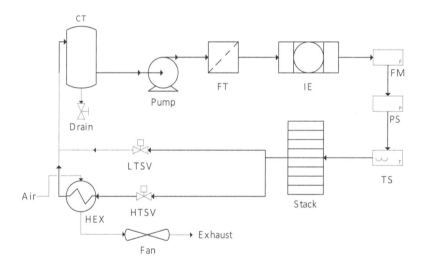

FIGURE 1.21
Schematic of a coolant loop. (CT = coolant tank, FT = filter, IE = ion exchanger, FM = flow meter, PS = pressure sensor, TS = temperature sensor, LTSV = low-temperature solenoid valve, HTSV = high-temperature solenoid valve, HEX = heat exchanger.)

through a filter (to remove particulates) and an ion exchanger (to remove ionic species). The flow rate, the pressure, and the temperature of the coolant can be monitored by a flow meter, a pressure sensor, and a temperature sensor, respectively. When the liquid coolant passes through the stack along the flow channels on the coolant plates, it removes some heat from the stack. Then the hotter coolant flows through a heat exchanger to become cooler before entering the coolant tank. (Please note that some designs do not send the coolant into the coolant tank.) The metallic radiating fins of the heat exchanger dissipate some heat and the fan mounted on the heat exchanger sends the heat out of the system. The heat dissipated by the heat exchanger can be simply calculated by using the difference between the coolant temperatures at the point of leaving the heat exchanger and at the point of entering the heat exchanger, the coolant flow rate, and the coolant heat capacity. When more heat needs to be dissipated, the fan will spin faster. If the fan is already at the maximum spin speed and the coolant inlet or outlet temperatures are still higher than desired, the heat exchanger is inadequate, and a larger one is required. A larger one can be one with a higher radiating fin surface area, or one with a faster fan, or both. For heat exchangers with the same total geometry (or dimension), the one with the larger total radiating fin surface area and faster fan can dissipate more heat. The fan should be mounted on the heat exchanger in such a way that it withdraws air from the radiating fins and sends the hot air out of the fuel cell system directly.

As shown in Figure 1.21, there is a draining valve at the bottom of the coolant tank for draining the coolant when needed. Typically, the amount of coolant will decrease with use, and fresh coolant needs to be added to the coolant tank when the coolant level drops to a preset value. If possible, a flow meter and a pressure sensor should be avoided in order to lower the cost. Also, their presence will increase the resistance to the flow of the coolant, leading to an increase in parasitic power losses.

Figure 1.21 shows two coolant paths around the heat exchanger. In order to facilitate the startup rate of the stack, the coolant leaving the stack bypasses the heat exchanger via one solenoid valve (LTSV). After the temperature of the coolant reaches a preset value, this solenoid valve is closed and the other solenoid valve (HTSV) is opened to direct the coolant through the heat exchanger.

Although we are accustomed to using the term *stack temperature*, it is not accurate because the stack temperature is not uniform. The stack temperature at the coolant *in* side will be lower than that at the coolant *out* side, and the temperatures in between the two sides change gradually. The temperatures at the coolant *in* side, the coolant *out* side, and their difference are all important. If the temperatures are higher than the desired values, the MEA lifetime can be shortened. If the difference between the two temperatures is too small, the stack can be flooded, while if it is too high, the stack can be dried up. More heat dissipation by the heat exchanger will lower both the stack coolant *in* side and the coolant *out* side temperatures. The coolant flow rate can more effectively impact the temperature difference, with a faster coolant flow rate leading to a smaller temperature difference.

Besides the stack, there are other components within a fuel cell system that may need heat removal. These components include the DC–DC converter, the control boards, the diodes, and some electrical or electronic components. About 10% of the electrical power from the stack turns into heat in the DC–DC converting process, and this heat needs to be dissipated; otherwise, the DC–DC converter will be overheated. Normally, fans are used to send the heat out of the DC–DC converter. A fuel cell system may have several control boards, and they generate some heat when in operation. If the heat generated is small, metallic radiating fins can be enough; if the heat generated is large, metallic radiating fins along with fans should be used. If diodes are used in the fuel cell system to prevent the current from flowing in a wrong direction, the heat generated by the diodes should also be removed using metallic radiating fins, or fans, or both.

Another major function of the HMM is to supply heat to maintain the temperature of certain components and modules. In order to prevent water from freezing within the stack at subzero temperatures while the fuel cell system is in the idling state, a simple approach is to warm up the liquid coolant to a preset temperature higher than 0°C and allow it to circulate through the stack. The needed energy can come from several sources, such as the internal power supply module, external grid power, and through burning of the fuel. Since the power output of a battery is severely affected by subzero temperatures, the battery bank used as the internal power supply may also need to be kept above 0°C in order for it to be able to supply the needed power during the startup of the fuel cell system.

For a fuel cell system incorporating fuel (e.g., CH_4) processing the HMM becomes much more complicated. The temperatures of the hydrodesulfurization unit, the reformer, the water-gas shift reactors, the preferential oxidizer, and the stack are all need to be properly controlled and maintained. There will be a heat exchanger between every two adjacent units mentioned above. For steam reforming, heat is needed to generate vapor from liquid water for the steam reforming reaction and the subsequent water-gas shift reactions. The conversion of liquid water to steam is typically carried out in one or more of the heat exchangers mentioned above.

1.4.10.6 Water Management Module (WMM)

There are several modules within a fuel cell system that need water. The function of a water management module (WMM) is to meet those needs. If all the water need is satisfied by the water produced by the stack, a water-neutral situation is achieved. Under such a situation, no external water needs to be carried by the fuel cell system, which in turn lowers the operational cost.

The WMM first humidifies the gaseous reactants, either air, or H_2, or both. It is achieved by allowing the incoming drier gas to pass through a humidifier.

For a fuel cell system that incorporates steam reforming, water needs to be provided to the steam reformer and the water-gas shift reactors. The best water source is that produced by the stack. Through a water-gas separation technique the water within the cathode exhaust is collected and stored as liquid. If the water is more than the fuel cell need, the excess water is auto-matically sent out of the fuel cell system. However, if the water is not enough the water produced by the catalytic burner and the preferential oxidizer should also be collected, or water is supplied from external sources.

Some liquid coolant is slowly and gradually lost when it circulates through the stack because the graphite plates are not completely impermeable to water. The collected liquid water from the cathode exhaust can be used to refill the coolant tank if water is used as the coolant.

1.4.10.7 Internal Power Supply Module (IPM)

An internal power supply module provides the power needed by certain components within a fuel cell system. The components include sensors, con-trol boards, pumps, fans, blowers, compressors, solenoid valves, contactors, switches, and so on. The IPM also provides the power to start the fuel cell system and helps carry some load when the fuel cell stack is inadequate to handle a sudden load jump. There are many types of sensors in a fuel cell system, such as the H_2 concentration sensors, the H_2 pressure sensors, the fluid flow rate sensors, the coolant-level sensors, the temperature sensors, the current sensors, the voltage sensors, the door-open sensors, the vibration sensors, and the flooding sensors. These sensors monitor the correspond-ing parameters to indicate the situation of the entire fuel cell system. The control boards may include a main board for controlling the system and sev-eral sub-boards for controlling various modules discussed in this chapter. Pumps, fans, blowers, compressors, solenoid valves, contactors, and switches all require power to perform the corresponding functions.

The power of the IPM may arise from three sources. One source is batteries and supercapacitors, which are typically integrated within the fuel cell sys-tem. When the fuel cell system is in the idling state, this source will provide all of the fuel cell component power needs, such as the sensors, the control boards, the valves, and the switches. A second source is the stack itself. This source only provides power when the fuel cell system is in operation. After the DC–DC conversion, the majority of the converted power is used by the load, and a small portion of the converted power is used by some fuel cell components. Depending on the fuel cell system design, a third portion of the converted power may be used to recharge the batteries or the supercapaci-tors within the system. A third source is the grid power. There is typically accessibility to the grid power for a backup power fuel cell system. During the idling of the fuel cell system, the grid power can be used to charge the batteries and the supercapacitors within the system. It can also be used to keep the stack and the batteries above 0°C on cold winter days.

The batteries and/or supercapacitors should be able to provide enough power for both the load and the fuel cell system's internal power needs at the moment that the grid power is down. This requires them to have enough discharging capacity and discharging rate. The capacity of a battery is expressed by ampere-hour (Ah) and is labeled on the battery pack by the battery manufacturer. But the Ah is normally measured by the battery manufacturer when the battery discharges at a low discharging rate. For example, for Pb-acid batteries, the discharging rate during the factory test is normally 0.05 C, which means that for 100-Ah batteries, the discharging current during the test is 5 A and the discharging lasts for 20 hours at room temperature before the battery voltage drops to the low cutoff value. At higher discharging rates, the actual discharging capacity will be lower than 100 Ah. For the same Pb-acid batteries, the discharging capacity can drop to 60 Ah when the discharging rate is 1 C (i.e., 100 A). At the same time, at a lower temperature such as −10°C, the same battery may only have a discharging capacity of 20 Ah at 1 C. Another important parameter is the achievable discharging rate of the battery. Batteries with higher discharging rates are favored in order to reduce the battery size. In addition to Ni-Cd and Ni-MH$_x$ batteries, lithium-ion batteries can provide relatively high discharging rates, and a 2-C discharging rate can be achieved by many battery manufacturers. Some battery manufacturers claim 5-C or 10-C discharging rates, but the reality (e.g., safety and durability) needs to be confirmed by users.

The discharging capacity and the discharging rate of a battery pack must be able to meet the fuel cell system needs when the battery pack is in the worst conditions (e.g., the lowest temperature and the highest discharging rate). First, the power output of the battery must meet the combined need of the load and the fuel cell system's internal power. Second, the battery pack must be able to last for the power need duration that is targeted by the fuel cell manufacturer. For example, if the total load is 1 kW and the fuel cell manufacturer wants a 50-V Li-ion battery discharging at 2 C, we can estimate what the Ah of the battery pack should be and how long it could last. Giving out 1 kW at 50 V means that the discharging current of the battery is 20 A. This means that the battery pack's capacity must be 10 Ah at the lowest temperature when it discharges at a 2-C rate. This battery pack can last for 0.5 hours maximum (= 10 A × 1 h/20 A).

During the start-up of a fuel cell system, the batteries provide all the power needs only for a few seconds. Afterward, the fuel cell stack begins supplying some power, and the output power from the batteries reduces correspondingly. With time, the stack supplies more and more power and the batteries supply less and less power, and finally, all the power is supplied by the stack. The duration of the load-sharing process by the stack and the battery depends on the discharging capacity of the batteries, the output voltage setup of the DC–DC converter, and how fast the stack can reach its nominal power output. For an air-cooled stack, the stack can reach the nominal power output in about 1 minute. For a liquid-cooled stack, the stack can reach the nominal power output in less than 5–10 minutes. But this does not necessarily mean that the load will be fully powered by an air-cooled stack within 1 minute, because the battery will continue giving out power until its voltage becomes lower than the DC–DC output voltage setup of the fuel cell system (around 53 V for telecommunications backup power applications). A battery with smaller Ah and a higher discharging rate will get out of the load-sharing process faster. Of course, if the battery completely discharges but the stack still cannot give out enough power for all the load needs, the system fails and it will shut down.

Components such as sensors, control boards, pumps, fans, blowers, compressors, solenoid valves, and switches very often need DC power at different input voltages. Some need 5 V, some need 12 V, and some may need 24 or 48 V. If the needed DC voltage is around that from the fuel cell system's main DC–DC converter, the latter can supply the DC power to the components directly. Otherwise, the voltage from the main DC–DC converter has to be converted to the desired DC voltages by additional DC–DC converters. Those additional DC–DC converters are normally much smaller than the main DC–DC converter. The output voltage setup of the system's main DC–DC converter is based on the need of the external load.

1.4.10.8 Power Conditioning Module (PCM)

A power conditioning module converts the DC power output from the stack to the input power needed by the load. One conversion is from DC power to DC power through a DC–DC converter. This conversion is needed because the load typically requires a stable input voltage within a certain range, while the stack output power changes frequently and its output DC voltage may not be in the range required by the load. Another conversion is from DC power to AC power through a DC–AC inverter when the load needs AC power. Both conversion processes will waste some power, and thus it is important to make the converter or inverter with a high converting or inverting efficiency. A conversion efficiency of 90% is an achievable good number.

Although there are numerous DC–DC converters in the electronic and electrical markets, there are nearly no off-the-shelf DC–DC converters that are suitable for fuel cells with kW power outputs. The reasons are twofold—profitability and technical challenges. First, the fuel cells are not currently produced at large scale, which discourages DC–DC converter manufacturers from developing DC–DC converters for fuel cells for profit reasons. Some DC–DC converter manufacturers may agree to develop DC–DC converters for fuel cells with sizeable developmental fees paid by the fuel cell developers, and then only sell the converters to the developers who pay the developmental fees. Second, DC–DC converters for fuel cells have larger technical challenges than regular DC–DC converters with respect to the input power, current, and voltage ranges. For kW-level fuel cell systems, the current will be tens to hundreds of amperes, far higher than that of regular DC–DC converters. The manufacturers have to develop and evaluate new components for such high currents, and figure out ways to dissipate the heat. Also, a stack

FIGURE 1.22
Picture of a DC–DC converter. Courtesy of Shanghai Jiaotong University and Beihang University.

output voltage can theoretically range from 0 to the OCV. Although in reality the lower end voltage will be far higher than 0 V, it still can be as low as 1/3 to 1/2 of the OCV. For example, for a stack with 100 unit cells, the output voltage during operation can range from 40 to 100 V under different operating conditions. Therefore, the DC–DC converter must be able to handle a wide input voltage range. For the above stack, if the DC–DC output voltage is set at 53 V for telecommunications backup power applications, the DC–DC converter has to be able to change the voltage-changing direction in two ways, up from 40 to 53 V, and down from 100 to 53 V. In addition, the DC–DC converter in a fuel cell system for telecommunications backup power applications must meet the low electromagnetic ripple and electrical noise–level requirements at different frequencies. Figure 1.22 is a picture of a DC–DC converter jointly developed by Beihang University and Shanghai Jiaotong University for 5-kW fuel cell systems.

1.4.10.9 Communications Module (COM)

A communications module handles both internal communications and external communications. The communications are dominated by the system's main control board, and the central processing unit (CPU) is a key component on this board. The internal communications are between the main control board and all the subcontrol boards such as those for pumps, fans, blowers, and solenoid valves. The external communications are between the fuel cell system and the load as well as remote monitoring devices.

The communications are based on selected and mutually understandable agreements, so that the entities at the communications ends can understand each other; otherwise, communications cannot proceed.

Remote communications allow a distant user to know the situation of the fuel cell system through a monitoring device. If the monitoring device is a computer, the user can remotely control and operate the fuel cell system. Remote communications can be through wireless, telephone line, or Internet. Broadband Internet can transport a large amount of data at faster rates. Wireless connections will be the choice when there is no telephone line or Internet port at the location where the fuel cell system is used.

A fuel cell system often has interfaces for connecting to a computer or to other electronic monitoring or controlling devices, and at least one port is for connecting to a Universal Serial Bus (USB) for data transfer or other purposes.

1.4.10.10 Control Module (CM)

A CM is the "brain" of a fuel cell system and is responsible for controlling the entire fuel cell system during its idling and operating time periods. The CM is composed of software and hardware, and the software is embedded in the hardware.

The CM is crucial for a fuel cell system, and it reflects how well the system integrator understands the entire fuel cell system. Without a sound understanding of the fuel cell system and all its components, a good CM cannot be made. The important things that the CM should cover include when and how to start the fuel cell system, how to respond when abnormal events happen, what actions to take to avoid component damage, and when and how to shut down the fuel cell system.

The component that needs the most care is the stack, the "heart" of the system, because it is probably the most easily damaged component in the fuel cell system. For example, a stack can be damaged in seconds when there is H_2 starvation (think of H_2 for a fuel cell as oxygen for a human being). In order to protect the stack, the voltage of each unit cell in the stack is normally monitored. When the voltage of a unit cell drops to a low value (such as 0.4 V), which is set beforehand by the system integrator, the CM will take actions to try to restore the voltage of this unit cell to the normal value (i.e., near the average cell voltage). If one action is ineffective, other actions will be taken in sequence. All these actions are programmed within the CM beforehand. If none of the preprogrammed actions can change the situation significantly, the CM should shut down the system to prevent the stack from being damaged. An experienced integrator is able to not only foresee most of the potential problems, but also program the CM to take the right actions to correct those problems automatically. A picture of the front of a mother control board is shown in Figure 1.23; the black square in the middle is the CPU.

FIGURE 1.23

Picture of a mother control board. Courtesy of Wuhan Intepower Fuel Cells Co., Ltd.

1.5 Summary

A fuel cell converts chemical energy directly into electrical energy with high efficiency. Based on the electrolytes used, there are proton exchange membrane fuel cells (including those directly using liquid fuels such as methanol and formic acid), phosphoric acid fuel cells, alkaline fuel cells, molten carbonate fuel cells, and solid oxide fuel cells. Since a fuel cell has the combined advantages of batteries and heat engines and many other unbeatable advantages, it will become one of the most critical technologies in the distant future.

A PEMFC uses a solid membrane that conducts protons as the electrolyte. Since it can start at ambient temperatures instantly, it is ideal for backup, portable, and motive power applications. The most important technologies concern the stack (like the heart of a human being) and the system controls (like the brain of a human being). The key components in a stack include the catalyst, PEM, GDM, plates, and gasket, while the controls include the operation algorithm, software, and electronic circuits. A fuel cell also needs various auxiliary components such as fans, blowers, compressors, pumps, heat exchangers, humidifiers, converters, valves, sensors, and batteries to work. A fuel cell system involves multidisciplinary skills and knowledge, and therefore it requires a team effort to develop.

2

Thermodynamics and Kinetics

2.1 Theoretical Efficiency

Everyone wishes that heat energy could be 100% converted to electrical energy or mechanical energy. But in reality, a 100% conversion efficiency can hardly be achieved by any device no matter how hard we try or how perfect the design is. One reason concerns theoretical limitations, and other reasons are due to unavoidable heat loss in the conversion processes.

Consider heat engines such as the internal combustion engine (ICE) used in vehicles. An ICE changes the chemical energy of a fuel such as gasoline into heat energy through combustion with oxygen from air. The heat leads to volume expansion of the gases in the reaction chambers, which apply forces to the pistons to cause them to move. So, in the entire process, the chemical energy first changes to heat energy, which in turn changes to mechanical energy. The mechanical energy drives the wheels of a vehicle to move. How much heat energy is converted to mechanical energy? According to the second law of thermodynamics, that is, Carnot's Theorem, the theoretical efficiency of such a heat engine, which converts heat energy to mechanical energy, is

$$\eta_{ICE} = (T_1 - T_2)/T_1 \qquad \text{(Eq. 2.1)}$$

where T_1 and T_2 are the absolute temperatures of the reaction chamber and the exhaust from the reaction chamber, respectively. The absolute temperature (in Kelvin) is simply obtained by adding 273 to the temperature in degrees Celsius (°C): $T_K = T_C + 273$. If we assume that T_1 and T_2 are 2373 and 1173 K (i.e., 2100 and 900°C), respectively, the theoretical efficiency is 50.6% (1200 K/2373 K). In other words, maximally 50.6% of the chemical energy can be converted into mechanical energy by the ICE in this case. In practice, due to various energy losses such as mechanical transfer, friction and incomplete fuel combustion, the actual efficiency is much lower than 50.6% achievable by cars. As discussed in Chapter 6, only about 14% of the chemical energy is converted to mechanical energy based on the driving distance of a common car. If we want to convert the mechanical energy further into electrical energy, additional energy will be lost in the conversion process. Obviously, a heat engine is far from ideal for converting chemical energy into electrical energy.

It is not difficult to understand why not all the chemical energy is converted to mechanical energy. Only a portion of the heat generated by combustion is transformed into mechanical energy, and the remaining heat is contained within the exhaust and is thus wasted. Only when the temperature of the exhaust is at 0 K will the theoretical efficiency reach 100%, which is not realistic. In practice, the theoretical efficiency can be increased by increasing the combustion temperature (and lowering the exhaust temperature), but this imposes challenges to materials and engineering.

One advantage of a fuel cell is its high theoretical efficiency of conversion of chemical energy to electrical energy. The chemical energy of a material can be regarded as its *enthalpy of formation* ($\Delta_f H$). The enthalpy change of a reaction (ΔH) represents the total chemical energy change, that is, the total chemical energy released by the reaction. A fuel cell produces electrical energy, and when the electrons pass through a load, electrical work is done. What corresponds to work is the Gibbs free energy of formation ($\Delta_f G$), because it is defined as the energy available to do external work, excluding any work associated with the changes in volume and pressure. The Gibbs free energy change of a reaction (ΔG) represents the total work that the reaction can do. The ratio of ΔG to ΔH is therefore the theoretical efficiency of a fuel cell, as shown by Equation 2.2:

$$\eta_{FC} = \Delta G / \Delta H \qquad \text{(Eq. 2.2)}$$

Under the standard condition (i.e., 298 K and 1 bar), the enthalpy of formation of H_2, O_2, and H_2O (gas form) is 0, 0, and -241.8 kJ mol^{-1}, respectively; the Gibbs free energy of formation of H_2, O_2, and H_2O (gas form) is 0, 0, and -228.6 kJ mol^{-1}, respectively (Gurvich et al. 1994). Unless otherwise specified, throughout this book all the cited thermodynamic data are taken from Lide (1994) for consistency. So, for Reaction 2.1

$$H_2 + 0.5O_2 = H_2O \text{ (g)} \qquad \text{(R. 2.1)}$$

the enthalpy change and the Gibbs free energy change with the formation of 1 mole of H_2O (g) are, respectively,

$$\Delta H^\circ = \Delta_f H^\circ{}_{H_2O} - (\Delta_f H^\circ{}_{H_2} + 0.5\Delta_f H^\circ{}_{O_2}) = -241.8 \text{ (kJ mol}^{-1}) \qquad \text{(Eq. 2.3)}$$

and

$$\Delta G^\circ = \Delta_f G^\circ{}_{H_2O} - (\Delta_f G^\circ{}_{H_2} + 0.5\Delta_f G^\circ{}_{O_2}) = -228.6 \text{ (kJ mol}^{-1}) \qquad \text{(Eq. 2.4)}$$

Then the theoretical efficiency for this case is

$$\eta = (-228.6)/(-241.8) = 94.5\% \qquad \text{(Eq. 2.5)}$$

Under the same condition, but water is in the liquid form, as shown in Reaction 2.2,

$$H_2 + 0.5O_2 = H_2O \text{ (l)} \tag{R. 2.2}$$

the enthalpy and the Gibbs free energy changes are −285.8 and −237.1 kJ mol^{-1}, respectively. Then the theoretical efficiency is

$$\eta = (-237.1)/(-285.8) = 83\% \tag{Eq. 2.6}$$

Both the enthalpy and the Gibbs free energy changes of formation are affected by the temperature. The temperature should be indicated when the theoretical efficiency is calculated.

The difference between the gaseous and liquid forms of water in the enthalpy of formation is due to the energy needed to evaporate water from liquid form to vapor form, which is called the *enthalpy of vaporization*. It is 44 kJ mol^{-1} at 298 K. When water is in liquid form, the enthalpy change is called the *higher heating value* (HHV) because it releases more heat energy. When water is in the vapor form, the enthalpy change is called the *lower heating value* (LHV) because it releases less heat energy. Although a proton exchange membrane fuel cell (PEMFC) typically runs at temperatures lower than the boiling point of water (100°C), the LHV is often used to calculate the theoretical efficiency. This is not done to get a better-looking, higher efficiency number, but because the water formed at the stack cathode initially is in the vapor form, and most of it exits the stack in this form along with the other gases in the exhaust.

Besides using the enthalpy change of formation, the chemical energy change of a reaction can also be obtained by using the bond strength (also called the *bond energy*). Fundamentally, they are the same thing. At 298 K, the bond strength of O–O for O_2, H–H for H_2, and HO–H and O–H for H_2O (gas form) is 436, 498.4, 498 (±4), and 427.6 kJ mol^{-1}, respectively (Kerr 1994). Therefore, the chemical energy change for Equation 2.1 is

$$\Delta H^\circ = (D^\circ_{HO-H} + D^\circ_{O-H}) - (D^\circ_{H-H} + 0.5D^\circ_{O-O}) = -240.4 \ (\pm4) \text{ kJ mol}^{-1} \tag{Eq. 2.7}$$

which is basically the same as −241.8 kJ mol^{-1}, obtained by using the enthalpy change of formation within the experimental errors.

Thermodynamically, ΔH and ΔG follow the relationship

$$\Delta H = \Delta G + T\Delta S \tag{Eq. 2.8}$$

where ΔS is the entropy change of the reaction. Entropy represents the degree of disorder of a system, and a more highly chaotic system has higher entropy. For Reaction 2.1, 1.5 moles of gaseous molecules turn into 1 mole of gaseous molecules, so the system becomes more ordered, that is, the entropy of the product is smaller than that of the reactants, and thus ΔS is negative. Since ΔH and ΔG are negative for this reaction, ΔH is more negative than ΔG.

For a reaction with positive ΔS, ΔH will be less negative than ΔG, and thus the ratio of ΔG to ΔH will be higher than 1, that is, the theoretical efficiency will be higher than 100%. For example, for Reaction 2.3

$$CH_3OH \ (g) + 1.5O_2 = CO_2 + 2H_2O \ (g) \qquad\qquad (R. \ 2.3)$$

2.5 moles of gaseous molecules turn into 3 moles of gaseous molecules, the entropy change is positive. ΔH° and ΔG° are −675.6 and −689 kJ, respectively, when 1 mole of CH_3OH reacts; therefore, the theoretically efficiency of this reaction is

$$\eta = (-689)/(-675.6) = 102\% \qquad\qquad (Eq. \ 2.9)$$

For a reaction with zero ΔS, ΔH will be equal to ΔG, and the theoretical efficiency will be 100%. For example, for Reaction 2.4

$$C \ (s) + O_2 = CO_2 \qquad\qquad (R. \ 2.4)$$

1 mole of gaseous molecules turns into 1 mole of gaseous molecules, the entropy change is 0. ΔH° and ΔG° are −393.5 and −394.4 kJ, respectively, when 1 mole of O_2 reacts, and then the theoretical efficiency is

$$\eta = (-394.4)/(-393.5) = 100.2\% \qquad\qquad (Eq. \ 2.10)$$

TΔS is thermal energy. Equation 2.8 can be understood as that the chemical energy ΔH consists of work ΔG and thermal energy TΔS. When TΔS is positive (i.e. system becomes less ordered), the reaction takes in some heat from the environment, resulting in ΔG being more negative than ΔH, and thus ΔG/ΔH is larger than 100%. In other words, some heat from the environment is used by the reaction to do work. When TΔS is negative (i.e. system becomes more ordered), the reaction releases some heat to the environment, resulting in ΔG being less negative than ΔH, and thus ΔG/ΔH is smaller than 100%. In either case, the conservation of energy is obeyed without energy being created or destroyed.

2.2 Voltage

By definition, electrical work is charge times voltage, as shown in Equation 2.11:

$$W = Q \times E \qquad\qquad (Eq. \ 2.11)$$

The charge of 1 mole of electrons is called the Faraday constant, F, which is 96,485 coulombs. For Reaction 2.1, 1 H_2 molecule provides 2 electrons, meaning that the charge is 2F for each mole of H_2 to react. As the Gibbs free energy change of a reaction also represents the work that can be done by the reaction, ΔG should be equal to 2FE in absolute value:

$$\Delta G = -2FE \text{ or } \Delta G^\circ = -2FE^\circ \qquad \text{(Eq. 2.12)}$$

A negative sign is put before 2FE because E is positive when ΔG is negative by definition. From the above equation,

$$E = -\Delta G/(2F) \qquad \text{(Eq. 2.13)}$$

At 25°C, when water is in liquid form, ΔG° is −237.1 kJ mol^{-1}; thus E° is 1.23 V. When water is in the gas form, ΔG° is −228.6 kJ mol^{-1}; thus E° is 1.18 V. Since the voltage here is calculated from ΔG, a thermodynamic parameter, it is called the *thermodynamic voltage*.

Reaction 2.1 is composed of an anode half reaction:

$$H_2 = 2H^+ + 2e^- \qquad \text{(R. 2.5)}$$

and a cathode half reaction:

$$0.5O_2 + 2H^+ + 2e^- = H_2O \qquad \text{(R. 2.6)}$$

The thermodynamic voltage is the difference in the standard reduction potentials between the two half reactions under the standard condition (i.e., 298 K, 1 bar). The anode is a standard hydrogen electrode, and its potential is 0 V by definition. Checking reference books, the above cathode electrode has a standard reduction potential of 1.229 V with water in liquid form (Vanýsek 1994). The potential difference between these two half reactions is therefore 1.229 V, the same as that obtained by using ΔG°. Clearly, the thermodynamic voltage should be the same regardless of the method used, either the standard Gibbs free energy change or the standard reduction potential.

Since the voltage is directly related to the Gibbs free energy change, the theoretical efficiency can also be expressed using the ratio of voltages. Using the LHV of −241.8 kJ mol^{-1}, E° is 1.25 V. The ratio of 1.18 to 1.25 is 94.4%, the same as that obtained by using the ratio of ΔG° to ΔH°.

Table 2.1 lists ΔG, ΔH, efficiency limits, and thermodynamic voltage E° at different temperatures. Water is in the gaseous form at all the temperatures for this table. Although there are only two temperature points below 100°C, the theoretical efficiency and the thermodynamic voltage are quite close to those at 127°C. So for PEMFCs that typically operate at temperatures lower than 80°C, using the theoretical efficiency of 94.5% and the thermodynamic voltage of 1.18 V should not cause significant errors at temperatures from ambient to 80°C. The data at 227°C, 627°C, and 927°C can be used for phosphoric

TABLE 2.1

Theoretical Electrical Efficiency and Thermodynamic Voltage of an
H_2/O_2 Fuel Cell at Different Temperatures

T (°C)	T (K)	ΔH (kJ/mol)	ΔG (kJ/mol)	Theoretical Efficiency (%)	Thermodynamic Voltage E° (V)
25	298	−241.826	−228.582	94.5	1.18
27	300	−241.844	−228.500	**94.5**	**1.18**
127	400	−242.845	−223.900	92.2	1.16
227	500	−243.822	−219.050	89.8	1.14
327	600	−244.751	−214.008	87.4	1.11
427	700	−245.620	−208.814	85.0	1.08
527	800	−246.424	−203.501	82.6	1.05
627	900	−247.158	−198.091	80.1	1.03
727	1000	−247.820	−192.603	77.7	1.00
827	1100	−248.410	−187.052	75.3	0.97
927	1200	−248.933	−181.450	72.9	0.94
1027	1300	−249.392	−175.807	70.5	0.91
1127	1400	−249.792	−170.132	68.1	0.88
1227	1500	−250.139	−164.429	65.7	0.85

Source: Calculated from L. V. Gurvich et al., "Thermodynamic Properties as a Function of Temperature," in *CRC Handbook of Chemistry and Physics (1913–1995), 75th Edition (Special Student Edition)*, editor-in-chief, David R. Lide, 5-48–5-71 (Boca Raton, FL: CRC Press, 1994).

acid, molten carbonate, and solid oxide fuel cells, respectively, because they normally operate at temperatures around these values. It can be clearly seen from Table 2.1 that both the theoretical efficiency and the thermodynamic voltage decrease with an increase in temperature.

Figure 2.1 shows the theoretical efficiency and the thermodynamic voltage versus temperature. Both of them decrease nearly linearly with temperature.

A linear curve fitting for E° versus T data points results in the following relationship: $E° = -0.0003T + 1.1966$, where E° is in unit V and T in degree Celsius (°C). So, the thermodynamic voltage E° decreases for about 0.3 mV for every degree Celsius increase in temperature.

2.3 Polarization

A thermodynamic voltage is the voltage under thermodynamic equilibrium conditions, which means that the current is zero. At open circuit, there is no external current flowing between the anode and the cathode of a cell, and the voltage should be equal to the thermodynamic voltage, but often they are

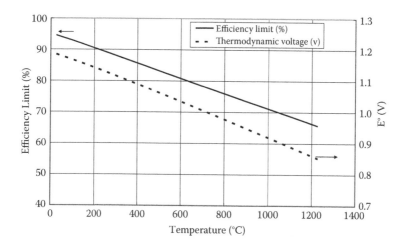

FIGURE 2.1
Theoretical electrical efficiency and thermodynamic voltage of an H_2/O_2 fuel cell at different temperatures.

not equal, as discussed below. The cell voltage at open circuit is called the *open-circuit voltage* (OCV).

For a PEMFC, the OCV is always smaller than the thermodynamic voltage for several reasons. One reason is that there are some side reactions occurring at the electrodes, especially at the cathode. The participants in the side reactions can be some impurities or a component of the catalyst layer. For example, when the cathode catalyst is Pt/C, besides Pt being oxidized to Pt O_x ($E° = 0.98$ V), the carbon catalyst support could be oxidized by the following reaction:

$$C + xH_2O = CO_x + 2xH^+ + 2xe^- \qquad \text{(R. 2.7)}$$

The standard reduction potential of this reaction is around 0.21 V, much lower than the standard reduction potential of oxygen (R. 2.6); therefore, a mixed potential between 0.21 and 1.18 V is produced. Since the oxygen reduction reaction and Pt oxidation reaction dominate at the cathode due to kinetic reasons, the mixed potential is near the higher end. The other reason is hydrogen crossover through the PEM from the anode to the cathode. This is like an internal current flow in the cell and thus leads the electrodes (mainly the cathode) away from the 0 current thermodynamic equilibrium conditions. Due to the slow kinetics of Reaction 2.6, this internal current flow significantly lowers the cathode potential. Detailed estimation is given later in this chapter. For these two reasons, the OCV of a PEMFC is typically between 0.95 and 1.0 V.

The voltage of a PEMFC will drop further when it generates power due to three types of power losses. A schematic V versus I curve is shown in Figure 2.2.

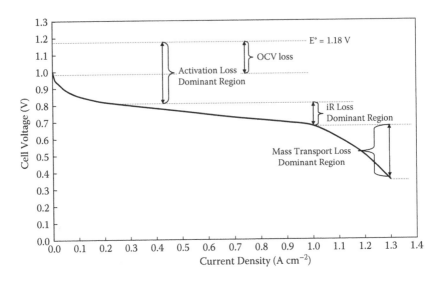

FIGURE 2.2
Typical voltage losses.

Unless otherwise specified, the current density is based on the geometric surface area of the electrode throughout this book.

In the low current density region, the loss that is quite steep is dominated by the slow reaction kinetics. This loss is called the *activation loss*, or the *activation overpotential*. This loss is mainly caused by the reaction at the cathode because the oxygen reduction reaction (ORR) is more sluggish than the hydrogen oxidation reaction (HOR) at the anode. This loss, ΔE_A, normally dominates at voltages higher than about 0.8 V. At lower voltages, the loss becomes less steep and is dominated by the ohmic resistance, that is, iR losses, or the *resistance overpotential*, ΔE_R. R refers to the sum of all the ionic and the electronic resistances in the electron and the proton flow paths. The ionic losses arise from the PEM and the anode and cathode catalyst layers when H^+ transports within the anode catalyst layer, through the PEM, and within the cathode catalyst layer. The electrical losses come from the anode current collector, the anode plate, the anode gas diffusion medium (GDM), the anode catalyst layer (CL), the cathode CL, the cathode GDM, the cathode plate, the cathode current collector, all the interfaces between neighboring components mentioned above, and the external connections. In the high current-density region, another steep decline occurs, which is dominated by the mass transport resistance, or the *concentration overpotential*, ΔE_M. The species involved in the mass transport mainly include the anode reactant (H_2), the cathode reactant (O_2), and the product (H_2O). Most of the time, the mass transport loss mainly comes from inadequate O_2 transport at the cathode, especially when air instead of pure O_2 is used as the reactant. Slower mass transport of O_2 causes its concentration within the catalyst layer to be lower. Due to the losses caused by these three reasons, the actual voltage when a fuel cell is in operation is

$$E = E^\circ - \Delta E_A - \Delta E_R - \Delta E_M \qquad \text{(Eq. 2.14)}$$

where ΔE_A, ΔE_R, and ΔE_M are the voltage losses due to activation, ohmic resistance, and mass transport resistance, respectively. Although losses, they are taken as positive numbers in Equation 2.14. Please note that the OCV loss is included in ΔE_A.

Although the voltage loss in the kinetic region is dominated by slow reaction kinetics, there are also some minor losses due to ohmic resistance and mass transport. Similarly, in the resistance (mass transport) region, most of the voltage loss arises from the ohmic resistance (mass transport), but losses due to slow kinetics and mass transport (ohmic resistance) also exist.

The thermodynamic voltage is related to the activities of the reactants and the products and is described by the Nernst equation, as shown in Equation 2.15:

$$E = E^\circ + [RT/(2F)] \cdot \ln(a_{H_2} a_{O_2}^{0.5}/a_{H_2O}) \qquad \text{(Eq. 2.15)}$$

where a_{H_2}, a_{O_2}, and a_{H_2O} are the activities of H_2, O_2, and H_2O, respectively. Since most water in a fuel cell exists in the vapor form, all the above species are in the gaseous form. Also, because the pressures of the gases are not high (normally less than a few bars), they can be treated as ideal gases without causing significant errors. In a mixture of ideal gases, the activity of each gaseous component is its partial pressure, defined as

$$a_{H_2} = P_{H_2}/P^\circ \qquad \text{(Eq. 2.16)}$$

$$a_{O_2} = P_{O_2}/P^\circ \qquad \text{(Eq. 2.17)}$$

$$a_{H_2O} = P_{H_2O}/P^\circ \qquad \text{(Eq. 2.18)}$$

where P° is the standard pressure, which is 1 bar (or 100 kPa). Then Equation 2.15 becomes the following using the bar as the pressure unit:

$$E = E^\circ + [RT/(2F)] \cdot \ln(P_{H_2} P_{O_2}^{0.5}/P_{H_2O}) \qquad \text{(Eq. 2.19)}$$

Also, the actual pressure of each gas is just its partial pressure when the bar is used as the unit of standard pressure.

According to the ideal gas law,

$$P = nRT/V \qquad \text{(Eq. 2.20)}$$

P is in direct proportion to n/V, which is the molar concentration of the gas. Therefore, a lower concentration of a gaseous reactant within the catalyst layer will result in a lower partial pressure and then a lower electrode potential.

We can estimate the voltage difference between using saturated air and O_2 based on the Nernst equation at 57°C, where the saturated vapor pressure is 0.17 bars. With pure O_2, its partial pressure is 0.83 bars, and with air having 21% O_2, the O_2 partial pressure is 0.17 bars (0.83 × 0.21). Therefore,

$$\Delta E = [RT/(2F)]\ln(P_{O_2}^{0.5}/P_{air}^{0.5}) = [(8.31 \times 330)/(4 \times 96485)] \times \ln(0.83/0.17) = 0.011 \text{ V}$$

Apparently, the change from O_2 to air does not significantly reduce the cell voltage based on the Nernst equation. But in reality, the difference is normally higher due to kinetic and mass transport reasons.

In the high current-density region, the reactant is consumed quickly; therefore, its concentration within the catalyst layer is far lower than the bulk concentration of the reactant in the flow-field channels. The higher the current density is, the lower the reactant concentration will be, leading to a decrease in the electrode voltage. When all the reactant is consumed within the catalyst layer, its concentration approaches 0. The current density at this moment is the largest that the electrode can generate and is called the *limiting current density*. The limiting current density depends on how fast the reactant can transport from the flow-field channels, through the GDM, into the CL to the catalyst sites. The major driving force for the reactant to transport here is through diffusion that is driven by the reactant concentration gradient between the bulk concentration and the concentration at the catalyst sites.

The highest power density normally appears at the current density just before the mass transport region starts. A well-designed cell should push the mass transport region to higher current densities as much as possible. The gas transporting ability of the GDM and the CL plays a crucial role. If they are easily flooded, the reactant will encounter more resistance to pass through because the transporting paths are severely blocked by liquid water. If this happens, a higher bulk reactant concentration or pressure helps.

In addition to liquid water, water vapor may also decrease the limiting current by lowering the reactant pressure in the catalyst layer and by potentially impeding the diffusion of oxygen toward the catalyst sites. When a stack is in operation, the water vapor in both the anode and the cathode is around the saturation pressure of the stack temperature. This is important for the stack to operate normally in order to avoid drying up of the membrane electrode assemblies (MEAs). Table 2.2 lists the saturated vapor pressures at different temperatures. A PEMFC normally operates at temperatures between 50°C and 70°C, which means that the saturated vapor pressure is between 0.1 and 0.3 bars. For a stack operating at a temperature near or higher than 100°C, the reactants must be pressurized.

TABLE 2.2

Saturated Vapor Pressure at Different Temperatures

T (°C)	T (K)	Pressure (kPa)	Pressure (bar)	Vapor% Vol.
7	280	0.99	0.010	1.0
17	290	1.92	0.019	1.9
27	300	3.54	0.035	3.5
37	310	6.23	0.062	6.2
47	320	10.54	0.104	10.4
57	330	17.20	0.170	17.0
67	340	27.17	0.269	26.9
77	350	41.65	0.412	41.2
87	360	62.14	0.615	61.5
97	370	90.45	0.896	89.6

Source: "IUPAC Recommended Data for Vapor Pressure Calibration," in *CRC Handbook of Chemistry and Physics (1913–1995), 75th Edition (Special Student Edition)*, editor-in-chief, David R. Lide, 6–109 (Boca Raton, FL: CRC Press, 1994).

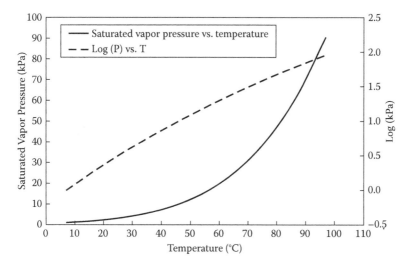

FIGURE 2.3
Saturated vapor pressure at different temperatures (total pressure at 1 bar).

The relationship between saturated vapor pressure and temperature is plotted in Figure 2.3. The saturated vapor pressure increases exponentially with temperature, and at temperatures higher than 40°C, the vapor pressure becomes significant. This implies that humidifying gaseous reactant becomes more effective when the humidifying temperatures are higher than 40°C, but the molar percentage of the reactant will drop more significantly in accordance. When the saturated vapor pressure is taken in the log form, log (P) is almost linearly related to the temperature, as shown in Figure 2.3.

Nitrogen may also lower the reactant pressure in the catalyst layer and impede the diffusion of oxygen toward the catalyst sites in the catalyst layer. At the cathode, O_2 in air is consumed within the catalyst layer, but N_2 is not. Therefore, the N_2 concentration or partial pressure within the catalyst layer may be much higher than that in fresh air, which in turn could reduce the partial pressure of O_2 in the catalyst layer or impede the diffusion of O_2 in the catalyst layer.

2.4 Tafel Equation

In the kinetic region, when the activation overpotential ΔE_A and current density (j) are plotted in the form of ΔE_A versus logj, a straight line exits. This was found in 1905 and is called the *Tafel plot*:

$$\Delta E_A = a + b \log j \qquad \text{(Eq. 2.21)}$$

Through experiments, both a and b can be determined.

Based on the thermodynamics, the Tafel plot was later found to be a simplified Buttler–Volmer equation when ΔE_A is larger than about 116 mV/n_r (Bard and Faulkner 1980), where n_r is the number of electrons involved in the rate-determining step (not necessarily equal to the number of electrons involved in one reactant molecule) and it is 1 for the ORR. For the ORR, the OCV is already more than 180 mV lower than the thermodynamic voltage; the activation overpotential ΔE_A will be even higher when there is a current flow. So, the activation overpotential ΔE_A for the ORR should follow the Tafel relationship. Based on the thermodynamics, the Tafel relationship is

$$\Delta E_A = [RT/(\alpha n_r F)] \ln j^\circ - [RT/(\alpha n_r F)] \ln j \qquad \text{(Eq. 2.22)}$$

$$= [2.3RT/(\alpha n_r F)] \log j^\circ - [2.3\ RT/(\alpha n_r F)] \log j$$

$$= -[2.3RT/(\alpha n_r F)] \log(j/j^\circ)$$

where j° is called the exchange current density, and α is called the charge transfer coefficient. By comparing Equation 2.21 with Equation 2.22, it is clear that

$$a = [2.3RT/(\alpha n_r F)] \log j^\circ \qquad \text{(Eq. 2.23)}$$

$$b = -2.3RT/(\alpha n_r F) \qquad \text{(Eq. 2.24)}$$

The exchange current density is the current density when an electrode is at thermodynamic equilibrium with zero net current flowing through. Any cathode or anode half reaction involves a forward reaction and a backward reaction, and at thermodynamic equilibrium, the rates of the forward and the backward half reactions are equal, and therefore, there is no net current. The forward or the backward half reaction rate is the exchange current density. A higher j° means that the reaction can get away from the thermodynamic equilibrium state with a smaller "push"; in other words, the energy barrier is lower, and therefore, the voltage loss due to activation will be smaller. In an acidic environment with Pt as the catalyst, j° for the HOR and the ORR is on the order of 0.1–1 mA cm^{-2} Pt and 0.1–1 µA cm^{-2} Pt, respectively, according to various studies. For example, it was reported that for the ORR at the Pt–Nafion interface and 30°C, j° and α are 0.28 µA cm^{-2} Pt and 0.48, respectively (Parthasarathy et al. 1992). Since the rate of a reaction is largely affected by the temperature, j° is temperature dependent.

The charge transfer coefficient represents what a portion of the input electrical energy is used to change the electrochemical reaction rate. When the linear plot of ΔE_A versus $\log j$ is extrapolated to $\Delta E_A = 0$, the intercept at the $\log j$ axis is $\log j^\circ$; then j° can be obtained. The slope of the plot is $2.3RT/(\alpha n_r F)$, from which α can be obtained. For the HOR on Pt, α is around 0.5. For the ORR on Pt, α is normally between 0.1 and 0.7, depending on the reaction environment; for general purposes, 0.5 can be used for estimation.

The Tafel plot is valid only when ΔE_A is large enough (e.g., >116 mV/n_r). If ΔE_A is very small leading to $\Delta E_A/(RT/n_r F)$ far smaller than 1

$$\Delta E_A = -[RT/(n_r F\, j^\circ)]\, j \qquad\qquad (Eq.\ 2.25)$$

that is, the activation overpotential is linearly related to the current density. Large ΔE_A means j° is smaller, while smaller ΔE means j° is larger.

Due to the extremely small j° for the ORR, the Tafel plot is valid whenever there is a current flowing through the cell. Since there is one electron involved in the rate-determining step for the ORR, i.e., $n_r = 1$, the Tafel slope at 298 K is around 118 mV per decade $[(1000 \times 2.3 \times 8.31 \times 298)/(0.5 \times 1 \times 96{,}485)]$.

In actual tests a Tafel slope of ca. 60 mV per decade also exists in the voltage region close to the OCV (i.e., in the low overpotential region). This slope corresponds to $\alpha n_r = 1$ (most likely $\alpha = 1$ and $n_r = 1$) when calculated using Equation 2.24. The presence of this smaller slope is because that the surface of Pt is in the oxidized state in this high potential region, and Pt oxide has lower catalytic activity toward the ORR than Pt. The exchange current density estimated using this smaller slope is smaller than that estimated using the larger Tafel slope (i.e., ca. 120 mV per decade). During the operation of a fuel cell the potential of the cathode is typically at 0.7 V or lower, which enables most of the Pt surface in the oxide-free form, and therefore, the exchange current density calculated based on the larger Tafel slope more accurately reflects the actual situation. Please note that the two Tafel slope regions are not clearly separated or well defined, and thus the slopes are affected by the points taken to draw the two straight lines

If the dissociation of an H_2 molecule into two adsorptive H atoms on the catalyst surface is the rate-determining step, the HOR will have two electrons involved in the rate-determining step, i.e., $n_r = 2$. Since the HOR has a much higher j^o (~0.5 mA cm^{-2} Pt), it may not follow the Tafel plot in PEMFCs. We can have a look at a situation where the electrode consists of Pt nanoparticles and the Pt loading is 0.5 mg cm^{-2} electrode geometric area.

The surface area for spherical Pt particles is given by

$$S = 3/(rd)$$

where r is the radius of the Pt particle (cm), d is the density of Pt (21.45 g cm^{-3}), and S is the surface area (cm^2 g^{-1}). Table 2.3 lists the j_{ap}^o based on the total Pt surface area within a 1-cm^2 electrode area consisting of different sized Pt nanoparticles. The case with r = 0.15 nm represents Pt atoms and if they are deposited on a support to form a monolayer, the total Pt surface area will be 4662 times of the geometric surface area of an electrode consisting of 0.5 mg cm^{-2} Pt. Here we use j_{ap}^o to represent the exchange current density per unit electrode geometric area and call it the *apparent exchange current density*. Please note that since not all of the catalyst surface can participate in the reactions, the actual j_{ap}^o will be smaller than the results shown in Table 2.3,

TABLE 2.3

j_{ap}^o for HOR and ORR with Pt Loading of 0.5 mg cm^{-2}

| Pt Radius r (nm) | Pt Surface Area S (cm^2 g^{-1}) | ΣS (cm^2 Pt per cm^2 electrode) | HOR | | | ORR |
			j_{ap}^o (mA cm^{-2} electrode)	RT/(n$_r$Fj$_{ap}^o$) (Ohm cm^2)		j_{ap}^o (mA cm^{-2} electrode)
0.15	9324009	4662	2331	0.006		1.399
0.25	5594406	2797	1399	0.009		0.839
0.5	2797203	1399	699	0.018		0.420
1	1398601	699	350	0.037		0.210
1.5	932401	466	233	0.055		0.140
2	699301	350	175	0.073		0.105
2.5	559441	280	140	0.092		0.084
3	466200	233	117	0.110		0.070
3.5	399600	200	100	0.128		0.060
4	349650	175	87	0.147		0.052
4.5	310800	155	78	0.165		0.047
5	279720	140	70	0.183		0.042

Note: HOR: j^o = 0.5 mA cm^{-2}, n_r = 2; ORR: j^o = 0.3 μA cm^{-2}.

which are obtained by simply using the particle sizes. It is more accurate to use the electrochemically active surface area (ECSA) based on either the H adsorption–desorption peak areas or the CO oxidative stripping peak area obtained through cyclic voltammetry to estimate j_{ap}°. In Table 2.3, the j° for the HOR and ORR is taken at 0.5 mA cm^{-2} and 0.3 µA cm^{-2}, respectively.

With 3-nm Pt particles, the j_{ap}° is 233 mA cm^{-2} electrode geometric area for the HOR. Only when the current density is much higher than this value will the Tafel plot be valid. But at such current densities, the iR drop becomes dominant. Therefore, for the HOR of a PEMFC with a reasonable Pt loading, the Tafel plot is rarely applicable. In contrast, a linear relationship between ΔE_A and j shown in Equation 2.25 will exist. Therefore, in both the activation and the resistance regions, ΔE is linearly related to j. For a PEM single cell, the cell resistance R is around 0.05–0.2 Ohm cm^2, depending on the thickness of the PEM, which is close to $[RT/(n_r F\, j_{ap}^{\circ})]$, that is, the slope of ΔE_A versus j, when the Pt particle radius is between 1 and 5 nm shown in Table 2.3 for the HOR. The above two kinds of resistance are additive, and the ΔE vs. j plot is a straight line.

The j° represents the intrinsic activity of a catalyst toward a reaction the most straight forward and unambiguous. We know that a cathode giving a higher V-I curve is more active than one giving a low V-I curve (please visualize two V-I curves in Figure 2.2), but often we do not know how to quantify the difference in the activity between the two electrodes. Using Equation 2.22 the quantification can be easily obtained in term of exchange current densities for the ORR. Assuming the apparent exchange current densities of the two electrodes are j_{ap1}° and j_{ap2}°, and the corresponding activation voltage losses are ΔE_1 and ΔE_2, then the difference between the two ΔE is

$$\Delta E = \Delta E_2 - \Delta E_1$$

$$= -[2.3RT/(\alpha n_r F)]\log(j/j_{ap2}^{\circ}) + [2.3RT/(\alpha n_r F)] \log(j/j_{ap1}^{\circ})$$

$$= [2.3RT/(\alpha n_r F)] \log(j_{ap2}^{\circ}/j_{ap1}^{\circ})$$

Rearranging the above equation results in

$$j_{ap2}^{\circ}/j_{ap1}^{\circ} = 10^{[\Delta E\alpha n_r F/(2.3RT)]} \qquad \text{(Eq. 2.26)}$$

The two V-I curves should be more or less parallel in the linear region (i.e., the iR loss region when the same PEM is used), and thus ΔE can be easily obtained. Also, the iR loss should be corrected from the V-I curves when obtaining ΔE. Theoretically, the activity ratio depends on the temperature, but is independent of the current density at where ΔE is measured. However, ΔE can often not be accurately measured in the kinetic region because it varies with the current density in this region; so it should be measured in the linear iR drop region.

Table 2.4 shows $j_{ap2}^{\circ}/j_{ap1}^{\circ}$ at different cathode ΔE. It can be seen that a 50, 100, and 150 mV difference in voltage, the better cathode is 2.35, 5.53, and

TABLE 2.4

Ratio of Apparent Exchange Current
Densities versus Differences in Activation
Voltage Losses (T = 340 K; α = 0.5; n_r = 1)

ΔE (mV)	$\log(j_{ap2}{}^\circ/j_{ap1}{}^\circ)$	$j_{ap2}{}^\circ/j_{ap1}{}^\circ$
5	0.037	1.09
10	0.074	1.19
20	0.148	1.41
30	0.223	1.67
40	0.297	1.98
50	0.371	2.35
60	0.445	2.79
70	0.520	3.31
80	0.594	3.93
90	0.668	4.66
100	0.742	5.53
110	0.817	6.56
120	0.891	7.78
130	0.965	9.23
135	1.002	10.05
140	1.039	10.95
150	1.114	12.99
175	1.299	19.91
200	1.485	30.53

12.99 times as active as the worse cathode, respectively. If a monolayer of
Pt atoms with a radius around 0.15 nm is made on a support in a core/shell
structure (or other thin layered structure), the specific Pt surface area and
the $j_{ap}{}^\circ$ will increase by 10 times (Table 2.3) compared with that of Pt particles with 1.5 nm radius that are commonly used today. It can be seen from
Table 2.4 that the activation loss with the former electrode will be ca. 135 mV
lower than that with the latter electrode.

The above analysis can also be used in rotating disc electrode (RDE) test,
where the difference in half wave potentials is used as ΔE; or for the positive
electrode of an electrolysis cell; or for both the anode and the cathode of an
alkaline fuel cell and direct methanol fuel cell, where the anode also suffers
high overpotential losses.

For HOR Equation 2.25 should be used to estimate the ratio of apparent
exchange current densities of two anodes. It is clear that the sought ratio is
the ratio of the two slopes, and can also be expressed as

$$j_{ap2}{}^\circ/j_{ap1}{}^\circ = \Delta E_1/\Delta E_2 \qquad \text{(Eq. 2.27)}$$

ΔE_1 and ΔE_2 are the anode overpotentials of the two anodes at the same current density (again iR loss should be corrected). The ratio does not depend on the temperature.

The activity difference between the two electrodes based on j_{ap}° can arise from difference either in the catalyst (e.g., different catalysts, or same catalyst but with different crystalline facets, or different particle size and geometry) or in the loading. What represents the intrinsic activity of a catalyst is the j°. With known catalyst loading and the total electrochemical active surface area (ECSA), j_2°/j_1° can be obtained with respect to specific mass or surface activities of the two catalysts used in the two electrodes. This ratio represents the intrinsic activity difference of the two catalysts.

For electrodes made with the same catalyst and the same method the apparent exchange current density should theoretically increase with the catalyst loading in proportion assuming the catalyst surface utilization is the same. For example, when Pt loading is doubled, j_{ap}° should be doubled, and thus the activation overpotential loss should decrease for about 40 mV based on the data in Table 2.4. However, the catalyst utilization often decreases with increase in the catalyst layer thickness accompanying the increase in Pt loading, and thus the increase may be less than 40 mV. Anyway, using Equation 2.26 to quickly estimate the difference in the activation voltage losses offers a good guide.

2.5 Voltage Loss due to H_2 Crossover at an Open Circuit

From $\Delta E_A = [2.3RT/(\alpha n_r F)]\log(j/j^\circ)$, the activation loss due to H_2 crossover can be estimated. After H_2 crosses through the PEM to reach the cathode catalyst layer, it will dissociate into protons and electrons according to Reaction 2.5 and combine with O_2 to form H_2O according to Reaction 2.6. Since the j° for the HOR is much larger than the j° for the ORR (over 1000 times), the activation overpotential loss is dominated by Reaction 2.6, that is, the ORR. If the Pt loading is 0.5 mg cm^{-2}, the Pt particle's diameter is 3 nm, and j° for the ORR is 0.3 μA/cm^2 Pt, then the apparent j_{ap}° for the ORR is 466 × 0.0003 = 0.14 electrode geometric area. With these assumptions, the activation overpotential loss ΔE_A (V) due the H_2 crossover can be calculated and the results are listed in Table 2.5 at different H_2 crossover current densities. It is striking to see that with an H_2 crossover current as small as 2 mA cm^{-2}, the electrode potential is decreased from 1.18 V to 1.04 V, a drop of 140 mV. The rate of the potential drop becomes smaller with a further increase in the current density. H_2 crossover current density is affected by conditions. Thinner

TABLE 2.5

OCV at Different H_2 Crossover Current Densities

j (mA cm^{-2})	j/j$_{ap}°$	log(j/j$_{ap}°$)	(2.3RT/αn$_r$F)log(j/j$_{ap}°$) (V)	E (V)
0.5	4	0.55	0.07	1.11
1	7	0.85	0.10	1.08
2	14	1.15	0.14	1.04
3	21	1.33	0.16	1.02
4	29	1.46	0.17	1.01
5	36	1.55	0.18	1.00
6	43	1.63	0.19	0.99
7	50	1.70	0.20	0.98
8	57	1.76	0.21	0.97
9	64	1.81	0.21	0.97
10	71	1.85	0.22	0.96
20	143	2.15	0.25	0.93
30	214	2.33	0.28	0.90
50	357	2.55	0.30	0.88
100	714	2.85	0.34	0.84
150	1071	3.03	0.36	0.82
200	1429	3.15	0.37	0.81

Note: T = 298 K, E° = 1.18 V, α = 0.5, n$_r$ = 1, j$_{ap}°$ = 0.14 mA cm^{-2} electrode.

membrane, higher H_2 inlet pressure, higher temperature, and higher relative humidity lead to higher H_2 crossover current density.

The relationship between the OCV and the H_2 crossover current density is plotted in Figure 2.4. It can be seen that most of the OCV loss occurs in the extremely low crossover current density region (e.g., <10 mA cm^{-2}).

It should be emphasized that the activation overpotential loss due to H_2 crossover should be estimated using the ORR not the HOR. Do not be misled into thinking that it is the H_2 that crosses through the PEM, and that it is the H_2 that is oxidized at the cathode, and thus the HOR should be used to estimate the activation overpotential loss.

Whether it is the H_2 crossover current or the output current generated by a PEMFC, the impact on the activation overpotential loss is the same because the same reaction process occurs at the cathode. In actual tests, a PEMFC normally can give out a current density of 150 mA cm^{-2} at around 0.80 V, which is in reasonable agreement with the result shown in Table 2.5.

In a direct methanol fuel cell (DMFC) since the voltage losses due to both ORR and the methanol oxidation reaction (MOR) are significant, the OCV loss is the summation of both.

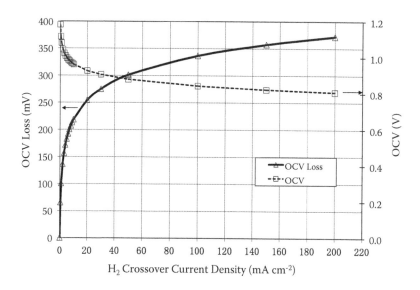

FIGURE 2.4
OCV versus H_2 crossover current density ($T = 298$ K, $j_{ap}^{\,\circ} = 0.14$ mA cm^{-2}).

2.6 Example

Figure 2.5 shows the actual results from a single cell test with a cathode Pt loading of 0.40 mg cm^{-2}. The OCV was 0.98 V. The resistance of the cell was found to be around 80 mΩ cm^2, measured by the impedance technique at current densities of 0.1 and 0.4 A cm^{-2}, as shown in Figure 2.6. It is clear from Figure 2.5 that even after the iR correction, the V–I curve still shows significant voltage decline with increase in the current density, implying that in this iR-dominated region there are still voltage losses due to activation and mass transport.

For the sole purposes of illustration, using three data points at current densities of 0.05, 0.067, and 0.10 A cm^{-2} ($\Delta E_A = 1.18 - V_{iR\text{-corrected}}$), ΔE_A versus log j is plotted in Figure 2.7. It is a straight line, and a curve fitting leads to the following correlation:

$$\Delta E_A = 0.404 + 0.093 \log j$$

So, the Tafel slope is 93 mV per decade. At $\Delta E_A = 0$, $\log j_{ap}^{\,\circ} = -4.344$; then, $j_{ap}^{\,\circ} = 0.045$ mA cm^2. This value is within the range of the ORR $j_{ap}^{\,\circ}$ estimated using the Pt particle sizes shown in the last column in Table 2.3. Using $2.3RT/(\alpha n_r F) = 0.404$, α is estimated to be 0.7. It must be noted that since only three data points were collected in the activation region during this test, errors must

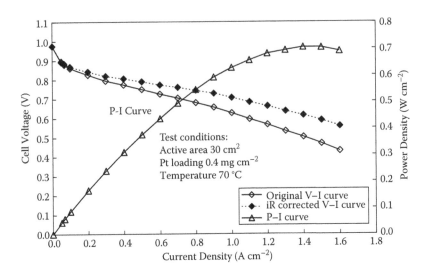

FIGURE 2.5
Single cell test data. Courtesy of Dalian Institute of Chemical Physics, Chinese Academy of Sciences.

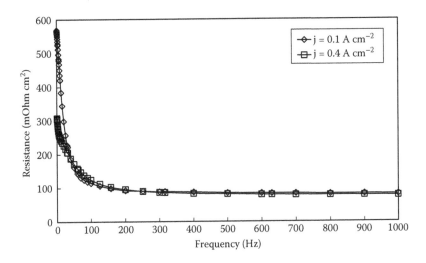

FIGURE 2.6
Total resistance measurement using the impedance technique. Courtesy of Dalian Institute of Chemical Physics, Chinese Academy of Sciences.

exist for the estimated $j_{ap}{}^{\circ}$ and α. In actual tests such as the one shown here, people normally do not take many points in the activation region because it is often not the aim of the tests. But for purposes of obtaining j° and α, more data points should be taken in the activation region in order to get more accurate results for $j_{ap}{}^{\circ}$ and α.

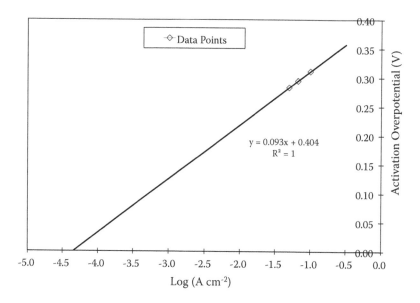

FIGURE 2.7
Activation overpotential versus log current density.

2.7 Limiting Current Density

2.7.1 Diffusion

The current density is proportional to the flux of the reactant arriving at the reaction sites (e.g., the catalyst surface). For neutral species such as H_2 and O_2, their transport involves both convection and diffusion. But near the catalyst surface, there is a stagnant layer, and the transport through this layer to the catalyst surface is solely via diffusion, as schematically shown in Figure 2.8a (neglecting the concentration gradient of the species within the GDM because the gradient is small). Subscripts A, R, and M mean the activation, resistance, and mass transport regions, respectively, and the superscript "o" refers to the bulk concentration.

It is obvious that the flux is determined by the diffusion coefficient of the reactant and its concentration gradient (Fick's first law):

$$J = -D\,dc/dx = D\,(c_{bulk} - c_s)/L \qquad \text{(Eq. 2.28)}$$

where D is the diffusion coefficient, dc/dx is the concentration gradient of the reactant (negative), c_{bulk} is the bulk concentration of the species, c_s is the surface concentration of the species, and L is thickness of the diffusion layer. For the second part of Eq. 2.28, it is assumed that c_{bulk} is larger than c_s, and $(c_{bulk} - c_s)/L$ has the opposite sign with respect to the concentration gradient. Multiplying the flux with charge results in the current density:

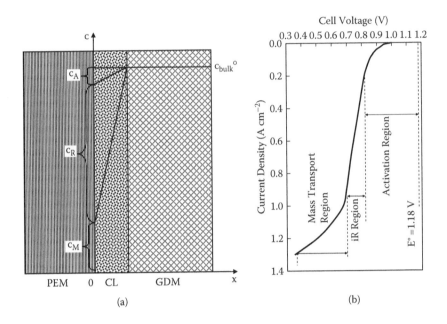

FIGURE 2.8
(a) Schematic of reactant diffusion through GDM and CL; (b) schematic of corresponding current density region.

$$j = -nFD \, dc/dx = nFD \, (c_{bulk} - c_s)/L \qquad \text{(Eq. 2.29)}$$

where n is the number of electrons given or received by each reactant molecule, and F is the Faraday constant.

At the limiting current, all the species arriving at the catalyst surface are consumed, that is, $c_s = 0$, then,

$$j_{lim} = -nFD \, dc/dx = nFD \, c_{bulk}/L \qquad \text{(Eq. 2.30)}$$

Dividing j by j_{lim} gives the following relationship:

$$j = j_{lim} \, (1 - c_s/c_{bulk}) \qquad \text{(Eq. 2.31)}$$

Then

$$c_s = c_{bulk} \, (1 - j/j_{lim}) \qquad \text{(Eq. 2.32)}$$

Referring between Figures 2.8a and 2.8b, the reactant concentration c_A in the activation region is very close to $c_{bulk}{}^o$, in the mass transport region c_M is near 0, and in the iR loss region c_R is between c_A and c_M.

According to the Nernst equations shown in Equations 2.15 and 2.19, for H_2, the voltage loss at current density j due to mass transport is

$$\Delta E_M{}^{H_2} = -[RT/(2F)]\ln(1 - j/j_{lim}) \qquad \text{(Eq. 2.33)}$$

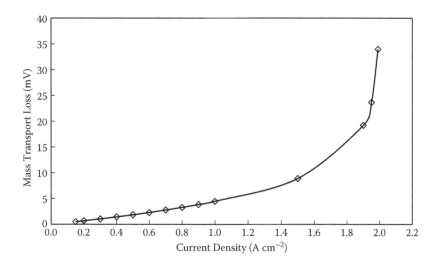

FIGURE 2.9
Voltage loss due to mass transport resistance (T = 298 K, j_{lim} = 2 A cm^{-2}).

and for O_2, the voltage loss due to mass transport is

$$\Delta E_M^{O_2} = - [RT/(4F)]\ln(1 - j/j_{lim}) \qquad \text{(Eq. 2.34)}$$

Taking j_{lim} = 2 A cm^{-2}, the mass transport loss due to O_2 calculated according to Equation 2.34 is shown in Figure 2.9. At current densities of 1.5 A cm^{-2} or less, $\Delta E_M^{O_2}$ is less than 10 mV. Even at a current density of 1.99 A cm^{-2}, $\Delta E_M^{O_2}$ is only 34 mV. It appears that the mass transport loss should not be significant as long as the current density is about 99% or less of the limiting current density. The results of this analysis contradict actual observations of PEMFCs, and the major reason is due to electrode flooding that significantly reduces the porosity of the GDE and thus the effective diffusion coefficients of the gaseous reactants, as discussed later in this chapter.

When the activation voltage loss, the ohmic resistance loss, and the mass transport loss are all taken into consideration, the cell voltage will be

$$E = E^\circ - \Delta E_A - \Delta E_R - \Delta E_M$$

$$= E^\circ - [RT/(\alpha n_r F)]\ln(j/j_{ap}^\circ) - iR + [RT/(nF)]\ln(1 - j/j_{lim}) \qquad \text{(Eq. 2.35)}$$

where n_r is 2 and 1, and n is 2 and 4 for H_2 and O_2, respectively. For a PEMFC using pure H_2, nearly all the activation loss comes from the ORR, and the mass transport loss is often dominated by O_2 at the cathode as well therefore, the above equation can be simplified as

$$E = E^\circ - [RT/(0.5F)]\ln(j/j_{ap(ORR)}^\circ) - iR + [RT/(4F)]\ln(1 - j/j_{lim(O_2)}) \qquad \text{(Eq. 2.36)}$$

From Equation 2.30, the diffusion layer thickness at the limiting current can be estimated:

$$L = nFD \, c_{bulk}/j_{lim} \qquad \text{(Eq. 2.37)}$$

Let us use the cathode side as an example to have some discussion because O_2 diffusion has a much larger impact on fuel cell performance than H_2 diffusion. If we assume that the current density is 1.5 A cm^{-2}, the O_2 diffusion coefficient in air is 0.27 cm^2 s^{-1} (at 60°C), the cathode temperature is 57°C (vapor pressure is 0.17 bars), the total gas pressure is 1 bar, the O_2 molar fraction in the gas mixture is then 17.4% (0.83 × 21%), and the O_2 concentration is 7.8 μmol cm^{-3} [(0.174 × 10^6 μmol mol^{-1})/(22.4 × 10^3 cm^3 mol^{-1})]; therefore, the diffusion layer thickness is

$$L = 4 \times 96485 \times 0.27 \times 7.8 \times 10^{-6}/1.5 = 0.54 \text{ cm}$$

However, since O_2 needs to diffuse through the GDM and the catalyst layer, its effective diffusion coefficient is affected by the porosity, and can be approximated by the Bruggeman expression:

$$D_{eff} = D\varepsilon^{1.5} \qquad \text{(Eq. 2.38)}$$

where D_{eff}, D, and ε are the effective diffusion coefficient, bulk diffusion coefficient, and porosity. GDM normally has a porosity of nearing 80%, and after being treated with PTFE (e.g., 20% wt.), the porosity is still higher than about 70%. With ca. 30% thickness reduction within a fuel cell under the clamping pressure, the porosity of the GDM will be around 60%. The porosity also depends on how many pores are blocked by liquid water during the fuel cell operation. Table 2.6 lists D_{eff} and L (at 1.5 A cm^{-2}) for O_2 with different porosities.

2.7.2 Porous Electrode

Normally, the catalyst layer (CL) is thinner than 20 μm, the microporous layer (MPL) thinner than 30 μm, and the GDM thinner than 250 μm, making a total thickness of a gas diffusion electrode (GDE) less than 300 μm. So, if the combined porosity is not less than 15% (normally higher than 50%), diffusion itself should be enough to supply O_2 for the cathode reaction at 1.5 A cm^{-2} because O_2 can diffuse for 315 μm as shown by Table 2.6. Therefore, there is no need to create convection in the GDM in order to supply O_2 more sufficiently to the CL, although there have been numerous patents and studies aiming at creating convection. Also, a PEMFC fuel cell should work fine with ambient pressure air. Another interesting implication from this simple analysis is that when starting a fuel cell under subzero temperatures, if the overall porosity of the gas diffusion electrode GDE (GDM+MPL+CL)

TABLE 2.6

Estimation on Cathode O_2 Diffusion Layer Thickness at Different Porosities (j_{lim} = 1.5 A cm^{-2}, D_{bulk} = 0.27 cm^2 s^{-1}, T = 57°C, fully saturated air)

Porosity (%)	D_{eff} (cm^2 s^{-1})	L (μm)
100	0.270	5419
90	0.231	4626
80	0.193	3877
70	0.158	3173
60	0.125	2518
50	0.095	1916
40	0.068	1371
30	0.044	890
20	0.024	485
15	0.016	315
10	0.0085	171
5	0.0030	61
1	0.0003	5
0.5	9.55E-05	2

remains at about 15% without being filled by ice until the cell temperature increases to above 0°C, the fuel cell should be able to start at quite a high current density (e.g., 1.5 A cm^{-2}). If a fuel cell started at 20 mA cm^{-2} current density from subzero temperatures (i.e., the current density is 1/75 of 1.5 A cm^{-2}), then it should be able to start as long as the overall porosity of the GDE is larger than ca. 1%. These statements are based on assumption that O_2 diffuses evenly over the entire surface of the electrode. In practice, the open pores are most likely to locate in a much smaller area, and then the activation loss will become much higher.

We can also consider the GDM and the CL separately (also at current density of 1.5 A cm^{-2}). The carbon paper-type GDM typically has a porosity around 60% after being teflonated and compressed in a cell, the diffusion length shown in Table 2.6 (2518 μm) is more than 10 times of its thickness (~175 μm thick under compression), so O_2 diffusion is not a problem. The CL is around 5~20 μm, adding the MPL of ca. 30 μm, the total thickness is less than 50 μm. At their typically porosity of ca. 30% O_2 can diffuse for 890 μm, much longer than 50 μm.

We can estimate what a maximum limiting current density the cathode could support by assuming the thickness of the GDM, the CL, and the MPL being 250, 15, and 25 μm, and their porosity being 70, 30, and 30%, respectively. With 30% thickness reduction, the porosity and thickness of the GDM in the fuel cell are reduced to 60% and 175 μm, respectively. The thickness and porosity of the CL and MPL layers are expected change little under compression. Then, the overall thickness and porosity of the GDE (i.e., GDM+MPL+CL) are 215 μm and 54.4%, respectively. The limiting current density is

$$j_{lim} = nFD \, c_{bulk}/L = 4 \times 96485 \times 0.27 \times 0.54^{1.5} \times 7.8 \times 10^{-6}/0.0215$$

$$= 15 \text{ A cm}^{-2} \tag{Eq. 2.39}$$

If we assume that the CL is just a surface (i.e., without any thickness and without MPL), and the GDM has a porosity of 60% and a thickness of 175 μm, then the limiting current density is

$$j_{lim} = 4 \times 96485 \times 0.27 \times 0.6^{1.5} \times 7.8 \times 10^{-6}/0.0175 = 21.6 \text{ A cm}^{-2}$$

For the above cases if the total pressure of the cathode gaseous components is increased from 1 bar to 3 bars, the limiting current density will be 3 times as high; if pure O_2 is used, the limiting current density will be around 5 times as high. Therefore, the limiting current density can range from 15 to 225 A cm^{-2} based on the diffusion of oxygen through GDE. These are very exciting numbers and no batteries have the potential to generate such high current densities. Unfortunately, due to flooding and iR loss operating a fuel cell at a current density higher than 5 A cm^{-2} is rarely achievable.

2.7.3 Flooded Electrode

Let us consider an extreme case where the GDM and CL are completely flooded. In other words, all the pores in the GMD and CL are filled with water and O_2 has to diffuse through water. The diffusion coefficient of O_2 in water is 4.8×10^{-5} cm^2 s^{-1} (at 60°C). We first need to figure out the O_2 concentration in water.

The mole fraction of O_2 in water follows the following equation in the temperature range of 273–348 K (Gevantman 1994):

$$\ln X = A + B/T^* + C \ln T^* \tag{Eq. 2.40}$$

where X is the mole fraction of O_2 in water, $T^* = T/100$ K, A = −66.7345, B = 87.4755, and C = 24.4526. Table 2.7 lists O_2 concentrations in water at different temperatures.

The O_2 concentration in water at 60°C is 8.84×10^{-7} mol cm^{-3}. Then the diffusion layer thickness at current density of 1.5 A cm^{-2} is

$$L = 4 \times 96485 \times 4.8 \times 10^{-5} \times 8.8 \times 10^{-7}/1.5 = 10.9 \times 10^{-6} \text{ cm} = 0.11 \text{ μm} \tag{Eq. 2.41}$$

TABLE 2.7

O_2 Concentrations in Water at Different Temperatures

T (°C)	T (K)	ln(X)	X (mole fraction)	O_2 Conc. (mol cm^{-3})
5	278	−10.27	3.48E-05	1.93E-06
10	283	−10.39	3.08E-05	1.71E-06
15	288	−10.50	2.77E-05	1.54E-06
25	298	−10.68	2.30E-05	1.28E-06
40	313	−10.89	1.87E-05	1.04E-06
50	323	−10.98	1.70E-05	9.45E-07
60	333	−11.05	1.59E-05	8.84E-07
70	343	−11.09	1.52E-05	8.47E-07
75	348	−11.10	1.50E-05	8.36E-07

Note: $\ln X = A + B/T^* + C \ln T^*$, $T^* = T/100$ K, $A = -66.7345$, $B = 87.4755$, $C = 24.4526$)

Source: L. H. Gevantman, "Solubility of Selected Gases in Water," in *CRC Handbook of Chemistry and Physics (1913–1995), 75th Edition (Special Student Edition)*, editor-in-chief. D. R. Lide, 6-3–6-6 (Boca Raton: CRC Press, 1994).

So, under such a circumstance, only when the combined thickness of the GDM and the CL is around 0.11 μm will the diffusion of O_2 be enough to support a current density of 1.5 A cm^{-2}. Therefore, under a completely flooded condition, an actual fuel cell with a combined thickness of the GDM and the CL of around 300 μm cannot generate a current density of 1.5 A cm^{-2} at all. Even for a nanostructured thin film electrode structure developed by 3M with a catalyst layer thickness as thin as 0.5 μm and in the absence of any diffusion medium, it is not possible to provide such a current density if the electrode is completely flooded.

Table 2.8 lists the air diffusion layer thicknesses at three different porosities that enable a cell to achieve the different current densities listed in the first column. The 100% porosity case represents bulk water; the 60% and 30% porosity cases represent complete water flooding of GDM and CL with 60% and 30% porosity, respectively. The effective diffusion coefficient of O_2 in water that is confined in porous medium is modified according to Equation 2.38. Clearly, a completely flooded catalyst layer with a thickness of 10~20 μm can only generate current densities on the order of less than 5 mA cm^{-2} even in the absence of the GDM. A completely flooded GDM with a thickness of ca. 175 μm cannot generate a current density of 1 mA cm^{-2}. Another way to look at the results shown in Table 2.8 is that in order to generate a current density of 1.5 A cm^{-2} a bulk liquid water layer (existing at the interface between the GDM and the plate is possible) cannot be thicker than 0.11 μm, and a liquid water layer within the GDM and the CL cannot exceed 50 and 20 nm thick, respectively. By using the concentration and the diffusion coefficient of H_2 in water readers can similarly estimate the anode flooding situation.

TABLE 2.8

O_2 Diffusion Layer Thickness through Water at
Different Current Densities and Porosities

j (mA cm^{-2})	L (µm)		
	100% Porosity	60% Porosity	30% Porosity
1	163.02	75.77	26.79
5	32.60	15.15	5.36
10	16.30	7.58	2.68
20	8.15	3.79	1.34
50	3.26	1.52	0.54
100	1.63	0.76	0.27
200	0.82	0.38	0.13
500	0.33	0.15	0.05
1000	0.16	0.08	0.03
1500	0.11	0.05	0.02

It is known that a phosphoric acid fuel cell (PAFC) can generate a current density around 1 A cm^{-2} before getting to the mass transport–related steep voltage drop region, indicating that the catalyst layer is not completely flooded by the liquid phosphoric acid, although the liquid acid is used as the electrolyte imbedded in either SiC or polybenzimidazole (PBI), and the catalyst layers are filled with the liquid acid for proton transport.

Table 2.9 lists the liquid water layer thicknesses under different current densities when none of the water generated at the cathode is removed (assuming the porosity of the CL is 30%). It is amazing to see that the liquid water layer would be as thick as the commonly used 20-µm thick CL in 1 minute, even at a current density of 0.1 A cm^{-2}. It can be foreseen easily

TABLE 2.9

Estimation of Liquid Water Layer Thicknesses from Cathode Water Production
at Different Current Densities

Current Density j (A cm^{-2})	Water Produced at Cathode (mmol cm^{-2} s^{-1})	Liquid Water Volume (mm^3 cm^{-2} s^{-1})	Liquid Water Thickness (µm s^{-1})	Water Thickness at 30% CL Porosity (µm s^{-1})
0.1	0.0005	0.009	0.1	0.3
0.5	0.0026	0.047	0.5	1.6
1	0.0052	0.093	0.9	3.1
2	0.0104	0.187	1.9	6.2
3	0.0155	0.280	2.8	9.3
4	0.0207	0.373	3.7	12.4
5	0.0259	0.466	4.7	15.6
10	0.0518	0.933	9.3	31.1
20	0.1036	1.866	18.7	62.2

then, that if a small portion of the water produced is not removed, it will accumulate within the CL quickly with time; then fresh air has to diffuse through the liquid water to the catalyst sites. Fortunately, with around three times the air stoichiometric ratio, the water can be taken out by the air exhaust as vapor based on thermodynamics (i.e., the saturated vapor pressure at a typical fuel cell operational temperature range).

Since water can diffuse through the PEM to the anode side for a PEMFC, let us estimate how much water can get to the anode side through diffusion. From the literature, the water's Fickian diffusion coefficient through Nafion ranges from about 3×10^{-6} to 3×10^{-5} cm^2 s^{-1} (Motupally 2000). Also, the diffusion coefficient of water can decrease for about one order of magnitude from a fully hydrated state ($\lambda = 22$) to a dry state (such as $\lambda = 3$) of the membrane. In order to simplify the calculation, let us assume that the membrane at the membrane–cathode interface is fully hydrated ($\lambda = 22$), the membrane at the membrane–anode interface has a λ of 3, and that the average diffusion coefficient of water though the membrane is 3×10^{-6} cm^2 s^{-1}. For Nafion with an EW of 1100, the concentration of $-SO_3H$ is 1.8 M (1000 cm^3 L^{-1} × 2.0 g cm^{-3}/1100 g mol^{-1}), and thus the water concentrations within the membrane are 39.6 M (1.8 M × 22) and 5.4 M (1.8 M × 3) at the membrane–cathode and the membrane–anode interfaces, respectively. Using Equation 2.28, the water flux from cathode to anode via diffusion can then be estimated for the PEM with different thicknesses. The results are listed in Table 2.10. Studies by Motupally et al. (2000) showed a water flux of nearly 0.01 mmol cm^{-2} s^{-1} through the Nafion 115 membrane, which is very close to the corresponding result shown in Table 2.10.

Comparing Table 2.10 with Table 2.9, it can be seen that at a current density of 20 A cm^{-2} or less and using 10 μm thick PEM, all the water generated at the cathode could get to the anode side. Currently, the thinnest PEM is around 25 μm. It enables all the water produced at the cathode to diffuse to the anode when the current density is as high as nearly 10 A cm^{-2}. Please note that those analyses are based on the assumption that the water arriving at the anode is effectively and immediately removed; in other words, a

TABLE 2.10

Estimated Amount of Water Diffusing through PEM of Different Thicknesses

PEM Thickness (μm)	Water Flux (mmol cm^{-2} s^{-1})
10	0.1026
25	0.0410
50	0.0205
87	0.0118
125	0.0082
175	0.0059

high water concentration gradient from the cathode to the anode exists at all times. Of course, if too much water transfers to the anode, the anode will be flooded.

What makes the water situation more complex is the huge amount of water that is electroosmotically dragged by protons from the anode to the cathode during the operation of the fuel cell. When the PEM is fully hydrated, each proton can drag three water molecules. Assuming a situation where 1.5 water molecules are dragged by each proton, Table 2.11 lists the amounts of water dragged to the cathode from the anode by protons at different current densities. The amount of water dragged by protons can be six times as much as that generated at the cathode, potentially making the flooding situation at the cathode much worse. Of course, the amount of water available to be dragged by protons depends on how much water the anode can have. It is not possible for the anode to have too much water for the proton to drag even if the fuel is fully humidified during the operation of a PEMFC. Without humidifying the fuel, the maximum amount of water to be dragged by protons will be that amount that diffuses to the anode from the cathode. It is very likely that the net flux of water to the anode will be near zero, and therefore the cathode may not be significantly flooded as long as there is enough air exhaust to take out the water produced at the cathode as vapor. Based on the saturated vapor pressure, an air stoichiometric ratio of 2.48 will be just right to take all of the water out when dry air (e.g., relative humidity, or RH = 0%) is sent to the stack and the outlet temperature of the stack is 57°C. In practice, since the RH of air cannot be 0%, and it is typically humidified to around 70%–80% RH relative to the stack outlet temperature, the air stoichiometric ratio is normally in the range of 2.5 to 3.0 to achieve water balance. Additional discussion will be presented later in this chapter.

TABLE 2.11

Amount of Liquid Water Dragged by Protons from Anode to Cathode at Different Current Densities (assuming each H^+ drags 1.5 H_2O molecules)

j (A cm^{-2})	Water Dragged by Protons (mmol cm^{-2} s^{-1})	Water Volume (mm^3 cm^{-2} s^{-1})	Water Layer Thickness (μm s^{-1})	Water Layer Thickness at 30% Cathode CL Porosity (μm s^{-1})
0.1	0.0016	0.028	0.3	0.9
0.5	0.0078	0.140	1.4	4.7
1	0.0155	0.280	2.8	9.3
2	0.0311	0.560	5.6	18.7
3	0.0466	0.840	8.4	28.0
4	0.0622	1.119	11.2	37.3
5	0.0777	1.399	14.0	46.6
10	0.1555	2.798	28.0	93.3
20	0.3109	5.597	56.0	186.6

For high-temperature fuel cells, such as molten carbonate fuel cells (MCFCs) and solid oxide fuel cells (SOFCs), where there is no liquid water at the fuel cell operating temperatures, the limiting current density can be much higher than that from a PEMFC, implying again that accumulation of liquid water plays a key role in limiting current.

From all the above simple analyses, it may be reasoned that the mass transport resistance lowers the cell performance with two different dominant ways. In the activation and the ohmic resistance regions, it is the presence of inactive species such as water vapor and nitrogen that lowers the oxygen concentration and hinders the diffusing of fresh air through the GDM and the CL toward the reaction sites. The hindrance does not affect the oxygen diffusion significantly, and thus, the voltage loss due to mass transport resistance is very small, as shown in Figure 2.9. In the limiting current density region, where a steep decline in voltage happens, it is mainly due to the accumulation of liquid water within the CL (and potentially the GDM) that reduces its porosity to less than 15%. Also, in a working fuel cell, the CL is never completely flooded by water; otherwise, it can only generate a current density on the order of mA cm^{-2}.

2.7.4 Achievable Current Density and Power Density

What current density could be achieved if liquid water accumulation did not occur in the GDE at all, as in MCFCs and SOFCs? It will be wrong if we expect the current densities to be on the order of 10's A cm^{-2} simply based on the achievable air flux through the GDE. What becomes dominating is the ohmic resistance, that is, the iR drops. Currently, the total resistance due to the PEM (or the electrolyte used in other types of fuel cells), GDM, plate, CL, and the contact resistance is around 0.1 ohm cm^2. As technology advances with time, it is possible that the total resistance will be reduced to 0.05 ohm cm^2. Table 2.12 displays the cell voltages when ΔE_A and ΔE_R are taken into consideration at different total resistances without considering the voltage loss due to mass transport (using apparent $j_{ap}^\circ = 0.14$ mA cm^{-2}, $E^\circ = 1.18$ V). Without any ohmic resistance (i.e., with only the activation loss), the fuel cell could reach a current density of 10 A cm^{-2} at a voltage of 0.61 V. When iR loss is involved, the voltage at a current density of 1 A cm^{-2} drops to 0.68 and 0.63 V for R being 0.05 and 0.10 ohm cm^2, respectively (Table 2.12). In order to maintain a reasonable fuel cell efficiency, people normally keep the individual cell voltage higher than 0.5 V (more often higher than 0.60 V for a PEMFC). Therefore, even without considering the loss due to mass transport, it would be excellent if a fuel cell reached a current density of 1 A cm^{-2} at 0.65 V or 2 A cm^{-2} at 0.55 V in actual use. When the ohmic resistance is reduced to 0.01 ohm cm^2, it would be possible to achieve a current density of 5 A cm^{-2} at a cell voltage of around 0.6 V. So, in order to achieve high current densities, one of the most important things to do is to significantly reduce the ohmic resistance.

TABLE 2.12

Hypothetical Voltage at Different Total Ohmic Resistances

		ΔE_R (V)		E (V)		
j (A cm^{-2})	ΔE_A (V)	R = 0.05 (Ω cm^2)	R = 0.1 (Ω cm^2)	R = 0.0 (Ω cm^2)	R = 0.05 (Ω cm^2)	R = 0.1 (Ω cm^2)
0.0005	0.07	0.000025	0.000050	1.11	1.11	1.11
0.001	0.10	0.000050	0.000100	1.08	1.08	1.08
0.002	0.14	0.000100	0.000200	1.04	1.04	1.04
0.003	0.16	0.000150	0.000300	1.02	1.02	1.02
0.004	0.17	0.000200	0.000400	1.01	1.01	1.01
0.005	0.18	0.000250	0.000500	1.00	1.00	1.00
0.006	0.19	0.000300	0.000600	0.99	0.99	0.99
0.007	0.20	0.000350	0.000700	0.98	0.98	0.98
0.008	0.21	0.000400	0.000800	0.97	0.97	0.97
0.009	0.21	0.000450	0.000900	0.97	0.97	0.97
0.01	0.22	0.000500	0.001000	0.96	0.96	0.96
0.02	0.25	0.001000	0.002000	0.93	0.92	0.92
0.03	0.28	0.001500	0.003000	0.90	0.90	0.90
0.05	0.30	0.002500	0.005000	0.88	0.88	0.87
0.1	0.34	0.005000	0.010000	0.84	0.84	0.83
0.15	0.36	0.007500	0.015000	0.82	0.81	0.81
0.2	0.37	0.010000	0.020000	0.81	0.80	0.79
0.3	0.39	0.015000	0.030000	0.79	0.77	0.76
0.4	0.41	0.020000	0.040000	0.77	0.75	0.73
0.5	0.42	0.025000	0.050000	0.76	0.74	0.71
0.6	0.43	0.030000	0.060000	0.75	0.72	0.69
0.7	0.44	0.035000	0.070000	0.74	0.71	0.67
0.8	0.44	0.040000	0.080000	0.74	0.70	0.66
0.9	0.45	0.045000	0.090000	0.73	0.69	0.64
1	0.45	0.050000	0.100000	0.73	0.68	0.63
1.5	0.48	0.075000	0.150000	0.70	0.63	0.55
2	0.49	0.100000	0.200000	0.69	0.59	0.49
5	0.54	0.250000	0.500000	0.64	0.39	0.14
10	0.57	0.500000	1.000000	0.61	0.11	−0.39

Note: $j_{ap}^{\circ} = 0.14$ mA cm^{-2}, $E^{\circ} = 1.18$ V.

Table 2.13 lists the expected highest power densities of a PEMFC at different current densities. Expecting power densities higher than those shown in the table is not realistic. For example, the expected highest power density is less than 0.68 and 1.18 W cm^{-2} at a cell voltage of 0.68 and 0.59 V, respectively, with gas pressure of 1 bar. The power density can be slightly higher when the gaseous reactants are pressurized to several bars.

TABLE 2.13

Highest Possible Power Densities of a PEMFC
at Different Resistance

j (A cm⁻²)	E (V)		P (W cm⁻²)	
	R = 0.05 (Ω cm²)	R = 0.1 (Ω cm²)	R = 0.05 (Ω cm²)	R = 0.1 (Ω cm²)
0.15	0.81	0.81	0.12	0.12
0.2	0.80	0.79	0.16	0.16
0.3	0.77	0.76	0.23	0.23
0.4	0.75	0.73	0.30	0.29
0.5	0.74	0.71	0.37	0.36
0.6	0.72	0.69	0.43	0.41
0.7	0.71	0.67	0.50	0.47
0.8	0.70	0.66	0.56	0.53
0.9	0.69	0.64	0.62	0.58
1.0	0.68	0.63	0.68	0.63
1.5	0.63	0.55	0.94	0.83
2.0	0.59	0.49	1.18	0.98

2.8 Impact of the Ionomer on the Three-Phase Boundary

A popular way of making the CL is using an ionomer such as Nafion as the binder and the provider of proton transport through the CL. The weight percentage of ionomer normally ranges from 15% to 35%. A question is raised that if the Pt nanoparticles are coated by a thin layer of ionomer film that prevents the reactant, such as O_2, from getting into direct contact with the Pt surface, how seriously will the ionomer film affect the electrode performance?

When the Pt surface is coated by a Nafion film, O_2 has to diffuse through the Nafion film to reach the Pt below. The O_2 diffusion coefficient and solubility in Nafion are around 0.6×10^{-6} cm² s⁻¹ and 1.6×10^{-5} mol cm⁻³, respectively (Haug and White 2000). The O_2 diffusion layer thicknesses at different current densities are calculated according to Equation 2.37 and are listed in Table 2.14. It can be seen that the Nafion thin film has little impact on the fuel cell performance at current densities achievable by a fuel cell. For example, at a current density of 1.5 A cm⁻², the thickness of the diffusion layer is 0.023 μm. In other words, when the Nafion film on Pt particles is not thicker than 23 nm, the film does not significantly affect fuel cell performance. If the Pt particle size is normally less than 5 nm in diameter, it is unimaginable that a Nafion film coating that is about 10 times its radius does not significantly affect its participation in the electrochemical reaction. Figure 2.10 shows an

TABLE 2.14

O_2 Diffusion Layer Thicknesses in
Nafion at Different Current Densities

j (mA cm⁻²)	L (μm)
1	34.740
5	6.948
10	3.474
20	1.737
50	0.695
100	0.347
200	0.174
500	0.069
1000	0.035
1500	0.023
2000	0.017
3000	0.012

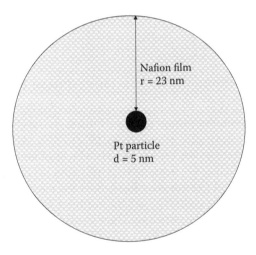

FIGURE 2.10
Illustration of Nafion film about 5 times the diameter of a Pt particle.

illustration of the two dimensions. The black dot in the middle represents a Pt particle with a diameter of 5 nm. The larger sphere represents a Nafion film about 23 nm thick. Again it is striking that such a "thick" Nafion coating on catalyst particles has little impact on the fuel cell performance.

Please note the current density mentioned above is based on the surface area of Pt particles. Referring to Table 2.3 the total Pt surface area within an electrode made of 2-nm radius Pt particles with a Pt loading of 0.5 mg cm⁻² is 350 times of the electrode geometric surface area. In other words, at a current density of 450 A cm^{-2} electrode (350 × 1.5 A cm^{-2} Pt) the coating of Nafion on the Pt nanoparticles within the electrode at a thickness of 0.023 μm has little impact to their participation in the electrochemical reaction. Of course, a fuel cell cannot generate a current density of 450 A cm^{-2} at all. Another way to look at this is that at a current density of 1.5 A cm^{-2} electrode, the current density at the surface of Pt catalyst is 4.3 mA cm^{-2} Pt (1.5 × 1000/350); referring to Table 2.14 it can be seen that the reaction is little affected even the Pt is covered with a 7 μm thick Nafion film

From this simple analysis we can conclude that very little catalyst surface is wasted due to Nafion coating on the catalyst particles within the catalyst layer. In other words, we do not need to worry about the covering of Pt particles by Nafion film within the catalyst layer.

Since Pt particles are normally deposited on a support, we can estimate the average Nafion film thickness that could coat the catalyst support. We assume that graphite particles with a diameter r_1 = 25 nm are used as the catalyst support, the densities (d) of graphite and Nafion are equal (around 2.0 g cm^{-3}), and after Nafion coating, the entire particle's diameter is r_2. Then, d(4/3) πr_1^3/d(4/3)πr_2^3 is the weight percentage of the graphite particles. Table 2.15 lists the average Nafion film thicknesses with different Nafion weight percentages in the CL. Even with 50% Nafion with respect to carbon its average coating thickness is only 3.6 nm (an illustration is shown in Figure 2.11), and therefore it should have negligible impact on oxygen diffusion. Most CLs use a Nafion weight percentage of around 25%, and thus the average Nafion film thickness is around 1.9 nm, that is, the Nafion film thickness is around one tenth of the support particle's diameter. Of course, in the CL Nafion will not form such a homogeneous coating on the primary catalyst support particles, which typically connect to form larger secondary particles.

We know that the catalyst surface utilization within a CL commonly made with Pt/C and Nafion is lower, and sometimes far lower, than 100%. The

TABLE 2.15

Average Nafion Film Thicknesses on Carbon Support with Different Nafion Weight Percentages with respect to Carbon in the CL

Nafion wt.%	Radius of Uncoated Particle r_1 (nm)	Radius of Coated Particle r_2 (nm)	Thickness of Nafion Coating r_2–r_1 (nm)
5	25	25.4	0.4
10	25	25.8	0.8
15	25	26.2	1.2
20	25	26.6	1.6
25	25	26.9	1.9
30	25	27.3	2.3
35	25	27.6	2.6
40	25	28.0	3.0
50	25	28.6	3.6

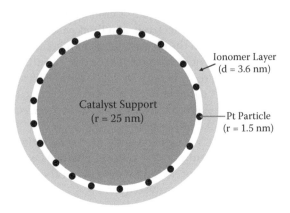

FIGURE 2.11
Illustration of Nafion film–coated catalyst support with 50% wt. Nafion in the catalyst layer.

previous analysis implies that the major reason for this is that some particles either have no ionomer nearby or lack a continuous proton transport path to the PEM. A thorough mixing of the ionomer with the catalyst particles is therefore very important in order to enhance the catalyst surface utilization.

Higher ionomer content in the CL was found to be able to aid the fuel cell startup process from subzero temperatures. A very minor reason is that the ionomer can absorb more liquid water so that the clogging of pores by ice in the CL is slowed down. The dominant reason is that O_2 can still transport through the ionomer to reach some catalyst sites even if all of pores were clogged by ice (Hiramitsu 2010).

However, it is not beneficial to add too much ionomer (e.g., more than 40% wt.) in the CL because it will reduce the electrical conductivity and the porosity of the CL, and make the CL more readily to be flooded under normal environmental temperatures with currently used porous GDE.

2.9 System Efficiency

The system electrical efficiency of a fuel cell is defined as the ratio of the electrical power sent out by the fuel cell system to the chemical energy (i.e., the enthalpy) of the fuel (e.g., H_2) received by the fuel cell system. This is the efficiency a user cares about the most, and a fuel cell developer tries every means to achieve the highest number.

The system efficiency is affected by the fuel utilization η_F, the stack efficiency η_S, the system parasitic power losses η_P, and the DC–DC converting efficiency η_{DC}. The system efficiency is

$$\eta_{Sys} = \eta_F \times \eta_S \times (1-\eta_P) \times \eta_{DC} \qquad \text{(Eq. 2.42)}$$

As described in Chapter 1, not all the fuel can be used during the operation of a fuel cell because some fuel will be wasted during the purging time period. How much hydrogen is lost depends on the purging frequency and the duration of purging. Around 5% loss of the fuel during purging is a reasonable number. In other words, the fuel utilization is normally around 95% at the current fuel cell technology level.

The stack efficiency is determined by the average cell voltage. Normally, the average cell voltage is around 0.69 V. Then, the stack efficiency is around 55% (0.69/1.25).

The DC–DC converting efficiency may range from 85% to 95%, depending on the skill of the manufacturer. A good estimate is 90% at present.

Parasitic power loss is due to the power needs of some fuel cell components, such as sensors, control boards, pumps, fans, blowers (or compressors), solenoid valves, and switches, and due to the power losses when currents pass through certain components such as the diodes and wires. Typically, the sensors, the control boards, the solenoid valves, the switches, the diodes, and the wires consume very little power. The pump for driving the liquid coolant does not consume too much power either. It is the fans (or blowers and compressors) that consume most of the parasitic power. For an air-cooled stack, the total parasitic power loss can be controlled to less than 5% of the stack output power, and for a liquid-cooled stack, the total parasitic power loss can be controlled to less than 10% of the stack output power.

Therefore, a fuel cell system with an air-cooled stack will have an electrical efficiency of around 45% ($0.95 \times 0.55 \times 0.95 \times 0.90$), and a fuel cell system with a liquid-cooled stack will have an electrical efficiency of around 42% ($0.95 \times 0.55 \times 0.90 \times 0.90$).

If the heat generated by a PEMFC is also harnessed for heating purposes, the thermal efficiency based on the coolant is around 40% (100% total − 55% stack electrical efficiency − 5% fuel loss). Therefore, the combined thermal and electrical efficiency will be 82% (42% electrical + 40% thermal) and 85% (45% electrical + 40% thermal) for liquid-cooled and air-cooled PEMFC systems, respectively. If a non-H_2 raw fuel is used, the total efficiency will be at least 5% lower due to the energy losses during the fuel processing steps. Such a system is called a *combined heat and power* (CHP) system. A CHP system typically uses liquid-cooled stacks and the heat is for hot water generation. Since not all of the thermal energy from the coolant can be harnessed, the actual combined thermal and electrical efficiency will be slightly lower than the above numbers. If part of the heat losses in the parasitic power (5–10% of the stack electric power output) and through the DC–DC conversion (10% of the stack electric power output) is harnessed for space heating, the total efficiency will be higher. Also, the 5% fuel that is lost from the stack during purging should also be utilized for heating purposes.

The H_2 consumed by the stack to generate electrical power can be easily calculated by using the current because the current is in proportion to the reactant flow rate (the Faraday law). Current (amperes) is the charge (coulombs) flow per unit time (second). One mole of H_2 will provide 2F coulombs of charge because each H_2 molecule provides 2 electrons. Then, the H_2 consumption rate by each unit cell is

$$n_{H_2} = I/(2F) \text{ (mole s}^{-1}) \tag{Eq. 2.43}$$

If the stack contains N unit cells, the total H_2 consumption rate is

$$n_{H_2} = NI/(2F) \text{ (mole s}^{-1}) \tag{Eq. 2.44}$$

Each mole of H_2 is 22.4 liters, and so the total H_2 consumption rate per minute is

$$n_{H_2} = 22.4 \times 60 \times NI/(2F) \text{ (L min}^{-1}) \tag{Eq. 2.45}$$

Similarly, each O_2 molecule involves 4 electrons, and therefore, the O_2 consumption rate per minute by a stack is

$$n_{O_2} = 22.4 \times 60 \times NI/(4F) \text{ (L min}^{-1}) \tag{Eq. 2.46}$$

There is about 21% oxygen in air; therefore, the air flow rate at 1 stoichiometric ratio is

$$n_{air} = (22.4 \times 60 \times NI)/(0.21 \times 4F) \text{ (L min}^{-1}) \tag{Eq. 2.47}$$

In practice, the air flow rate is typically at a stoichiometric ratio of between 2.5 and 3. The total air flow rate will be the stoichiometric ratio times the result from Equation 2.47.

By assuming that the water vapor also follows the ideal gas law, its production rate at the cathode is (each H_2O molecule involves 2 electrons)

$$n_{H_2O(g)} = 22.4 \times 60 \times NI/(2F) \text{ (L min}^{-1}) \tag{Eq. 2.48}$$

or

$$n_{H_2O} = 18 \times 60 \times NI/(2F) \text{ (g min}^{-1}) \tag{Eq. 2.49}$$

where 18 is the molar mass of water molecules. Equation 2.49 applies to water in either vapor or liquid form.

In order to estimate the fuel utilization, the amount of fuel being sent to the stack must be known. If there is a hydrogen flow meter, the flow rate given by this meter can be used. During purging, the flow rate is likely to be different from that when the fuel cell is in operation at a fixed load. However, since the purging duration is very short (100–500 ms), the flow difference caused

TABLE 2.16

Compressibility Factor of H_2 at Different Pressures and Temperatures

T (°C)	Pressure (MPa)						
	0.1	1	5	10	30	50	100
−150	1.0003	1.0036	1.0259	1.0726	1.3711	1.7167	—
−125	1.0006	1.0058	1.0335	1.0782	1.3231	1.6017	2.2856
−100	1.0007	1.0066	1.0356	1.0778	1.2880	1.5216	2.1006
−75	1.0007	1.0068	1.0355	1.0751	1.2604	1.4620	1.9634
−50	1.0007	1.0067	1.0344	1.0714	1.2377	1.4153	1.8572
−25	1.0006	1.0065	1.0329	1.0675	1.2186	1.3776	1.7725
0	1.0006	1.0062	1.0313	1.0637	1.2022	1.3462	1.7032
25	**1.0006**	**1.0059**	**1.0297**	**1.0601**	**1.1879**	**1.3197**	**1.6454**
50	1.0006	1.0056	1.0281	1.0567	1.1755	1.2969	1.5964
75	1.0005	1.0053	1.0266	1.0536	1.1644	1.2770	1.5542
100	1.0005	1.0050	1.0252	1.0507	1.1546	1.2596	1.5175
125	1.0005	1.0048	1.0240	1.0481	1.1458	1.2441	1.4852

Source: Adapted from NIST Reference, Fluid Thermodynamic and Transport Properties Database, http://www.nist.gov/srd/nist23.htm.

by purging can be neglected. If there is not a flow meter, the gas pressure change of the H_2 tank can be used to do the estimation by using the ideal gas law, $P_1V = n_1RT$, $P_2V = n_2RT$, where V is the tank volume, T is the environment temperature, both of which do not change, P_1 and P_2 are the gas pressures before and after use, and n_1 and n_2 are the moles of H_2 within the tank before and after use, respectively. The H_2 amount consumed is $n_1 - n_2$.

H_2 is not an ideal gas, especially when its pressure is high. Table 2.16 lists the compressibility factor (Z) of H_2 at different temperatures and pressures. n_1 and n_2 should be corrected by the compressibility factors corresponding to the pressures (and temperatures):

$$n_1 = P_1V/RTZ_{P_1} \qquad \text{(Eq. 2.50)}$$

$$n_2 = P_2V/RTZ_{P_2} \qquad \text{(Eq. 2.51)}$$

where Z_{P_1} and Z_{P_2} are the compressibility factors at pressures P_1 and P_2, respectively. Without using the compressibility factor, the amount of H_2 used will be overestimated, leading to lower (than actual) H_2 utilization estimations. For temperatures that are not given in Table 2.16, the compressibility factor at the nearest temperature can be used. A curve can be drawn using the data at 25°C and the compressibility factor at any pressure less than 100 MPa can be estimated.

2.10 Water Balance

Achieving water balance within a stack is the most important task for the stack to be able to operate smoothly and continuously. If there is more water than the stack needs, the stack will be flooded; if there is not enough water, the stack will dry up. Due to the many difficulties in balancing water, some earlier players gave up the development of PEMFCs.

A stack has two different water sources. One source is from the humidification of the reactants, either H_2 or air or both. The other source arises from the water produced at the cathode of the stack. Basically, reactant humidification is done by passing the reactant through a water source to be saturated with water vapor at the humidification temperature. The amount of water taken in by the reactant depends on the temperature of the water source. For example, at a total pressure of 1 bar, if the temperature of the water source is 47°C, the saturated vapor pressure will be 0.104 bars (please refer to Table 2.2) in the gas mixture as air passes through the water source. If the dry gas flow rate is 500 L min^{-1}, about 58 L of water vapor is taken into the stack per minute (500 × 0.104/0.896). If the stack contains 100 unit cells and operates at a current of 100 A, then the water (vapor) production rate is

$$n_{H_2O(g)} = 22.4 \times 60 \times 100 \times 100/2/96485 = 69.6 \ (L \ min^{-1})$$

So, the total water vapor at the stack cathode is 127.6 liters per minute (58 + 69.6).

The amount of oxygen consumed at the cathode is

$$n_{O_2} = 22.4 \times 60 \times 100 \times 100/4/96485 = 34.8 \ (L \ min^{-1})$$

Therefore, there will be 465.2 liters of air (500 − 34.8) leaving the stack cathode per minute. In this case, the air stoichiometric ratio is 3.0 (500 × 21%/34.8).

If the stack outlet temperature is at 57°C, that is, it is 10°C higher than the inlet temperature at saturation with a vapor pressure of 0.170 bars, there will be 95.3 liters per minute of water vapor leaving the stack (465.2 × 0.17/0.83). This number is smaller than 127.6 L min^{-1}, and thus the excess water may accumulate at the cathode, causing flooding when the net amount of water transporting through the PEM is assumed to be 0 (scenario 1).

In order to remove the excess water, an operator can either lower the humidification temperature, increase the stack outlet temperature, or increase the air flow rate. For example, for the above case, if the dry gas flow rate is increased to 1000 L min^{-1}, 116 L min^{-1} water vapor will be taken into the stack (1000 × 0.104/0.896) from humidification, leading the total water vapor at the stack cathode to 185.6 L min^{-1} (116 + 69.6). The total air leaving the stack will be 965.2 L min^{-1}, taking away water vapor at 197.7 L min^{-1}, which is higher than 185.6 L min^{-1}; therefore, flooding will not occur. In this case, the air stoichiometric ratio is 6.0 (1000 × 21%/34.8), which will impose a larger burden on the air supply device and results in more parasitic power losses (scenario 2).

In the above scenarios, it is assumed that the stack inlet temperature is the same as the humidification temperature of 47°C, and thus fully saturated air is sent to the stack. In practice, the stack inlet temperature is often controlled higher than the humidification temperature. In other words, air is only partially humidified when it enters the stack. For example, if the humidification temperature is set at 37°C, the RH of air will be 59.6% (0.062/0.104) relative to the 47°C stack inlet temperature. Water vapor taken into the stack will be 33 L min^{-1} at a dry air flow rate of 500 L min^{-1} (500 × 0.062/0.938). The total amount of water vapor at the cathode will be 102.6 L min^{-1} (33 + 69.6), still higher than the 95.3 L min^{-1} that is taken out by the air exhaust, but will be less severe than the situation presented in scenario 1 (scenario 3).

If the stack outlet temperature is increased to 67°C while still keeping the temperature difference between the stack inlet and outlet temperatures at 10°C, the humidification temperature the same as the stack inlet temperature, and the inlet air flow rate at 500 L min^{-1}, the total amount of water from humidification and production at the cathode will be 172 L min^{-1} (102.4 + 69.6), while the water taken out by air exhaust will be 171.2 L min^{-1}, achieving (Scenario 4) a water balance.

The four scenarios discussed above are summarized in Table 2.17. It can be seen that a water balance situation can be achieved at a lower air stoichiometric ratio when the stack outlet temperature is higher while keeping the temperature difference between the stack inlet temperature and the humidification temperatures the same.

Although all the methods mentioned above can be used to balance the water, the methods that are based on controlling temperatures are more practical and easier to achieve than the method based on the air flow rate in an actual fuel cell system. The stack inlet and outlet temperatures are mainly controlled by the heat exchanger and the coolant flow rate, and the amount of heat removed is mainly controlled by the fan mounted on the heat exchanger.

TABLE 2.17

Hypothetical Water Balance Scenarios (100 Unit Cells at 100 A)

Dry Air Flow Rate (L min^{-1})	Humid. Temp. (°C)	Stack Outlet T (°C)	Vapor Taken in by Air (L min^{-1})	Vapor Produced (L min^{-1})	Vapor Taken out by Exhaust (L min^{-1})	Out-Vapor Minus In-Vapor (L min^{-1})	Hypothetical Stack Symptom
500	47	57	58	69.6	95.3	−32.3	Flooding
1000	47	57	116	69.6	197.7	12.1	Drying up
500	37	57	33	69.6	95.3	−7.3	Flooding
500	57	67	102.4	69.6	171.2	−0.9	Balanced

The water situation at the cathode will be more complicated than the above analysis because some water will transport through the PEM between the two electrodes. The driving forces include diffusion due to the water concentration gradient between the two electrodes and the electroosmotic drag of water molecules by protons from the anode to the cathode. Since water is produced at the cathode and humidified air can take more water into the cathode than humidified H_2 because the air flow rate is much higher than the H_2 flow rate, water transport through the PEM driven by its concentration gradient is from the cathode to the anode. The number of water molecules electroosmotically dragged by protons from the anode to the cathode is affected by the hydration level of the PEM, with higher hydration levels allowing more water molecules to be dragged. For a fully hydrated PEM, about three water molecules will be dragged by each proton.

In addition, some liquid water will be mechanically pushed out of the stack by the exhaust gases. The amount of water being pushed out as liquid depends on the operation condition of the stack and can hardly be accurately estimated. Therefore, if the above simple analysis shows that the stack is not humidified sufficiently, its drying condition may actually be worse than is indicated by the analysis. In contrast, if the above simple analysis shows that the stack is slightly overhumidified, the stack may not actually experience flooding.

Water balance needs to be achieved at both the anode and the cathode. If H_2 is not humidified (a dead-ended operation does not humidify H_2), and by assuming that the amount of water sent out of the anode during purging is not significant because of its short duration, and by assuming that the net water transport through the PEM is zero, that is, water transport due to the water concentration gradient is counterbalanced by the electroosmotic drag, the entire cell water balance can be approximated by balancing water at the cathode. Then, the simple analyses presented above can be used as a guide or starting point. If this is followed, minor adjustments to the humidification temperature, the stack outlet temperature, or the temperature difference between the stack inlet and outlet may be able to meet the water balance requirement.

If the stack operates in a load-following fashion and the load (i.e., the current) changes frequently, balancing water becomes even more difficult. The cathode water-balancing parameters (e.g., humidification temperature, stack outlet temperatures) can be estimated beforehand using a fixed air stoichiometric ratio for several representative load levels using the simple analysis mentioned above. The fuel cell system can automatically use these parameters corresponding to each load level during operation, in conjunction with some minor adjustments based on the actual situation, to maintain the water balance.

The minor adjustments should be carried out by the fuel cell system automatically without intervention from an operator when a fuel cell is in operation. When water is not balanced, some symptoms will show up in the stack; typically its performance will decline continuously with time. The most important thing for the system to determine is whether this decline is due to flooding or the MEA drying up. It is best for the fuel cell developer to store the knowledge needed to make this judgment in the controls module beforehand, based on experience. Normally, if the decline proceeds smoothly without voltage spikes, the stack is determined to be drying up. Then the system should slightly increase the humidification temperature or decrease the stack outlet temperature. In contrast, if the decline proceeds with frequent voltage fluctuations and/or spikes, the stack is determined to be flooded. Then the system should slightly decrease the humidification temperature or increase the stack outlet temperature. If the situation improves after a minor adjustment is made, the system made the right judgment. Further adjustment can be made gradually until the situation is corrected to satisfaction. The satisfaction can be that the voltage decline rate per hour is less than a preset value. On the other hand, if the situation gets worse after a minor adjustment is made, the system made the wrong judgment, and then the system should automatically try the opposite adjustment.

If a few unit cells are flooded while the majority of the unit cells have just the right amount of water, the stack can be operated by periodically purging the stack with higher reactant flow rates to temporarily restore the voltage of the flooded unit cells. If a few unit cells are drying up while the majority of the unit cells have just the right amount of water, the stack may be usable if the humidification temperature is periodically increased or the stack outlet temperature is reduced by increasing the coolant flow rate, enabling the stack to maintain its basic function.

The worst case is that within a stack some cells are flooded, some cells are drying up, and the other cells have the right amount of water. This could be due to a poor stack design leading to uneven reactant or coolant distribution among cells, higher heat dissipation rates by the end cells, and nonuniform properties of the GDMs, the CLs, and the flow-field plates. It is very difficult to operate such a stack, and its life is very likely to be short.

2.11 Thoughts on Ultra Thin Catalyst Layers

The use of porous catalyst layer (CL) is based on the consideration that pores are needed for transporting gaseous reactants and water because a material diffuses much faster through empty space than through liquid or solid materials, and both hydrophobic and hydrophilic pores are needed for transporting gases and water, respectively. Numerous work has been done in order to create and maintain both kinds of pores within the CL, but an ideal situation is not yet achieved. A major problem encountered with such a CL with a thickness around 10~20 μm is that flooding appears to be unavoidable. In order to better utilize the catalyst surface, CLs are typically made with iono-mer as the binder and proton conductor since 1997, but such CLs are more prone to be flooded due to their lower hydrophobicity.

It is advantageous to make CLs as thin as possible in order to maximize the catalyst utilization and to reduce the iR loss and the mass transport loss. However, it becomes tricky regarding mass transport loss because if the CL is not severely flooded, it can reduce the mass transport loss, but if a proper water management situation is not achieved thin CL often can be severely or even completely flooded quickly because there is much less space to store water produced at the cathode (we can think the CL as a water reservoir, and a larger one takes longer time to be filled and has more time to send water out to the GDM, and thus it is less likely to be completely flooded). It can be seen from Table 2.9 that a water layer can become as thick as the CL in 1 minute even at a current density as low as 0.1 A cm^{-2} if none of the water produced at the cathode is removed. Therefore, water management becomes even more crucial and difficult for thin CLs.

What would happen if the CLs are made ultra thin? As shown in Table 2.8 oxygen can transport through 20 nm thick liquid water if the CL has a poros-ity of 30% to support a current density of 1.5 A cm^{-2} Pt. This current density is based on the surface area of Pt. Since the total Pt surface area is hundreds to thousands times larger than the electrode geometric surface area, the cur-rent density on the Pt surface will be correspondingly smaller than that on the electrode geometric surface area. In other words, if the current density is 1.5 A cm^{-2} electrode, the current density on the Pt surface is hundreds to thousands times smaller, i.e. on mA cm^{-2} levels. Checking Table 2.8 it can be seen that at such level of current densities, oxygen can transport for several μm. Please be warned however that this kind of correlation is too much sim-plified, but the Author has not figured out how to more accurately link them. Therefore, in the following discussion we completely ignore the correlation and just assume that the water later cannot be thicker than 20 nm. Since the CL is so thin, it is critical to increase the total catalyst surface area for reac-tion kinetics.

An ultra thin CL with the structure shown in Figure 2.12 is first postu-lated for the sake of discussion. The oriented or aligned supports (conduct-ing or insulating) are coated by a monolayer of closely packed catalyst (e.g., Pt) atoms. It will be better if the support can enhance the activity of the cata-lysts (e.g., through electronic or geometric effects). The space between those aligned structures (columns) will be filled with water completely during the operation of the fuel cell. The electrical connection is achieved by the contact between the Pt monolayer and the gas diffusion medium (GDM) (the sup-port will participate in electron conduction if it is a conducting material). The interfacial electron conductance between the columns and the GDM can be enhanced by several methods when required (e.g., making the Pt layer at the column base thicker, coating the entire column base with a good elec-tron conductor). The protons coming from the PEM transport through liquid

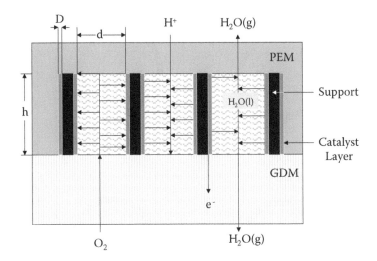

FIGURE 2.12
Side view of an ultra thin catalyst layer composed of catalyst-coated nano-columns.

water to reach the surface of every Pt atoms because water conducts protons in nano-sized space. Water formed within the water-filled CL will be in the liquid form, and transports as liquid to either the PEM (then the anode CL) or the GDM, where it is taken out by the reactant exhaust primarily as vapor. The flux of oxygen should also be adequate to diffuse through this ultra thin water layer to the catalyst surface. Therefore, such an ultra thin CL structure achieves enough transports of protons, electrons, reactants and water achieving a ~100% catalyst surface utilization.

Since the height (h) of the aligned columns (more accurately the final thickness of the CL after hot-bonding) should be short enough so that O_2 can diffuse to the tip of them at the operating current density in order to fully use the catalyst surface, the distance between the columns should then be controlled as short as possible for achieving dense package of the columns (and thus maximizing the total catalyst surface area) as long as it does not negatively affect the O_2 diffusion.

Let us consider a case as an example with the help of Figure 2.13. Columns coated with Pt atoms on the outer surface are aligned on 1 cm by 1 cm geometric GDM surface. The radius of each column is 5 nm, the distance between each column is 10 nm, and the height of each column is 20 nm. Table 2.18 shows the steps of calculation. It can be seen that 1 cm² GDM geometric surface area hosts 250 billion columns, and the total catalyst surface area is 5480 cm². Since the total surface area of 0.15 nm radius Pt atoms is 9324009 cm² g⁻¹ (Table 2.3), the Pt loading of this ultra thin CL is then 0.59 mg cm⁻² electrode (5480 cm² Pt cm⁻² electrode × 1000 mg g⁻¹/9324009 cm² g⁻¹). Considering the facts that the currently used conventional CL made with ca. 1.5 nm radius Pt nanoparticles needs a Pt loading of around 0.4 mg cm⁻² for a

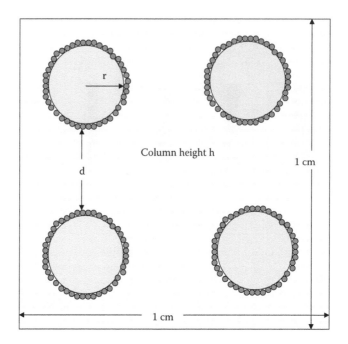

FIGURE 2.13
Top view of an ultra thin catalyst layer composed of catalyst-coated nano-columns.

good performance, and the surface area of those particles are about 10 times as low per unit mass as Pt atoms with 0.15 nm radius (Table 2.3), a Pt loading of around 0.04 mg cm^{-2} should therefore be adequate for the ultra thin CL presented here. This loading is about 15 times as low as the Pt loading presented in Table 2.18, which implies that the thickness of the ultra thin CL can still be reduced for more than 10 times. Therefore, flooding is not expected to be a problem for a 2 nm thick CL at all. Also, the DOE Pt target (0.1 g kW^{-1}) is achieved.

Another interesting implication from the above result is that if the space between all the columns is filled with ionomer completely to form a nonporous CL (NPCL) as shown in Figure 2.14, even though the oxygen permeability in Nafion is about 4 times as low as in water, oxygen should also have no problem to diffuse through a few nm thick ionomer film (Table 2.14). In such an ionomer-filled NPCL water and protons will transport through the ionomer film easily to achieve 100% catalyst utilization. Actually, the NSTF electrode made by 3 M (Debe 2006) is similar to such a CL, but it is about 250 nm thick.

Since the 3M CL works and is more than 10 times thicker than the ultra thin CL, the ultra thin CL presented here should work and probable be better.

If hollow columns are used with both of their outside and inside surfaces being coated with catalyst atoms as shown in Figure 2.15, the total catalyst

TABLE 2.18

Calculations for Ultra Thin Catalyst Layer Shown in Figure 2.13

Parameters	Equations	Results
Radius of aligned column (cm)	r	5.00E-07
Surface taken by each column (cm²)	πr^2	7.85E-13
Distance between columns (cm)	d	1.00E-06
No. of columns on 1 cm² GDM area	$[1/(2r+d)]^2$	2.50E+11
Height of each column (cm)	h	2.00E-06
Peripheral length of each column (cm)	$2\pi r$	3.14E-06
Surface area of each column (cm²)	$2\pi rh$	6.28E-12
Radius of catalyst atom (cm)	R	1.50E-08
No. of catalyst atoms in column height direction	$h/(2R)$	6.67E+01
No. of catalyst atoms in peripheral direction	$2\pi r/(2R)$	1.05E+02
No. of catalyst atoms on each column surface	$[h/(2R)]\times[2\pi r/(2R)]$	6.98E+03
Each catalyst atom surface area (cm²)	$4\pi r^2$	3.14E-12
Total catalyst surface area on each column (cm²)	$[h/(2R)]\times[2\pi r/(2R)]\times4\pi r^2$	2.19E-08
Total catalyst surface area on 1 cm² electrode (cm²)	$[1/(2r+d)]^2\times[h/(2R)]\times[2\pi r/(2R)]\times4\pi r^2$	5.48E+03

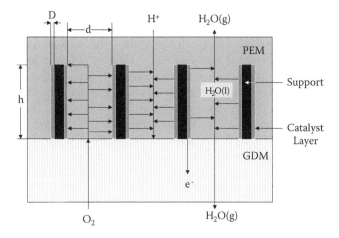

FIGURE 2.14

Side view of a non-porous ultra thin catalyst layer composed of catalyst-coated columns.

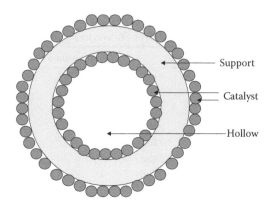

FIGURE 2.15
Top view of a hollow column with both surfaces coated with catalyst.

surface area per geometric GDM surface area can be further increased; then, the height of the columns can be further reduced. Please note that if the inner diameter of the hollow columns is too small (e.g., a few nm), ionomer may not be able to penetrate but liquid water can, and thus all the catalyst surfaces can be utilized.

Ultra thin CL can also be made with catalyst coated nanoparticles. A schematic representation is shown in Figure 2.16. A catalyst (e.g., Pt) monolayer is coated on the support electrically conducting or insulating. It is again better if the support can enhance the activity of the catalyst through electronic or geometric effects. An ultra thin CL less than 10 nm thick consisting of those particles is first applied onto GDM (better with a microporous layer). Good electrical connections among particles and between the CL and the GDM are achieved due to their intimate contact via the conducting catalyst layer. Subsequently, an ionomer solution is applied into the CL, and the ionomer should be able to penetrate into the space among the clusters (i.e., secondary particles) of those supported catalysts. Ionomer may not be able to penetrate into the space among the primary supported catalyst particles, but it will be filled with liquid water during the operation of the fuel cell. The presence of water within those nano-sized pores will provide proton transport even though there is no ionomer in it, and therefore, 100% catalyst utilization is achievable. The total catalyst surface area per GDM geometric surface area is expected to be even higher than that shown in Figure 2.12 because the primary and secondary particles shown in Figure 2.15 can be more closely packed than the columns shown in Figure 2.12.

Preventing flooding of the GDM becomes the focus for the ultra thin CLs described above. The GDM must be able to transport water out effectively; otherwise, if it is flooded, the entire gas diffusion electrode (GDE) will not be able to work properly. Fortunately, since the flooding problem changes from within both the CL and the GDM for the conventional CL to just within the GDM for the ultra thin CL, the problem is simplified and should be easier

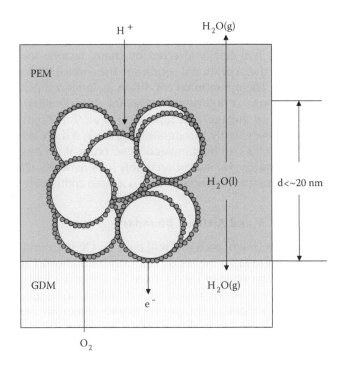

FIGURE 2.16
Side view of a non-porous ultra thin catalyst layer composed of catalyst-coated particles.

to be tackled. Optimizing the properties of the GDM such as its thickness, hydrophobicity, porosity and structure then becomes the focus.

Since the gas diffusion flux is proportional to the gas pressure (or concentration), operating the fuel cell with pressurized gases will enable the thickness of the ultra thin CL to be increased and thus the total catalyst surface area per geometric electrode area when needed.

Although oxygen has higher solubility in the currently available ionomers than in water but it has lower diffusion coefficient. Efforts to design ionomers with increased gas permeability by increasing either the gas diffusion coefficient or the solubility or both in it should be considered. The ionomer used in the ultra thin NPCL can be different from that for making the PEM in order to achieve the best performance and durability for both of their purposes.

2.12 Startup and Shutdown Strategies

Durability is most challenging in the development of all kinds of fuel cells. For commercialization fuel cells have to have lifetimes comparable to those of traditional technologies. The lifetime targets of a fuel cell for different applications can be found in Table 4.3.

The durability of a fuel cell is affected by many factors, including the materials themselves, the operational condition, the control strategy, and the design of the system. The operational condition includes temperature, relatively humidity, pressure, contaminants, reactant stoichiometric ratio, voltage cycling, open circuit voltage, and formation of air/fuel boundary at the anode. As mentioned repeatedly throughout this book, if there is H_2 (or other fuels) starvation the stack can be damaged in seconds. With careful design and proper control strategy, H_2 starvation is able to be avoided. However, the impacts of OCV and air/fuel boundary have not been completely resolved.

2.12.1 Impacts of OCV and Air/Fuel Boundary

After a fuel cell is shut down, each cell will have an OCV around 1 V. Air in the cathode chamber and H_2 in the anode chamber gradually diffuse through the PEM to the other chamber, where O_2 from air will react with H_2 to form water, which lowers the pressure inside both chambers accordingly. The diffusion rates are higher through thinner (or poorly manufactured) PEM, at higher temperatures, and with higher reactant pressures. Readers can easily calculate the fluxes of H_2 and O_2 by using the Fick's first law of diffusion (Equation 2.30) and their permeability (O_2: 9.6×10^{-12} mol cm^{-1} s^{-1}; H_2: 8.5×10^{-12} mol cm^{-1} s^{-1}) at different PEM thicknesses. The absolute pressures within the anode chamber and the cathode chamber (for closed cathode chamber) can drop much lower than the ambient pressure in 10s minutes. For example, the pressure of the anode chamber may drop to as low as 0.5 bara in 15 minutes. The lower pressures within the chambers will facilitate the diffusion of air from the environment into the stack. Finally, both the anode and the cathode chambers are filled with air, and the final pressure is the ambient pressure. Although the OCV between the cathode and the anode is 0 V, the potentials at the anode/PEM and the cathode/PEM interfaces are both around 1 V, as shown in Figure 2.17. Since higher potentials cause faster aging of the electrode components, the fuel cell may decay faster when not in operation than when in operation, which is a very disturbing fact.

What is worse is the formation of O_2/H_2 boundary at the anode during the shutdown and startup processes. This critical phenomenon was disclosed in 2005–2006 (Reiser 2005 and Tang 2006). As shown by Figure 2.18, during the startup when H_2 gets into the anode that is already filled with air, an O_2/H_2 boundary is formed at the anode. The dotted vertical line hypothetically represents the O_2/H_2 boundary in the figure, and it separates the single cell into I, II, III, and IV parts. These four parts form an internal circuit as indicated by the arrows for the flows of electrons and protons in the figure. The half reaction at Part I is the common hydrogen oxidation reaction with an electrode/ PEM interfacial potential around 0 V; the half reaction at Parts II and III is the

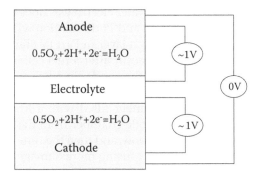

FIGURE 2.17
Typical OCV and the anode/PEM and the cathode/PEM interfacial potentials during shutdown time period.

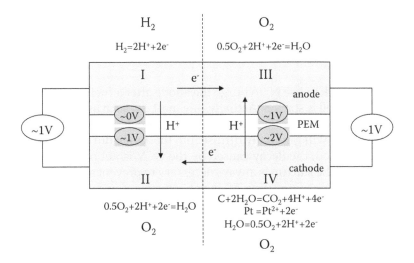

FIGURE 2.18
Voltage situation when an O_2/H_2 boundary forms at anode.

common oxygen reduction reaction with an electrode/PEM interfacial potential around 1 V. What is the electrode/PEM interfacial potential at Part IV? We can estimate it by the following simplified reasoning: since the overall potential difference between and cathode and the anode is around 1 V, and the potential difference between Part IV and Part III should be close to this potential difference, and Part III has an electrode/PEM potential around 1 V, then the electrode/PEM interfacial potential is around 2 V (1 + 1) at Part IV. In various tests, the electrode/PEM interfacial potential at Part IV is around 1.6 V. Under such a high potential, carbon corrosion, Pt oxidation and dissolution, and water electrolysis will occur as shown by the reactions listed under Part IV of Figure 2.18. Water electrolysis does not cause damage to the cathode, but carbon corrosion and Pt dissolution will quickly and significantly damage the cathode catalyst layer in Part IV (Tang 2006).

Similar process and damage can occur after a fuel cell is shut down if air from the environment diffuses into the anode chamber.

The O_2/H_2 boundary moves along the surface of the anode when a second gas (e.g., air) gets into the chamber filled with a first gas (e.g., H_2). If the second gas diffuses into the anode chamber from the environment, the movement of the boundary is quite slow, and then the time that Part IV experiences a voltage of ca. 1.6 V is long, causing more damage. If the second gas is purged into the anode that is filled with the first gas, the time will be short for the boundary to move over the entire anode, and therefore, the damage caused to Part IV will be much smaller.

2.12.2 Strategies

Methods presented below can be used individually or in combination.

2.12.2.1 N_2 Purge

People often think to use N_2 to purge the anode after a fuel cell is shut down or before a fuel cell is started. Actually, any inert gas such as water vapor and methane can be used for such a purging purpose. However, since air from the environment will gradually diffuse into the anode during the non-operational time period, the decay caused by the OCV is not prevented, and N_2 purging during the startup is always necessary to prevent air/fuel boundary formation. Also, purging using N_2 is not convenient because a N_2 cylinder must be carried for motive and portable applications and be installed on sites for stationary applications. If N_2 is not available purging the anode with air or cathode exhaust is also helpful because it can dramatically shorten the presence time of air/H_2 boundary at the anode.

2.12.2.2 H_2 Purge

Instead of using an inert gas to purge the anode, H_2 (from either the fuel tank or the anode exhaust) is used to purge the cathode after the fuel cell is shut down. With H_2 presence in the anode, the formation of O_2/H_2 boundary at the cathode will not cause any interfacial potential to go beyond ca. 1 V as shown by Figure 2.19. If H_2 can be maintained in both the anode and the cathode during a short shutdown time period (such as less than 10 minutes), restarting the fuel cell will not result in any potential higher than about 1 V. Therefore, both the potential damages caused by OCV and O_2/H_2 boundary are avoided. Before H_2 purging a dummy or auxiliary load can be used to

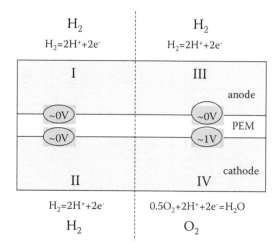

FIGURE 2.19
Voltage situation during startup of a fuel cell with both anode and cathode chambers filled with H_2.

diminish the concentration of O_2 in the cathode; in this process, the air inlet and outlet are closed but the H_2 inlet is kept open to prevent H_2 starvation.

However, for longer shutdown time period, air from the environment will diffuse into both the anode and the cathode chambers, potentially resulting in damage due to formation of air/H_2 boundaries, and when both chambers are filled with air, the OCV damage will start; also, during the subsequent startup, an O_2/H_2 boundary will form to cause further damage. In order to prevent air from diffusing into the chambers, H_2 needs to be maintained within the chambers. This can be done by keeping the H_2 inlet valves to both chambers opened and set up the H_2 inlet pressure slightly higher (such as at 0.05 barg) than the ambient pressure during the entire shutdown time period. Obviously some H_2 will be lost in the process for maintaining its pressure and presence, but the amount of loss should not be significant because the gas diffusion between the stack and the environment is expected to be very low.

2.12.2.3 H_2 Diffusion

A drawback of the above described H_2 purging is that a H_2 purging loop must be provided to the cathode, which slightly increases the system cost and the operational complexity. In order to overcome this drawback, we can rely on H_2 diffusion through the PEM from the anode to the cathode as shown in Figure 2.20. After the fuel cell is shut down, the anode outlet, the cathode inlet and outlet are closed but the anode inlet is kept open. The initial stage

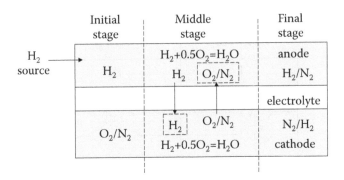

FIGURE 2.20
Diffusion of H_2 and air through PEM.

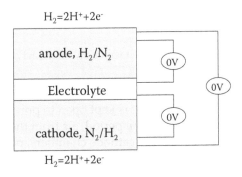

FIGURE 2.21
OCV and the anode/PEM and the cathode/PEM interfacial potentials when both anode and cathode chamber are filled with H_2.

shows that the anode and cathode have H_2 and air, respectively, immediately after the fuel cell is shut down. The middle stages shows that some H_2 diffuses to the cathode and some air diffuses to the anode; H_2 and O_2/N_2 in the dotted rectangles mean that they come from the opposite chamber. H_2 and O_2 will chemically react in both chambers once they meet, but such a minor chemical reaction is not expected to cause any concerns. In the final stage, both chambers are filled with a mixture of H_2 and N_2 because all of the O_2 originally present in cathode chamber is consumed. Both the OCV and the anode/PEM and the cathode/PEM interfacial potentials are around 0 V as shown in Figure 2.21. The H_2 inlet is kept open for the entire non-operational time period, and its pressure is set slightly higher than the ambient pressure.

If a dummy load is used to first diminish the concentration of O_2 in the cathode, the essence of the diffusion process does not change, but the cathode chamber can be filled with H_2 much faster. H_2 can also be quickly pumped from the anode to the cathode if a small external voltage is applied between the anode and the cathode (refer to Figure 4.14). However, relying on diffusion itself is simplest.

Please note that diffusion of O_2 to the anode through the PEM does not form an O_2/H_2 boundary. This is different from situations that O_2 from the environment diffuses into the anode from the flow field channels.

2.12.2.4 *Filling Stack Enclosure with H_2*

If H_2 loss is a concern for the above two approaches, especially if the seal of the stack is poor, the following method schematically represented by Figure 2.22 can be considered. The entire stack is placed in a gas-tight enclosure made of metal or plastics that is impermeable to gases. After the fuel cell is shut down the enclosure is filled with dry H_2 at a pressure slightly higher than the ambient pressure. Diffusion of H_2 and air through the PEM between the two chambers happens in the same way as described in the previous subsection. However, since the environment outside the stack now is H_2, both chambers will be finally filled predominantly with H_2; and afterward, no additional H_2 is needed because the enclosure is gas-tight. Dry instead of humidified H_2 should be used to prevent water condensation on the outer surface of the stack. It will be fine as long as the enclosure can endure a H_2 pressure slightly higher than the ambient pressure, but the material should be able to work against the H_2 embrittlement for durability and safety concerns. These needs can be easily met.

The enclosure also isolates the stack from the outside environment that will prevent cell shortening by foreign materials (e.g., liquid, metals). If some thermal insulating materials are incorporated inside or outside of the enclosure, it can also help keep the stack temperatures in cold days. Some desiccant can be placed inside the enclosure if water concentration is a concern.

This method can also be used for open cathode stacks as long as the enclosure can be sealed during the non-operational time period and air getting into and out of the enclosure is not obstructed during the operational time period.

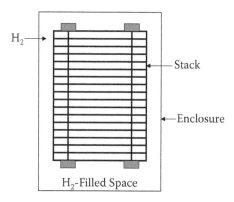

FIGURE 2.22
Illustration of a stack with its enclosure filled with H_2.

2.13 Summary

The theoretical electrical efficiency of a fuel cell is the ratio of the Gibbs free energy change to the enthalpy change of a reaction ($\Delta G/\Delta H$). The thermodynamic voltage of the cell is related to the Gibbs free energy change in the form of $E = -\Delta G/(nF)$. For an H_2/O_2 fuel cell operated at normal temperatures, the theoretical electrical efficiency/thermodynamic voltage is 94.5%/1.18 V and 83%/1.23 V when the product water is in the vapor and liquid forms, respectively. Both the theoretical electrical efficiency and the thermodynamic voltage decrease with an increase in temperature (the latter decreases for about 0.3 mV for every degree Celsius temperature increase). For example, when the temperature is at 927°C for SOFCs, they are at 72.9% and 0.94 V, respectively.

During operation, the cell voltage will be much lower than the thermodynamic voltage due to the activation loss ΔE_A, the iR loss ΔE_R, and the mass transport resistance loss ΔE_M, leading the voltage to

$$E = E^\circ - \Delta E_A - \Delta E_R - \Delta E_M$$

$$= E^\circ - [RT/(\alpha n_r F)]\ln(j/j_{ap}^\circ) - iR + [RT/(nF)]\ln(1 - j/j_{lim})$$

ΔE_A, ΔE_R, and ΔE_M dominate in the lower, middle, and higher current density regions, respectively.

Based on the diffusion of O_2 only, a fuel cell can generate a current density from 15 to 225 A cm^{-2} electrode. However, due to the activation loss and the iR loss, a fuel cell cannot generate a current density higher than ~5 A cm^{-2}. Based on Pt catalyst the fuel cell can maximally generate a current density of 1 and 2 A cm^{-2} at around 0.65 and 0.55 V, corresponding to maximum power densities of 0.65 and 1.10 W cm^{-2}, respectively. If the activity of the catalyst is increased for 10 times by using Pt alloys and core/shell particle structures, the voltage will increase for nearly 150 mV, corresponding to about 25% power density increase at the above current densities.

In the limiting current density region the significant mass transport loss is not due to consumption of the reactant but due to the flooding of the GDE. For example, a water layer as thin as 0.1 µm within the catalyst layer will cause the fuel cell not able to support a current density of 1.5 A cm^{-2}. Minimizing flooding by keeping at least 15% porosity is necessary for a normal fuel cell operation.

Based on the consideration that reactants, water and protons can transport through an ultra thin layer of either liquid water or ionomer at a thickness of less than ca. 20 or 5 nm to support the current densities during the typical operation of a fuel cell, an ultra thin catalyst layer concept is proposed. Such an ultra thin catalyst layer appears to be able to solve the CL flooding problem, and the catalyst loading per electrode geometric area can be reduced to about 0.04 mg cm^{-2} or less when supported catalyst monolayer is used.

Preventing drying up of such ultra thin CLs should be carefully considered during the operation of the fuel cell.

When a fuel cell is at the open circuit, its voltage is also far lower than the thermodynamic voltage, mainly due to the crossover of H_2 through the PEM from the anode to the cathode. At an H_2 crossover current density of as low as $2\ mA\ cm^{-2}$, the OCV drops from 1.18 V to around 1.04 V. But once the fuel cell is in operation, the voltage drop due to H_2 crossover becomes unimportant.

A fuel cell system will unavoidably encounter losses due to stack inefficiency (~55% stack efficiency), incomplete usage of fuel (95% utilization), parasitic power needs (5–10% of the stack power), and DC–DC converting loss (~10% of the stack power); the expected system electrical efficiency will be around 40%.

Coating of catalyst particles by an ionomer film is not likely to affect the availability of the catalyst surface for the electrochemical reaction. Only after the thickness of the coating exceeds about 23 nm, which is about 10 times the radius of the catalyst particle, will it affect the electrochemical reaction at a current density of $1.5\ A\ cm^{-2}$ Pt. With a typical Nafion loading of 15–35% wt. in the catalyst layer, the coating of Nafion on the catalyst support is around several nanometers thick and thus it is not expected to have any negative impact on the availability of the catalyst particles underneath. The less than 100% catalyst utilization mainly arises from the lack of ionomer around some catalyst particles.

Balancing water is the most difficult task for a PEMFC due to the high water production rate and the various water movement possibilities. Although a beforehand estimate is very helpful in determining the values of certain key parameters, such as the reactant humidification temperature, the stack inlet and outlet temperatures, the coolant flow rate, and the reactant stoichiometric ratio (mainly air), automatic adjustment by the fuel cell system itself is important in achieving the optimal operation conditions, especially if the fuel cell is in the load-following mode.

In order to avoid the accelerated cell damage caused by OCV and by the formation of an air/fuel boundary, keeping and maintaining the anode chamber and the cathode chamber in hydrogen environment when the fuel cell is in the idling state is probably the best strategy.

3

Hydrogen H$_2$

3.1 Properties of H$_2$

Hydrogen (H) is the third most abundant element on earth and exists almost exclusively in water and hydrocarbons. Its diatomic form, H$_2$, does not exist naturally on earth. The term *hydrogen* refers to H$_2$ throughout this book unless otherwise specified.

H$_2$ has found wide usage for various applications as a fuel, coolant, and reactant. Table 3.1 lists the physical and chemical properties of H$_2$. Even in liquid form, H$_2$ is about 13 times lighter than liquid water.

Extreme care should be taken when handling H$_2$ as well as other fuels. When its volume percentage in air is between 4 and 77, H$_2$ can combust when very low ignition energy is available. H$_2$ is the lightest gas, and when it leaks it can rise at a rate of around 20 m s^{-1}. Unless it is in a completely gas-tight environment, it is unlikely that H$_2$ will accumulate; this makes it difficult for H$_2$ to reach its lower combustion limit when there is minor leakage. Because gaseous fuels such as methane (CH$_4$) and propane (C$_3$H$_8$) have densities (~0.68 and ~1.91 g l^{-1} at 15°C, respectively) that are close to or denser than air (~1.23 g l^{-1} at 15°C) at ambient conditions, when they leak, they tend to accumulate in the environment, making the situation more dangerous. When H$_2$ combusts or explodes, the process finishes instantly, making rescue efforts easier than when other fuels combust or explode. Billions of people watched the H$_2$ explosions of the Japan nuclear power plants on television in 2011. The explosions finished instantly without any noticeable combustion to follow. An H$_2$ explosion dramatically contrasts with explosions caused by gasoline, where combustion will follow and continue until all the fuel and combustible materials are consumed. The combustion of gasoline also gives off dark and thick smoke, making it more difficult to locate potential victims.

3.2 Generation of H$_2$

H$_2$ can be generated by electrolysis of water, thermal splitting of water, reforming of hydrocarbons, reaction of certain chemicals with water, and

TABLE 3.1

Physical and Chemical Properties of H_2

Parameter	Unit	Value
Molar mass	g mol^{-1}	2.016
Gas density (NTP)	mg cm^{-3}	0.084
Melting point	°C	−259
Boiling point	°C	−253
Critical temperature	°C	−240
Critical pressure	MPa	1.293
Thermal conductivity at 300 K	mW m^{-1} K^{-1}	186.9
Liquid density	mg cm^{-3}	71
Combustion limits in air	% vol.	4–77
Combustion limits in O_2	% vol.	4–94
Heat of combustion at 298 K	kJ mol^{-1}	−241.8
Auto-ignition temperature	°C	560
Ignition energy	μJ	20

biological processes using enzymes and bacteria. Reforming is the widely used industry process, and large-scale reforming is quite mature.

3.2.1 Reforming of Hydrocarbons and Alcohols

There are several methods for reforming hydrocarbons to produce H_2, including partial oxidation, steam reforming, and autothermal reforming. We can use CH_4 as an example to illustrate the processes.

Under a suitable condition and using the right catalysts, CH_4 can go through the following chemical reaction with oxygen:

$$CH_4 + 0.5O_2 = CO + 2H_2 \qquad \Delta H^\circ = -247 \text{ kJ mol}^{-1} \qquad \text{(R. 3.1)}$$

This is a partial oxidation process, where each CH_4 molecule gives off two H_2 molecules and one CO molecule. The condition needs to be controlled to minimize the reaction between O_2 and H_2 or CO that is produced. This is an exothermal reaction, with 247 kJ of heat being released per mole of CH_4 in the process. Once the reaction starts, it can be sustained without input of additional heat.

In a steam reforming process, CH_4 reacts with steam through the following reaction:

$$CH_4 + H_2O \text{ (g)} = CO + 3H_2 \qquad \Delta H^\circ = 205.9 \text{ kJ mol}^{-1} \qquad \text{(R. 3.2)}$$

In order to achieve a thermal balance, the partial oxidation and the steam reforming can be combined, and the process is called *auto-thermal reforming*. If n fraction of CH_4 goes through Reaction 3.2, 1-n fraction of CH_4 goes through Reaction 3.1, then, $205.9n - 247(1-n) = 0$, $n = 0.54$. So, the auto-thermal reaction ($\Delta H° = 0$ kJ mol⁻¹) is

$$CH_4 + 0.54H_2O \text{ (g)} + 0.23O_2 + 0.86 N_2 = 0.86 N_2 + CO + 2.54H_2 \quad \text{(R. 3.3)}$$

Since air instead of pure O_2 is normally used as the O_2 source, the amount of inert gases in air (N_2, Ar, and CO_2) is represented by N_2 in Reaction 3.3. The partial oxidation will occur first near the inlet portion of the reactor, and the steam reforming occurs later in the remaining portion of the reactor.

The ratios of O_2 to CH_4 and H_2O to CH_4 should be slightly higher than those expressed in Reactions 3.1 and 3.2 in order to prevent the formation of carbon. Carbon can form through several reaction routes, and one of them is shown in Reaction 3.4:

$$CH_4 = C \text{ (s)} + 2H_2 \quad \text{(R. 3.4)}$$

Once C forms, it can form filaments quickly to cover the surface of the reforming catalyst. Such a coking process can quickly disable the reformer. When coking happens, adding extra steam or O_2 could rejuvenate the reforming catalyst to certain extent. Reaction between stream and C is shown in Reaction 3.5:

$$C + H_2O \text{ (g)} = CO + H_2 \quad \text{(R. 3.5)}$$

As CO is a severe poison to the anode catalyst of a PEM fuel cell it needs to be converted to H_2 through *water–gas shift* (WGS) process:

$$CO + H_2O \text{ (g)} = CO_2 + H_2 \qquad \Delta H° = -41.2 \text{ kJ mol}^{-1} \qquad \text{(R. 3.6)}$$

This is an exothermic process and 41.2 kJ heat is released when each mole of CO is reacted.

The overall reaction is

$$CH_4 + 2H_2O \text{ (g)} = CO_2 + 4H_2 \qquad \Delta H° = 164.7 \text{ kJ mol}^{-1} \qquad \text{(R. 3.7)}$$

A total of four H_2 molecules are generated from each CH_4 molecule. The theoretical volumetric H_2 concentration is 80%. The overall reaction is endothermic, requiring 164.7 kJ of heat for processing each mole of CH_4. Throughout the reforming process, additional heat input is needed in order to sustain the reaction. Steam reforming normally involves the reforming process and the following water–gas shift (WGS) process. Reaction 3.2 occurs exclusively in the reforming step, while Reaction 3.6 occurs partially in the reforming step. Reaction 3.6 is mainly carried out in the WGS reactor.

TABLE 3.2

Molar Gibbs Free Energy of Formation for CO, H_2O, CH_4, CO_2, H_2, and O_2 at Different Temperatures

T (°C)	T (K)	$\Delta_f G°_{CO}$ (kJ mol^{-1})	$\Delta_f G°_{H_2O}$ (kJ mol^{-1})	$\Delta_f G°_{CH_4}$ (kJ mol^{-1})	$\Delta_f G°_{CO_2}$ (kJ mol^{-1})	$\Delta_f G°_{H_2 \, or \, O_2}$ (kJ mol^{-1})
25	298	−137.2	−228.6	−50.5	−394.4	0
27	300	−137.3	−228.5	−50.4	−394.4	0
127	400	−146.3	−223.9	−41.8	−394.7	0
227	500	−155.4	−219.1	−32.5	−394.9	0
327	600	−164.5	−214.0	−22.7	−395.2	0
427	700	−173.5	−208.8	−12.5	−395.4	0
527	800	−182.5	−203.5	−2.0	−395.6	0
627	900	−191.4	−198.1	8.7	−395.7	0
727	1000	−200.3	−192.6	19.5	−395.9	0
827	1100	−209.1	−187.1	30.4	−396.0	0
927	1200	−217.8	−181.5	41.3	−396.1	0
1027	1300	−226.5	−175.8	52.3	−396.2	0
1127	1400	−235.2	−170.1	63.2	−396.2	0
1227	1500	−243.7	−164.4	74.2	−396.3	0

Source: L. V. Gurvich, V. S. Iorish, V. S. Yungman, and O. V. Dorofeeva, "Thermodynamic Properties as a Function of Temperature," in *CRC Handbook of Chemistry and Physics (1913–1995)*, 75th Edition (Special Student Edition), editor-in-chief, David R. Lide, 5-48–5-71 (Boca Raton: CRC Press, 1994).

Clearly, steam reforming can produce the largest amount of H_2 from CH_4, but the reaction needs heat to be sustained, and some CH_4 or the anode exhaust from the stack, which always contains some H_2 (and likely some unconverted CH_4) has to be burned to sustain the reaction temperature. Autothermal reforming can achieve the thermal balance, faster startup, quick response to load change, higher space velocity (e.g., more than 1 order of magnitude higher than steam reforming), but it is more complicated for controlling the amounts of both steam and air.

Since the proton exchange membrane fuel cell (PEMFC) anode catalyst can be poisoned by CO at ppm levels of concentration, it is necessary to estimate how much CO will exist at each of the fuel processing steps. Let us use the steam reforming as an example to do some analyses. Table 3.2 lists the molar Gibbs free energy of formation of the species involved, plus values for CO_2 and O_2 for later analysis.

From these data, the Gibbs free energy changes of Reaction 3.2 are calculated. Since

$$\Delta G = -RT \ln K \qquad (Eq. 3.1)$$

where R and K are the gas constant and the reaction equilibrium constant, respectively, then,

TABLE 3.3

ΔG_1 and K_1 for $CH_4 + H_2O$ (g) $= CO + 3H_2$ and ΔG_2 and K_2 for $CO + H_2O$ (g) $= CO_2 + H_2$ at Different Temperatures

T (°C)	T (K)	ΔG_1 (kJ mol⁻¹)	ΔG_2 (kJ mol⁻¹)	K_1	K_2
25	298	141.94	−28.62	0.00	104652.83
27	300	141.55	−28.55	0.00	93944.67
127	400	119.39	−24.42	0.00	1548.54
227	500	96.16	−20.45	0.00	137.31
327	600	72.22	−16.66	0.00	28.28
427	700	47.78	−13.04	0.00	9.41
527	800	23.00	−9.56	0.03	4.21
627	900	−2.00	−6.22	1.31	2.30
727	1000	−27.15	−2.98	26.25	1.43
827	1100	−52.39	0.15	308.38	0.98
927	1200	−77.67	3.20	2414.17	0.73
1027	1300	−102.97	6.17	13787.52	0.57
1127	1400	−128.25	9.07	61332.64	0.46
1227	1500	−153.51	11.91	223133.04	0.38

$$K = \exp(-\Delta G/RT) \tag{Eq. 3.2}$$

ΔG and K are shown in Table 3.3, where the subscripts 1 and 2 refer to Reaction 3.2 and Reaction 3.6, respectively.

The results clearly show that at temperatures lower than 800 K, the equilibrium constant is too small for the conversion of CH_4 to CO and H_2. At 900 K, the equilibrium constant becomes higher than 1, indicating that a significant portion of CH_4 can be converted. Higher conversion is achieved with an increase in the reforming temperature for R. 3.2.

Next, let us figure out the actual concentrations of all the species involved in the fuel processing steps. The following analysis neglects the impact of Reaction 3.6 on Reaction 3.2 during the reforming step. For Reaction 3.2, if we assume that there are initially 0.5 moles of CH_4, 0.5 moles of H_2O (i.e., the initial total number of moles of reactants is 1), and n moles and 3n moles of CO and H_2 formed, respectively, then at thermodynamic equilibrium, the amounts (equivalent to the molar or volumetric concentration, and the partial pressure) of CH_4, H_2O, CO, and H_2 are $0.5 - n$, $0.5 - n$, n, and 3n, respectively. Therefore,

$$K_1 = ([C_{CO}] [C_{H_2}]^3)/([C_{CH_4}] [C_{H_2O}]) = 27n^4/(0.5 - n)^2 \tag{Eq. 3.3}$$

Rearranging Equation 3.3 results in

$$n = [-K_1^{0.5} + (K_1 + 4 \times 27^{0.5} \times 0.5\ K_1^{0.5})^{0.5}]/(2 \times 27^{0.5}) \tag{Eq. 3.4}$$

TABLE 3.4

Species Contents at Equilibrium during CH_4 Steam Reforming Step

T (K)	K_1	C_{CH_4} (mole)	C_{H_2O} (mole)	C_{CO} (mole)	C_{H_2} (mole)	CH_4 Conversion Efficiency (%)
298	0.00	0.5000	0.5000	0.0000	0.0000	0.0
300	0.00	0.5000	0.5000	0.0000	0.0000	0.0
400	0.00	0.5000	0.5000	0.0000	0.0001	0.0
500	0.00	0.4990	0.4990	0.0010	0.0029	0.2
600	0.00	0.4918	0.4918	0.0082	0.0247	1.6
700	0.00	0.4618	0.4618	0.0382	0.1147	7.6
800	0.03	0.3853	0.3853	0.1147	0.3440	22.9
900	1.31	0.2606	0.2606	0.2394	0.7183	47.9
1000	26.25	0.1351	0.1351	0.3649	1.0948	73.0
1100	308.38	0.0578	0.0578	0.4422	1.3265	88.4
1200	**2414.17**	**0.0240**	**0.0240**	**0.4760**	**1.4281**	**95.2**
1300	13787.52	0.0106	0.0106	0.4894	1.4682	97.9
1400	61332.64	0.0051	0.0051	0.4949	1.4846	99.0
1500	223133.04	0.0027	0.0027	0.4973	1.4918	99.5

The amounts of CH_4, H_2O, CO, and H_2 in the reformer at equilibrium are listed in Table 3.4. Again, the units for the amounts can also be regarded as number of moles, molarity, volumetric concentration, or partial pressure (standard pressure is 1 bar).

Clearly, in order to maximize the conversion of CH_4 to CO and H_2, the reformer temperature should be kept as high as possible, but high temperatures impose challenges to materials and could accelerate the sintering rate of the reforming catalysts. The last column shows how much CH_4 is converted, that is, the conversion efficiency. For example, if the reformer temperature is maintained at 1200 K (927°C), there will be 0.024 moles of CH_4 unconverted, accounting for 4.8% of the original amount of CH_4 fed to the reformer.

The results shown in Table 3.4 are plotted in Figure 3.1. In order to achieve a conversion efficiency of 88% or higher, the reforming temperature should be kept at 827°C or higher.

Due to the high concentration of CO after the reforming step, the gas mixture is typically sent to WGS reactors. The reactors will significantly lower the CO concentration through Reaction 3.6. We assume that the reformer is at 1200 K (927°C), and another 0.5 moles of H_2O are added to react with CO according to Reaction 3.6; then the initial number of moles of CO, H_2O, and H_2 are 0.476, 0.524, and 1.428, respectively (refer to Table 3.4 at 1200 K). We further assume that n moles of CO_2 are formed in the WGS reactor. Then, at equilibrium, the number of moles of CO, H_2O, CO_2, and H_2 are 0.476 – n, 0.524 – n, n, and 1.428 + n, respectively. Therefore,

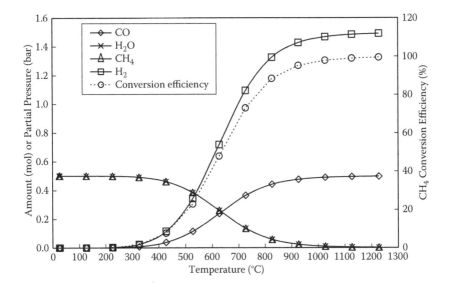

FIGURE 3.1
Amounts of species in reformer at equilibrium and CH$_4$ conversion efficiency.

$$K_2 = ([C_{CO_2}][C_{H_2}])/([C_{CO}][C_{H_2O}]) \qquad \text{(Eq. 3.5)}$$

$$= [n(1.428 + n)]/[(0.476 - n)(0.524 - n)]$$

Rearranging Equation 3.5 results in

$$(K_2 - 1)n^2 - (1.428 + K_2)n + 0.249K_2 = 0. \qquad \text{(Eq. 3.6)}$$

Then,

$$n = \{1.428 + K_2 - [(1.428 + K_2)^2 - 4 \times 0.249\, K_2(K_2 - 1)]^{0.5}\}/[2(K_2 - 1)] \quad \text{(Eq. 3.7)}$$

The results at different WGS temperatures are listed in Table 3.5.

In the above analysis it is assumed that 0.5 moles of water is added in the reforming step and another 0.5 moles of water is added in the water-gas shift process. In practice, (slightly) more than 1 mole of water may be added altogether in the reforming step in order to simplify the process and prevent cooking.

In contrast to reforming, lower WGS temperatures convert CO to CO$_2$ (and H$_2$) more effectively. For example, at 1200 K, only 17.4% of CO is converted, but at 700 K, 57.7% is converted, and at 500 K, 87.9% of CO is converted. But even at as low as 300 K (27°C), there is still 0.3% CO in the mixture, too high to be used directly in a PEMFC. The second to last column shows the percentage of CO in the gas mixture. The last column shows the percentage

TABLE 3.5

Equilibrium Species Contents in the WGS Reactor

T (K)	CO (mole)	H_2O (mole)	CO_2 (mole)	H_2 (mole)	CO Conversion Efficiency (%)	CO% in Mixture	H_2% in Mixture
298	0.008	0.056	0.468	1.896	98.4	0.3	78.1
300	0.008	0.056	0.468	1.896	98.4	0.3	78.1
400	0.016	0.064	0.460	1.888	96.7	0.6	77.8
500	0.057	0.105	0.419	1.847	87.9	2.4	76.1
600	0.127	0.175	0.349	1.777	73.2	5.2	73.2
700	0.201	0.249	0.275	1.703	57.7	8.3	70.1
800	0.265	0.313	0.211	1.639	44.4	10.9	67.5
900	0.313	0.361	0.163	1.591	34.2	12.9	65.5
1000	0.349	0.397	0.127	1.555	26.7	14.4	64.0
1100	0.375	0.423	0.101	1.529	21.3	15.4	63.0
1200	0.393	0.441	0.083	1.511	17.4	16.2	62.2
1300	0.406	0.454	0.070	1.498	14.6	16.7	61.7
1400	0.416	0.464	0.060	1.488	12.5	17.2	61.3
1500	0.424	0.472	0.052	1.480	10.9	17.5	61.0

of H_2 in the gas mixture, and it is 76% at 500 K, about 4% lower than the theoretical value (80%). In addition to the incomplete conversion of CO to H_2 through Reaction 3.6, the major contribution arises from the incomplete conversion of CH_4 to CO and H_2 in the reforming step through Reaction 3.2.

The equilibrium concentrations of the species are plotted in Figure 3.2. Please note that the CO concentration shown in Table 3.5 is based on the stoichiometric steam-to-CH_4 ratio (S/C) of 2 (R. 3.7). The CO concentration will be lower in practice because people normally use the S/C ratio higher than 2. This is good for lowering the CO concentration and for preventing formation of carbon, but it will lower the system's overall thermal efficiency because additional heat is needed for generating the additional amount of steam.

Although lower temperatures favor the conversion of CO, the reaction rate will be slower. So, the WGS process cannot be carried out at too low a temperature; if it is, either the reaction rate is not fast enough to meet the fuel cell need, or the WGS reactor will be too large. In practice, the WGS is likely to be carried out in two reactors; one is at a higher temperature such as 400°C (HTS, or high-temperature shift), and the other is at a lower temperature such as 200°C (LTS, or low-temperature shift). Also, the coolant flow should be better in a counter flow direction with respect to the flow of the reactants. In such a configuration the exit of the reformate will have the lowest temperature favoring the highest conversion of CO to CO_2.

In both the reforming (i.e., Reaction 3.2) and the WGS (i.e., Reaction 3.6) steps, if a material that can consume CO_2 exists, both processes will be shifted to the right side (e.g., the product side), helping achieve higher CH_4 conversion efficiency (or lower reforming temperature at the same CH_4 conversion

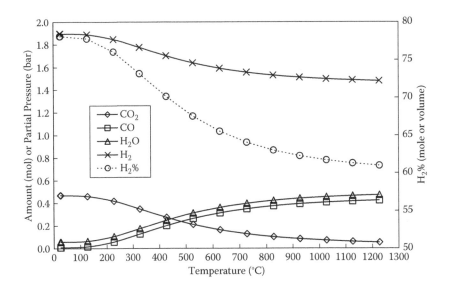

FIGURE 3.2
Equilibrium concentrations within the WGS.

efficiency) and lower CO concentration. CaO is a common material that can combine with CO_2 to form $CaCO_3$, and thus it is a potential candidate. CaO (or other chemicals that can react with CO_2) should be in a form that is stable in the reforming or the WGS process. Beaver et al. reported such a sorption enhanced steam reformer using hydrotalcite (promoted by K_2CO_3) as the sorbent that lowered the CH_4 reforming temperature to less than 550°C and achieved >99% CH_4 conversion with CO concentrations less than ca. 20 ppm (Beaver 2010). An inconvenience is that the sorbent needs to be frequently regenerated because it will be quickly saturated by CO_2, which seriously limits its applicability.

After the LTS, there will be around 2% CO in the gas mixture (Table 3.5). It is then removed by a preferential oxidizer (Prox) at temperatures slightly higher than 100°C through Reaction 3.8.

$$CO + 0.5O_2 = CO_2 \qquad\qquad \Delta H° = -283 \text{ kJ mol}^{-1} \qquad (R. 3.8)$$

The catalyst used should facilitate the oxidation of CO, but not H_2, in order to minimize the loss of H_2. The ratio of the moles of CO that are converted to CO_2 to 2 times the total moles of O_2 consumed in the Prox is the *reaction selectivity*. The selectivity is affected by the catalyst and the temperature. Higher temperatures do not favor selectivity because more H_2 will be oxidized by O_2. In addition, higher temperatures will lead to more CH_4 formation through the reverse reactions of R. 3.2 and R. 3.7 and the formation starts to accelerate at temperatures higher than ca. 200°C. Optimally, the selectivity can reach around 50%, that is, about the same amount of H_2 and CO will be oxidized

TABLE 3.6

ΔG_3 and K_3 for Prox Reaction

T (°C)	T (K)	ΔG_3	K_3
25	298	−257.21	1.28E + 45
27	300	−257.05	6.01E + 44
127	400	−248.32	2.78E + 32
227	500	−239.50	1.08E + 25
327	600	−230.67	1.24E + 20
427	700	−221.85	3.66E + 16
527	800	−213.06	8.30E + 13
627	900	−204.31	7.31E + 11
727	1000	−195.58	1.67E + 10
827	1100	−186.90	7.58E + 08
927	1200	−178.25	5.80E + 07
1027	1300	−169.64	6.60E + 06
1127	1400	−161.06	1.03E + 06
1227	1500	−152.52	2.06E + 05

by O_2 in the best-case scenario. Therefore, for converting 2% CO to CO_2 with the remaining CO at ppm levels, the amount of air added will be around 9.5% [(2% × 0.5)/(50% × 0.21)]. Table 3.1 shows that the combustion limits for H_2 in air are 4–77%, which means that the combustion limits for air in H_2 are 23–96%. Since 9.5% is out of this range, explosion should not be able to occur (the mixture also contains nearly 20% CO_2). Even so, the amount of air sent to the Prox should be carefully controlled to avoid the formation of an explosive gas mixture.

It can be seen from Table 3.6 that the equilibrium constant is huge for Reaction 3.8 at temperatures between 100°C and 200°C. In order to achieve a reasonable reaction rate and to minimize the amount of H_2 that is oxidized, the Prox temperature is normally kept between 110°C and 150°C.

From reforming to HTS, LTS, and Prox, the temperatures of the reactors are around 1200 K (927°C), 700 K (427°C), 500 K (227°C), and 400 K (127°C), respectively. There are heat exchangers between adjacent reactors to control the temperatures. The gas mixture from the Prox is higher than that suitable for a regular PEMFC, and thus there is another heat exchanger to lower the temperature. Clearly, a PEMFC system becomes very complicated when non-H_2 is used as the starting fuel.

Another popularly used hydrocarbon is C_3H_8. The steam reforming reaction is as follows:

$$C_3H_8 + 3H_2O \text{ (g)} = 3CO + 7H_2 \qquad \Delta H° = 497.7 \text{ kJ mol}^{-1} \qquad (R. 3.9)$$

The steam to carbon ratio is 1. Each C_3H_8 molecule produces seven H_2 molecules. The theoretical molar fraction of H_2 in the product mixture is 70%.

Gasoline is widely used nowadays, and it can be represented by octane (C_8H_{18}). The steam reforming reaction is

$$C_8H_{18} \text{ (g)} + 8H_2O \text{ (g)} = 8CO + 17H_2 \qquad \Delta H^\circ = 1300.5 \text{ kJ mol}^{-1} \quad \text{(R. 3.10)}$$

The steam to carbon ratio is 1. Each C_8H_{16} molecule produces 17 H_2 molecules. The theoretical molar fraction of H_2 in the product mixture is 68%.

Another potential alkane fuel is ethane, C_2H_6, and the steam reforming reaction is

$$C_2H_6 + 2H_2O \text{ (g)} = 2CO + 5H_2 \qquad \Delta H^\circ = 346.6 \text{ kJ mol}^{-1} \quad \text{(R. 3.11)}$$

Since reforming alkanes larger than CH_4 involves the breaking of the C–C bonds, avoiding the formation of C filaments becomes more challenging.

Methanol and ethanol can also be reformed to form H_2 according to Reactions 3.12 and 3.13, resulting in 75% theoretical molar fraction of H_2 in the product mixture.

$$CH_3OH \text{ (g)} + H_2O \text{ (g)} = CO_2 + 3H_2 \qquad \Delta H^\circ = 49.3 \text{ kJ mol}^{-1} \quad \text{(R. 3.12)}$$

$$C_2H_5OH \text{ (g)} + 3H_2O \text{ (g)} = 2CO_2 + 6H_2 \quad \Delta H^\circ = 173.3 \text{ kJ mol}^{-1} \quad \text{(R. 3.13)}$$

In order to figure out the suitable temperatures for reforming C_2H_6, C_3H_8, CH_3OH, and C_2H_5OH, let us calculate the equilibrium constants at different temperatures.

Table 3.7 lists the Gibbs free energy of formation of these four chemicals. Tables 3.8 and 3.9 list the molar Gibbs free energy changes and the equilibrium constants of Reactions 3.9, 3.11, 3.12, and 3.13, respectively.

Equilibrium constants reaches to the order of 10^4 at temperatures of 900 K (627°C), 800 K(527°C), 600 K (327°C), and 500 K (227°C) for Reactions 3.11, 3.9, 3.13, and 3.12, respectively. So, for the reforming of ethane, propane, ethanol, and methanol, the temperatures will be around 900 K (627°C), 800 K (527°C), 600 K (327°C), and 500 K (227°C), respectively, for nearly complete conversion of the chemicals.

For the three alkanes (methane, ethane, and propane) discussed here, the reforming temperature decreases with an increase in the carbon chain length, with propane having the lowest reforming temperature. For the two alcohols discussed here, the reforming temperature increases with an increase in the chain length. Since methanol can be effectively reformed at about 250°C, it is a preferred fuel for applications requiring quick startup.

Catalysts perform a crucial role during fuel processing. They should most catalyze the desired reactions (i.e., forming H_2 and reducing CO) and least catalyze any undesired reactions (e.g., formation of carbon, methane and any other hydrocarbon compounds, and consumption of H_2). Cheekatamarla

TABLE 3.7

Molar Gibbs Free Energy of Formation for C_2H_6, C_3H_8, CH_3OH, and C_2H_5OH at Different Temperatures

T (°C)	T (K)	$\Delta_f G°_{C_2H_6}$	$\Delta_f G°_{C_3H_8}$	$\Delta_f G°_{Methanol}$	$\Delta_f G°_{Ethanol}$
25	298	−32.0	−23.5	−162.3	−167.9
27	300	−31.7	−23.0	−162.1	−167.5
127	400	−13.5	15.0	−148.5	−144.2
227	500	5.9	34.5	−134.1	−119.8
327	600	26.1	65.0	−119.1	−94.7
427	700	46.8	96.1	−103.7	−69.0
527	800	67.9	127.6	−88.1	−43.0
627	900	89.2	159.4	−72.2	−16.8
727	1000	110.8	191.4	−56.2	9.5
827	1100	132.4	223.6	−40.1	36.0
927	1200	154.1	255.8	−23.9	62.5
1027	1300	175.9	288.0	−7.6	89.1
1127	1400	197.6	320.2	8.6	115.6
1227	1500	219.4	352.4	24.9	142.2

Source: L. V. Gurvich, V. S. Iorish, V. S. Yungman, and O. V. Dorofeeva, "Thermodynamic Properties as a Function of Temperature," in *CRC Handbook of Chemistry and Physics (1913–1995), 75th Edition (Special Student Edition)*, editor-in-chief, David R. Lide, 5-48–5-71 (Boca Raton: CRC Press, 1994).

TABLE 3.8

Molar Gibbs Free Energy Change for Steam Reforming C_2H_6, C_3H_8, Methanol, and Ethanol

T (°C)	T (K)	$\Delta G_{C_2H_6}$	$\Delta G_{C_3H_8}$	$\Delta G_{Methanol}$	$\Delta G_{Ethanol}$
25	298	214.8	297.7	−3.5	64.9
27	300	214.0	296.5	−3.8	64.2
127	400	168.6	217.6	−22.2	26.6
227	500	121.4	156.4	−41.8	−12.9
327	600	73.0	83.6	−62.0	−53.6
427	700	23.8	9.8	−82.8	−95.3
527	800	−25.9	−64.6	−104.0	−137.6
627	900	−75.9	−139.4	−125.4	−180.4
727	1000	−126.1	−214.5	−147.1	−223.5
827	1100	−176.4	−289.7	−168.9	−266.8
927	1200	−226.9	−364.9	−190.8	−310.3
1027	1300	−277.3	−440.1	−212.7	−354.0
1127	1400	−327.7	−515.3	−234.7	−397.7
1227	1500	−378.0	−590.4	−256.8	−441.4

TABLE 3.9

Equilibrium Constants for Steam Reforming C$_2$H$_6$, C$_3$H$_8$, Methanol, and Ethanol

T (°C)	T (K)	K$_{C_2H_6}$	K$_{C_3H_8}$	K$_{Methanol}$	K$_{Ethanol}$
25	298	2.10E − 38	6.18E − 53	4.10E + 00	4.20E − 12
27	300	5.19E − 38	2.26E − 52	4.63E + 00	6.55E − 12
127	400	9.39E − 23	3.66E − 29	8.07E + 02	3.34E − 04
227	500	2.06E − 13	4.49E − 17	**2.31E + 04**	2.21E + 01
327	600	4.41E − 07	5.20E − 08	2.52E + 05	**4.67E + 04**
427	700	1.67E − 02	1.84E − 01	1.52E + 06	1.30E + 07
527	800	4.90E + 01	**1.66E + 04**	6.22E + 06	9.71E + 08
627	900	**2.55E + 04**	1.25E + 08	1.92E + 07	2.97E + 10
727	1000	3.90E + 06	1.62E + 11	4.87E + 07	4.77E + 11
827	1100	2.42E + 08	5.79E + 13	1.06E + 08	4.75E + 12
927	1200	7.58E + 09	7.80E + 15	2.03E + 08	3.28E + 13
1027	1300	1.40E + 11	4.94E + 17	3.56E + 08	1.70E + 14
1127	1400	1.71E + 12	1.72E + 19	5.79E + 08	7.00E + 14
1227	1500	1.48E + 13	3.71E + 20	8.83E + 08	2.40E + 15

and Finnerty reviewed reforming catalysts in 2006 (Cheekatamarla 2006). Interest in reactor design can refer to (Önsan 2011).

3.2.2 Cracking of Ammonia

Since ammonia (NH$_3$) contains 75% atomic hydrogen, it can also be a good H$_2$ source for fuel cells after it goes through the following reaction in the presence of suitable catalysts.

$$NH_3 \text{ (g)} = 0.5N_2 + 1.5H_2 \qquad \Delta H° = 45.9 \text{ kJ mol}^{-1} \qquad \text{(R. 3.14)}$$

Table 3.10 lists the thermodynamic parameters of NH$_3$ at different temperatures.
    If we assume that the initial molar concentration of NH$_3$ is 1, and n moles of it cracks to form N$_2$ and H$_2$, then

$$K = ([C_{N_2}]^{0.5} [C_{H_2}]^{1.5})/[C_{NH_3}] = [(0.5n)^{0.5}(1.5n)^{1.5}]/(1-n) \qquad \text{(Eq. 3.8)}$$

Rearranging Equation 3.8 results in

$$1.3n^2 + Kn - K = 0 \qquad \text{(Eq. 3.9)}$$

Then

$$n = [-K + (K^2 + 5.2K)^{0.5}]/2.6 \qquad \text{(Eq. 3.10)}$$

It can be seen from Table 3.11 that at 700 K (427°C), 98.8% of NH$_3$ is converted to N$_2$ and H$_2$, and the H$_2$ percentage in the product mixture reaches

TABLE 3.10

Enthalpy Change, Gibbs Free Energy
Change, and Equilibrium Constant during
NH_3 Cracking at Different Temperatures

T (°C)	T (K)	ΔH_{NH_3}	ΔG_{NH_3}	K_{NH_3}
25	298	45.9	16.4	1.33E – 03
27	300	46.0	16.2	1.49E – 03
127	400	48.1	6.0	1.65E – 01
227	500	49.9	–4.8	3.15E + 00
327	600	51.4	–15.8	2.40E + 01
427	700	52.7	–27.2	1.07E + 02
527	800	53.7	–38.6	3.34E + 02
627	900	54.5	–50.2	8.26E + 02
727	1000	55.1	–61.9	1.72E + 03
827	1100	55.6	–73.6	3.15E + 03
927	1200	55.9	–85.4	5.24E + 03
1027	1300	56.1	–97.2	8.07E + 03
1127	1400	56.3	–109.0	1.17E + 04
1227	1500	56.3	–120.8	1.61E + 04

TABLE 3.11

Equilibrium Contents during NH_3 Cracking at Different Temperatures

T (°C)	T (K)	C_{NH_3}	C_{N_2}	C_{H_2}	NH_3 Conversion Efficiency (%)	H_2% in Mixture
25	298	9.69E – 01	0.016	0.047	3.1	4.6
27	300	9.67E – 01	0.017	0.050	3.3	4.8
127	400	7.01E – 01	0.149	0.448	29.9	34.5
227	500	2.39E – 01	0.380	1.141	76.1	64.8
327	600	4.90E – 02	0.476	1.427	95.1	73.1
427	700	1.19E – 02	0.494	1.482	98.8	74.6
527	800	3.86E – 03	0.498	1.494	99.6	74.9
627	900	1.57E – 03	0.499	1.498	99.8	74.9
727	1000	7.55E – 04	0.500	1.499	99.9	75.0
827	1100	4.13E – 04	0.500	1.499	100.0	75.0
927	1200	2.48E – 04	0.500	1.500	100.0	75.0
1027	1300	1.61E – 04	0.500	1.500	100.0	75.0
1127	1400	1.11E – 04	0.500	1.500	100.0	75.0
1227	1500	8.05E – 05	0.500	1.500	100.0	75.0

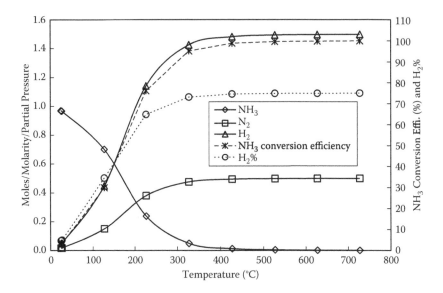

FIGURE 3.3
Equilibrium concentrations for NH$_3$ cracking.

74.6%. At higher temperatures, the conversion efficiency is even higher, achieving 100% at 1000 K (727°C). The data are plotted in Figure 3.3.

Since there is no CO formation in the NH$_3$ cracking process, WGS and Prox reactors are not needed, further simplifying the process. However, since NH$_3$ is basic and can react with Nafion in the catalyst layers and the PEM, a filter containing some acidic materials should be used to trap any unconverted ammonia before the product mixture enters the PEMFC.

Table 3.12 summarizes the processing of fuels discussed so far on the reforming temperature, HTS temperature, LTS temperature, Prox temperature, and the theoretical volumetric H$_2$% in the product mixtures after reforming and WGS processes. The data listed in Table 3.12 can be used as a guide for designing the fuel processors in the absence of any CO$_2$-consuming materials (in the presence of CO$_2$-consuming materials, the reforming and

TABLE 3.12

Comparison of Fuel Processing among CH$_4$, C$_2$H$_6$, C$_3$H$_8$, CH$_3$OH, C$_2$H$_5$OH, and NH$_3$

Parameter	CH$_4$	C$_2$H$_6$	C$_3$H$_8$	CH$_3$OH	C$_2$H$_5$OH	NH$_3$
Reforming Temp. (K)	1200	900	800	500	600	700
HTS Temp. (K)	700	700	N/A	N/A	N/A	N/A
LTS Temp. (K)	500	500	500	N/A	N/A	N/A
Prox Temp. (K)	400	400	400	400	400	N/A
H$_2$% Molar Limit	80	78	77	75	75	75

WGS temperatures can be lowered). Due to the incomplete conversion, addition of excess amount of water, and some potential losses, the actual $H_2\%$ will be a few percentages lower than those listed in the table. But if the $H_2\%$ volume from actual fuel processors is much lower than those numbers, the processors need improvement.

3.2.3 Electrolysis of Water

3.2.3.1 Efficiency

Water electrolysis is the opposite process to the fuel cell reaction

$$H_2O \text{ (l)} = 0.5O_2 + H_2 \qquad\qquad \text{(R. 3.15)}$$

with $\Delta H^\circ = 285.8$ kJ mol^{-1}, and $\Delta G^\circ = 237.1$ kJ mol^{-1}. The difference between the two is thermal energy $T\Delta S$, 48.7 kJ mol^{-1}. It needs electrical power to proceed. Electrolysis of liquid water is discussed in this section, although water vapor can be similarly electrolyzed. The thermodynamic voltage is 1.23 V at 298 K [(237.1 × 1000)/(2 × 96485)], which can be considered the theoretical voltage enabling the electrolysis of water to start. Based on ΔH°, the voltage is 1.48 V [(285.8 × 1000)/(2 × 96485)]. We can view Reaction 3.15 as indicating that 237.1 kJ mol^{-1} electrical energy input will result in 285.8 kJ mol^{-1} chemical energy. The theoretical efficiency of electrolysis under standard conditions is 121% (285.8/237.1). This may be surprising, but if we reverse the process by allowing O_2 and H_2 to form liquid water (i.e., the fuel cell mode), the electrical efficiency based on the HHV is 0.83. 0.83 times 1.21 results in 1, which means that the overall efficiency of the two opposite processes is 100%, obeying the law of energy conservation. The differences in the two efficiency numbers arise from the thermal or heat energy: a fuel cell releases heat to the environment while an electrolyzer takes in heat from the environment, and the total energy does not change in either process.

If an electrolyzer works at 100% efficiency (i.e., at 1.48 V under the standard condition), the heat generated from the electrolyzer will be equal to the heat needed for the electrolysis to proceed, and therefore, a thermoneutral situation is achieved. In other words, the electrolyzer neither releases nor absorbs heat to or from the environment. The voltage, 1.48 V under this condition, is called the thermoneutral voltage. Since ΔH increases with temperature, then the thermoneutral voltage increases slightly with temperature (please refer to Table 2.1 for water in vapor phase). Also, since the difference between ΔH and ΔG increases with temperature, the heat needed for the electrolysis to proceed increases with the temperature.

Reaction 3.15 is composed of two half reactions, as shown by Reactions 3.16 and 3.17, respectively.

Positive electrode: H_2O (l) $= 0.5O_2 + 2H^+ + 2e^-$ (R. 3.16)

Negative electrode: $2H^+ + 2e^- = H_2$ (R. 3.17)

Reaction 3.16 produces O_2 and is called the *oxygen evolution reaction* (OER). The protons transport through the PEM from the positive electrode to the negative electrode, where they combine with the electrons from the power source to form H_2.

Due to the activation, ohmic resistance, and mass transport voltage losses, the needed voltage for water electrolysis is much higher than 1.23 V. We can perform an estimate here on an electrolyzer using PEM as the electrolyte. The activation loss is dominated by the loss due to the OER, and the ohmic resistance loss is dominated by the PEM. In order to minimize the crossover of the products O_2 and H_2 through the PEM, an electrolyzer typically uses thicker membranes. When Nafion 117 is used as the PEM, the total resistance of a cell is around 0.2 ohm cm² (sum of ionic resistance, electronic resistance, and contact resistance), and if we assume that the apparent j_{ap}°, the charge transfer coefficient α, and the number of electrons in the rate-determining step for the OER are 0.14 mA cm⁻², 0.5, and 1, respectively, we can estimate the external voltage required to generate different current densities at 25°C due to the activation loss ΔE_A (using the Tafel equation) and the resistance loss ΔE_R ($= iR$). The results are listed in Table 3.13. Please note that the efficiency is based on ΔH° (corresponding to 1.48 V) and the activation loss is based on ΔG° (corresponding to 1.23 V). Figure 3.4 shows plots of j versus E and j versus η.

If we assume that the loss due to the mass transport resistance is negligible, then at an external voltage of 1.6–1.7 V, the electrolysis current densities will range from 50 to 500 mA cm⁻² with an efficiency around 85–95%.

Please note that j_{ap}° of 0.14 mA cm⁻² is adapted from the ORR with 0.5 mg cm⁻² Pt loading (Section 2.4). Since Pt is not the best catalyst for the OER, and the catalyst loading at the positive electrode is often at 2–4 mg cm⁻², the actual j_{ap}° for OER is about 10 times higher than 0.14 mA cm⁻², and thus the activation loss can be 100~150 mV lower than those shown in Table 3.13 (check Table 2.4 and refer to the dotted V-I curve in Fig. 3.4). Popular catalysts for OER include Ir, IrO_2, PtIr, and RuIr. For example, Jung et al. reported that the onset potential of OER is about 150 mV lower when using $Pt_{70}Ir_{30}$ catalyst than using Pt catalyst (Jung 2009).

If the H_2 and O_2 produced are pressurized, further efficiency loss will occur. By simply using the Nernst equation,

$$\Delta E_{H_2} = [RT/(2F)]\ln(P_{2H_2}/P_{1H_2})$$ (Eq. 3.11)

$$\Delta E_{O_2} = [RT/(4F)]\ln(P_{2O_2}/P_{1O_2})$$ (Eq. 3.12)

TABLE 3.13

External Voltage Required to Generate Different Current
Densities via Liquid Water Electrolysis (T = 298 K, E° = 1.23 V,
j_{ap}° = 0.14 mA cm^{-2}, α = 0.5, and n_r = 1)

j (mA cm^{-2})	j/j_{ap}°	ΔE_A (V)	ΔE_R (V)	E (V)	η (%)
0.5	4	0.07	0.0001	1.30	114
1	7	0.10	0.0002	1.33	111
2	14	0.14	0.0004	1.37	108
3	21	0.16	0.0006	1.39	107
4	29	0.17	0.0008	1.40	106
5	36	0.18	0.0010	1.41	105
6	43	0.19	0.0012	1.42	104
7	50	0.20	0.0014	1.43	103
8	57	0.21	0.0016	1.44	103
9	64	0.21	0.0018	1.45	102
10	71	0.22	0.002	1.45	102
20	143	0.25	0.004	1.49	99
30	214	0.28	0.006	1.51	98
50	357	0.30	0.01	1.54	96
100	714	0.34	0.02	1.59	93
150	1071	0.36	0.03	1.62	91
200	1429	0.37	0.04	1.64	90
250	1786	0.38	0.05	1.66	89
300	2143	0.39	0.06	1.68	88
400	2857	0.41	0.08	1.72	86
500	3571	0.42	0.10	1.75	85
600	4286	0.43	0.12	1.78	83
700	5000	0.44	0.14	1.81	82
800	5714	0.44	0.16	1.83	81
900	6429	0.45	0.18	1.86	80
1000	7143	0.46	0.20	1.89	79

where P_{2H_2}, P_{2O_2} are the H_2 and O_2 pressures after pressurization; P_{1H_2}, P_{1O_2} are the ambient pressures of H_2 and O_2 (1 bar), respectively; and ΔE_{H_2} and ΔE_{O_2} are the voltage losses at the H_2 and O_2 evolution sides, respectively. The voltage losses at 60°C and different levels of pressurization are listed in Table 3.14. Since the PEM cannot endure a pressure difference more than about 5 bars, it is best to make the volume of the H_2 evaluation chamber twice that of the O_2 evolution chamber; then the pressure difference between the PEM will be 0. For automobile applications, people normally pressurize H_2 to either 350 bars or 700 bars to reduce the space taken by the fuel storage tanks. Table 3.14 shows that a total of 10.2% and 11.5% losses occur at these two pressures, respectively.

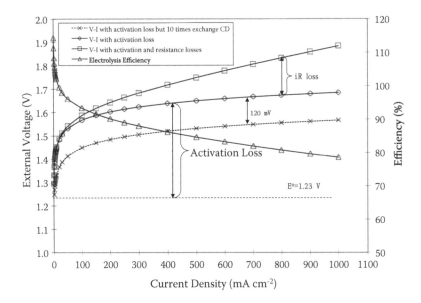

FIGURE 3.4
Electrolysis current density versus external voltage and efficiency.

TABLE 3.14

Voltage Losses from Pressuring H$_2$ and O$_2$ (60°C)

H$_2$/O$_2$ Pressure (bar)	H$_2$ Side Loss (V)	O$_2$ Side Loss (V)	Total Loss (V)	Percentage Loss (%)
5/5	0.023	0.012	0.035	2.8
10/10	0.033	0.017	0.050	4.0
50/50	0.056	0.028	0.084	6.8
75/75	0.062	0.031	0.093	7.6
100/100	0.066	0.033	0.099	8.1
200/200	0.076	0.038	0.114	9.3
300/300	0.082	0.041	0.123	10.0
350/350	0.084	0.042	0.126	10.2
400/400	0.086	0.043	0.129	10.5
500/500	0.089	0.045	0.134	10.9
600/600	0.092	0.046	0.138	11.2
700/700	0.094	0.047	0.141	11.5
750/750	0.095	0.047	0.142	11.6
800/800	0.096	0.048	0.144	11.7
900/900	0.098	0.049	0.146	11.9
1000/1000	0.099	0.050	0.149	12.1

FIGURE 3.5
LN H_2 station. Courtesy of Beijing Jonton Hydrogen Tech. Company.

The losses from pressurization listed in Table 3.14 are theoretical losses. In an electrolysis cell, it is very difficult to achieve a pressure higher than 100 bars due to engineering difficulties and the crossover of the gaseous products through the electrolyte. Typically, people use compressors driven by an electrical motor to pressurize the gases. If the combined efficiency of the electrical motor and the compressor is around 80% by pressurizing H_2 from 1 bar to 700 bars, then the overall efficiency of an electrolyzer due to activation loss, ohmic resistance loss, and pressurization loss will be slightly more than 70% (90% × 80%).

So, in order to achieve a higher overall efficiency, it is better to only slightly pressurize the gases within the electrolyzer, as long as the pressure is adequate for feeding the fuel cell, and not use electrical motors and compressors.

Figure 3.5 shows an H_2 generation and storage station built in 2005 in Beijing by LN Green Power. It uses two alkaline solution electrolyzers to generate 300 Nm^3 H_2 per hour. H_2 pressure was internally boosted to 16 bars in the electrolyzer, and then externally boosted to as high as 770 bars in one step.

3.2.3.2 Internal Pressurization

For a membrane electrolyzer, we can figure out how high the pressure can be achieved within the electrolyzer. By using the equation

$$j_{lim} = -nFD \, dc/dx = nFD \, c_{bulk}/L \qquad \text{(Eq. 3.13)}$$

and assuming Nafion 117 is being used, we can estimate the limiting current that will cross over the membrane with a thickness of 175 μm. At ambient pressure, the diffusion coefficient and solubility of O_2 in Nafion are around 0.6×10^{-6} cm² s⁻¹ and 1.6×10^{-5} mol cm⁻³, respectively (Haug and White 2000). If we assume that the O_2 solubility in Nafion is linearly proportional to its pressure, and the parameters for H_2 are the same as those for O_2, Table 3.15 lists the O_2 and H_2 crossover current densities at different pressures.

Although the crossover current densities are not significant (for example, at O_2/H_2 pressures of 100/200 bars, the crossover current density is 21.2 mA cm⁻²), potential formation of explosive gas mixtures must be considered and avoided. The last two columns in Table 3.15 list the ratios% of O_2 or H_2 crossover current density to the electrolysis current density (ECD) at 200 and 400 mA cm⁻², respectively. For a mixture consisting of H_2 and O_2, the combustion range is around 4–94% vol. H_2 (Wikipedia). Table 3.15 shows that the maximum H_2 pressure within an electrolyzer cannot be higher than about 70 and 140 bars at 200 and 400 mA cm⁻² electrolysis current densities, respectively, in order for the H_2 vol.% to be lower than 4% (assuming the O2 chamber also contains about 50% water vapor).

TABLE 3.15

O_2 and H_2 Limiting Crossover Current Density (CD) at Different Pressures and Ratio% of O_2 and H_2 Limiting Crossover CD to Electrolysis CD

O_2/H_2 Pressure (bar)	O_2 or H_2 Limiting Crossover CD (mA cm⁻²)	Ratio% of O_2 or H_2 limiting Crossover CD to Electrolysis Current Density (ECD)	
		200 mA cm⁻² ECD	400 mA cm⁻² ECD
1/2	0.2	0.11	0.05
5/10	1.1	0.53	0.26
10/20	2.1	1.06	0.53
15/30	3.2	1.59	0.79
20/40	4.2	2.12	1.06
25/50	5.3	2.65	1.32
30/60	6.4	3.18	1.59
35/70	7.4	**3.71**	1.85
40/80	8.5	4.23	2.12
50/100	10.6	5.29	2.65
70/140	14.8	7.41	**3.71**
100/200	21.2	10.59	5.29
150/300	31.8	15.88	7.94
200/400	42.3	21.17	10.59
300/600	63.5	31.76	15.88
350/700	74.1	37.06	18.53
400/800	84.7	42.35	21.17

From this simple analysis, it can also be concluded that the electrolyzer works best with the gaseous products near the ambient pressure. When hydrogen infrastructure is established in the future, most of the H_2 will be transported via pipelines to end users, and thus for stationary applications, there is probably no need to pressurize H_2 higher than several bars.

However, for automobile applications, H_2 has to be highly pressurized due to space limitations. If the overall electrolysis plus pressurization efficiency is 70% to generate 700 bars H_2, and the overall fuel cell system efficiency is 40%, then the entire electrolyzer–fuel cell system can achieve a well-to-wheel electrical efficiency of around 28%. This is still about twice of the tank-to-wheel efficiency achieved by conventional internal combustion engines (ICEs).

Converting electrical power to electrical power at an overall efficiency of around 28% seems to be discouraging and uneconomical. However, the power used to electrolyze water can be more or less free when renewable energy such as solar and wind energy is used. When fossil fuels are used up in the distant future, it is unavoidable that H_2 will become the major energy carrier, especially for use in the transportation sector because solar and wind power can hardly be used directly as motive power. Nowadays, only a small portion of the energy generated from wind is used by the grid in many regions because the power is not stable. But when it is used to electrolyze water, this is not a problem. It is expected that when fuel cells are commercialized, the H_2 economy will push solar and wind energy to much greater deployment.

3.2.3.3 Proton Onsite PEM Electrolyzer

Building a membrane electrolyzer is no less challenging than building a PEM fuel cell. Proton OnSite (formerly Proton Energy Systems, Inc.) is a world leader in developing and manufacturing PEM-based electrolyzers. The product portfolio includes the HOGEN® S Series, H Series, and C Series. The systems are fully automated with push-button start/stop, E-stop, onboard H_2 leak detection, remote shutdown capability, and automatic load-following, fault detection, and system depressurization. The S10, S20, and S40 systems in the S Series offer net H_2 generation rates of 0.265, 0.53, and 1.05 Nm^3 h^{-1}, respectively, at an H_2 delivery pressure of 13.8 barg. The H2m, H4m, and H6m systems in the H Series offer net H_2 generation rates of 2, 4, and 6 Nm^3 h^{-1}, respectively, at an H_2 delivery pressure of 15 barg (with a 30 barg option). The C10, C20, and C30 systems in the C Series offer net H_2 generation rates of 10, 20, and 30 Nm^3 h^{-1}, respectively, at an H_2 delivery pressure of 30 barg. Table 3.16 lists the major parameters of the C10, C20, and C30 systems. Figure 3.6 shows a picture of the Proton OnSite electrolyzer product family.

We can estimate the overall system efficiency of Proton OnSite electrolyzers based on the average electrical energy consumption of around 6.7, 7.0, and 6.0 kWh to produce 1 Nm^3 H_2 at 1 atm for the S, H, and C Series products, respectively. 1 Nm^3 H_2 at 1 atm (14.7 psia) and 0°C contains 44.6 moles of H_2

TABLE 3.16

Major Specifications of Proton OnSite HOGEN® C Series PEM Electrolyzers
(Courtesy of Proton OnSite)

Parameter	Unit	Electrolyzer Model		
		C10	C20	C30
Net H$_2$ rate (0°C, 1 bar)	Nm^3 h^{-1}	10	20	30
H$_2$ delivery pressure	barg		30	
Power consumption	kWh Nm^{-3}H$_2$	6.2	6.0	5.8
H$_2$ purity	%		99.9998	
			(water/N$_2$/O$_2$ < 2/2/1 ppm)	
Turndown range	%		0–100	
Field upgradeability	—		To 30 Nm^3 h^{-1} H$_2$	N/A
Maximum water consumption rate	L h^{-1}	9	17.9	26.9
Water temperature	°C		5–50	
Water pressure	barg		1.4–4.1	
Input water quality	µS cm^{-1}		< 1/0.1 (required/recommended)	
Liquid coolant temperature	°C		Electrolyzer: 5–50	
			Power Supply: 5–40	
Electrolyzer max. heat load	kW h^{-1}	28.7	58.4	88.1
Power supply max. heat load	kW h^{-1}	4.8	8.4	12.1
Recommended breaker rating	kVA	100	200	250
Electricity specification	—		342–456 Vac, 3 phases, 50 Hz	
			432–528 Vac, 3 phases, 60 Hz	
Dimension W/D/H	mm/mm/mm		Electrolyzer: 2388/914/2007	
			Power Supply: 1880/914/2007	
Product weight	kg	2041	2449	2812
Storage/transport temp.	°C		5–60	
Ambient temp. range	°C		5–40	
Altitude range	m		Up to 2000	
Maximum on-board H$_2$ inventory	Nm^3		0.08 @ 30 barg	
Audible noise at 1 meter	dB		<75	

Source: Modified from Proton Onsite Website, HOGEN C Series Specification Sheet, http://www.protononsite.com/pdf/HOGEN_C.pdf.

$[(1 \times 1000)/(0.0821 \times 273)]$. The energy content at a higher heating value (HHV) within such an amount of H$_2$ is 3.54 kWh ($44.6 \times 285.8/3600$). Therefore, the system efficiency of the S, H, and C Series products is 52.8% (3.54/6.7), 50.6% (3.54/7.0), and 59.0% (3.54/6.0), respectively. Please note that this efficiency is for converting AC electrical energy to chemical energy and includes all of the energy needed to produce the hydrogen and purify it to the stated purity specification, including AC–DC power conversion and hydrogen drying. Energy needed to dry hydrogen to the required purity specification is quite significant (>10%). Otherwise, the system efficiency is about 10% higher.

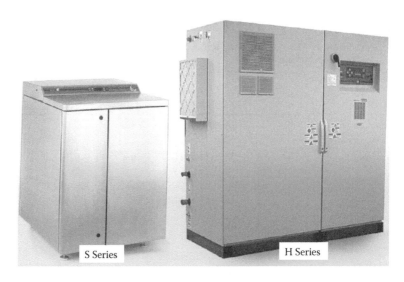

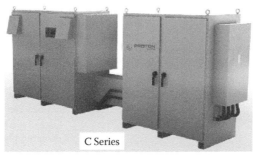

FIGURE 3.6
Proton OnSite HOGEN® electrolyzer product family. Courtesy of Proton OnSite.

3.2.4 Chemical Sources

3.2.4.1 Metals

Many metals can react with water to produce H_2 through Reaction 3.18.

$$M + 2xH_2O = M(OH)_{2x} + xH_2 \qquad \text{(R. 3.18)}$$

The hydrogen source comes from the water. Theoretically, any metal with "activity" higher than that of H_2 could react with water to form H_2. In other words, any metal with a standard reduction potential lower than 0 V (the standard hydrogen potential) could react with water to form H_2. Let us take the reaction between Na and water as an example:

$$2Na + 2H_2O \ (l) = 2NaOH + H_2 \qquad \text{(R. 3.19)}$$

We know from general chemistry classes that this reaction can proceed vigorously at room temperature without using any catalyst. Two Na atoms react with two water molecules to result in one H$_2$ molecule. The H$_2$ mass percentage from an H$_2$-generating device using this reaction can theoretically reach 2.4% wt. [(1 × 2)/(2 × 18 + 2 × 23)]. Considering the mass of the storage device for Na and water, and the connection piping, the H$_2$ mass percentage will be smaller than 2.4%. Since the standard reduction potentials are so much lower for Li, K, Ca, Na, and Mg than the standard hydrogen potential, they should be able to react with water easily. Table 3.17 lists some metals that could potentially be used for generating H$_2$ for fuel cell applications.

Without considering the weight of water, the largest theoretical H$_2$% weight will be 22.2% from Be, followed by Li (14.5%), Al (11.1%), and Mg (8.2%). When the weight of water that reacts with the metal to release H$_2$ is included, the largest theoretical H$_2$% weight is only 4.44% for Be, followed by Li (4.02%), Al (3.7%), and Mg (3.3%). Those numbers are still lower than the 5.5% gravimetric H$_2$ storage target set by the US Department of Energy (DOE) for 2017 (the ultimate target is 7.5%). Fortunately, since the fuel cell will generate water, if some of the water can be used, the situation may be changed. For example, if one third of the water needed for the reaction is collected, then the system only needs to carry the remaining two thirds of the water. As can be seen from Table 3.17, the H$_2$ weight percentage reaches 6.06% for Be, followed by Li (5.29%), Al (4.76%), and Mg (4.14%). If all the water produced by the fuel cell is collected, the system only needs to carry half of the water needed; then the H$_2$ weight percentage reaches 7.41% from Be, followed by Li (6.29%), Al (5.56%), and Mg (4.73%). Under such a scenario, Be approaches the DOE ultimate target, but Li, Al and Mg are still below the target. Considering the cost and abundance of metals and the difficulties in controlling their reactions with water, Al is the most preferred choice, followed by Mg, for generating H$_2$.

Typically, water is added to the metal storage tank at a carefully controlled rate according to the H$_2$ flow rate needed by the fuel cell. Obviously, if the addition rate of water is not enough, the fuel cell will suffer from H$_2$ starvation, which can destroy the stack in seconds. On the other hand, if water is added too quickly, H$_2$ will build up in the reaction vessel, which could cause the vessel to burst. There should be a sensor in the reaction vessel to monitor the H$_2$ pressure inside. It should be possible to quickly and automatically stop the addition of water once the pressure reaches a preset high-pressure limit. Also, the amount of either the metal or the water or both consumed or remaining should be monitored, especially during the operation of the fuel cell, to prevent running the fuel cell after either the metal or the water depletes. When either the metal or the water is found to be near depletion, it needs to be quickly added, or the fuel cell is shut down. Water is easier to monitor than metal. Alternatively, the cumulative charge generated by the stack can be used as a control factor. When it is near the designed capacity

TABLE 3.17

Metals That Could Be Used to Generate H_2 by Reacting with H_2O

Element	Molar Mass (g mol^{-1})	Reaction with H_2O	Standard Red. Potential (V)	H_2% with 100% Water (wt.)	H_2% with 2/3 Water (wt.)	H_2% with 1/2 Water (wt.)	H_2% without Water (wt.)
Li	6.9	$Li + H_2O = LiOH + 0.5H_2$	−3.04	4.02	5.29	6.29	14.49
Na	23	$Na + H_2O = NaOH + 0.5H_2$	−2.71	2.44	2.86	3.13	4.35
K	39.1	$K + H_2O = KOH + 0.5H_2$	−2.93	1.75	1.96	2.08	2.56
Be	9	$Be + 2H_2O = Be(OH)_2 + H_2$	−1.85	4.44	6.06	7.41	22.22
Mg	24.3	$Mg + 2H_2O = Mg(OH)_2 + H_2$	−2.37	3.32	4.14	4.73	8.23
Ca	40.1	$Ca + 2H_2O = Ca(OH)_2 + H_2$	−3.80	2.63	3.12	3.44	4.99
Mn	54.9	$Mn + 2H_2O = Mn(OH)_2 + H_2$	−1.19	2.20	2.53	2.74	3.64
Fe	55.9	$Fe + 2H_2O = Fe(OH)_2 + H_2$	−0.45	2.18	2.50	2.71	3.58
Co	58.9	$Co + 2H_2O = Co(OH)_2 + H_2$	−0.28	2.11	2.41	2.60	3.40
Ni	58.7	$Ni + 2H_2O = Ni(OH)_2 + H_2$	−0.26	2.11	2.42	2.61	3.41
Zn	65.4	$Zn + 2H2O = Zn(OH)_2 + H_2$	−0.76	1.97	2.24	2.40	3.06
Al	27	$Al + 3H2O = Al(OH)_3 + 1.5H_2$	−1.66	3.70	4.76	5.56	11.11

of the H_2-generating device based on the lower amount of either the metal or the water, a signal for either adding more fuel (metal, water, or both) or shutting down the fuel cell should be generated for the system to perform actions accordingly. Another approach is to use the pressure sensor mentioned previously. When it indicates that the H_2 pressure approaches a preset low-pressure limit, it signals the fuel cell system to take actions (adding fuel or shutting down the fuel cell). The H_2 pressure change from normal to alarmingly low could happen in seconds, so the fuel cell system must be able to act quickly.

Serious consideration should be given to any unreacted metal that remains after a fuel cell is shut down. For example, some remaining metal particles may be covered by a metal oxide layer that prevents the metal beneath from participating in H_2 generation effectively or efficiently. It is a technical challenge to maintain the remaining metal (such as Al) intact after being partially used.

3.2.4.2 Complex Metal Hydrides

Some metal hydrides, such as $LiBH_4$, $NaBH_4$, $Mg(BH_4)_2$, and $Ca(BH_4)_2$, have pretty high theoretical gravimetric hydrogen contents (18.4%, 10.6%, 14.8%, and 11.5%, respectively), but only a portion of the hydrogen can be reversibly released, making the actual H_2 amount released much lower. When they react with water and the mass of the water is included, their releasable H_2% weight is 13.9%, 10.8%, 12.7%, and 11.3%, respectively, all higher than the 7.5% DOE ultimate target, even when the weight of water is taken into consideration. Since the fuel cell reaction will produce twice the amount of the water needed for the metal hydride reactions, all the water needed can be recovered from the fuel cell reaction, and therefore, the H_2% weight can be as high as 36.9%, 21.2%, 29.7%, and 23.0% for $LiBH_4$, $NaBH_4$, $Mg(BH_4)_2$, and $Ca(BH_4)_2$, respectively. Those results are summarized in Table 3.18. Another group of complex metal hydrides contains AlH_4^- such as $LiAlH_4$, Li_3AlH_6, $Mg(AlH_4)_2$, $Ca(AlH_4)_2$, $LiMg(AlH_4)_3$, as well as AlH_3.

TABLE 3.18

Potential H_2% wt. Obtained from Metal Hydrides

Metal Hydride	Molar Mass (g mol⁻¹)	H_2% (wt.)	Reaction with H_2O	H_2% with Water (wt.)	H_2% without Water (wt.)
$LiBH_4$	21.7	18.4	$LiBH_4 + 2H_2O = LiBO_2 + 4H_2$	13.9	36.9
$NaBH_4$	37.8	10.6	$NaBH_4 + 2H_2O = NaBO_2 + 4H_2$	10.8	21.2
$Mg(BH_4)_2$	53.9	14.8	$Mg(BH_4)_2 + 4H_2O = Mg(BO_2)_2 + 8H_2$	12.7	29.7
$Ca(BH_4)_2$	69.7	11.5	$Ca(BH_4)_2 + 4H_2O = Ca(BO_2)_2 + 8H_2$	11.3	23.0

The most popular metal hydride is sodium borohydride, $NaBH_4$, which reacts with water according to Reaction 3.20.

$$NaBH_4 + 2H_2O \ (l) = NaBO_2 + 4H_2 \qquad \Delta H^\circ = -218 \ kJ \ mol^{-1} \qquad (R. \ 3.20)$$

One mole of sodium borohydride reacts with two moles of water to produce four moles of H_2 and one mole of sodium borate. The theoretical H_2% weight from such a system is 10.8% when water is externally added, or 21.2% when water is collected from the fuel cell reaction. Due to this high H_2 weight percentage, $NaBH_4$ is regarded as one of the best fuels for portable fuel cells. The major drawbacks of using $NaBH_4$ are its high cost and the inconvenience associated with handling the caustic $NaBO_2$ produced.

It is worth mentioning that $NaBH_4$ can be directly oxidized electrochemically at the anode of an alkaline fuel cell (AFC) with the following anode reaction:

$$NaBH_4 + 8OH^- = NaBO_2 + 6H_2O + 8e^- \qquad (R. \ 3.21)$$

The reduction of O_2 at the cathode is

$$O_2 + 2H_2O + 4e^- = 4OH^- \qquad (R. \ 3.22)$$

And the overall reaction is

$$NaBH_4 + 2O_2 = NaBO_2 + 2H_2O \qquad E^\circ = 1.64 \ V \qquad (R. \ 3.23)$$

But due to the cost and various technical reasons, people rarely use $NaBH_4$ as a direct fuel for AFCs.

Hydrazine (N_2H_4) has been proposed by some for use as an H_2 source because it contains 12.5% weight hydrogen. Its melting and boiling points are at 1.5°C and 113.7°C, respectively, so it is normally in either solid or liquid form. H_2 can be released easily because the reaction is exothermic:

$$N_2H_4 \ (l) = N_2 + 2H_2 \qquad \Delta H^\circ = -50.6 \ kJ \ mol^{-1} \qquad (R. \ 3.24)$$

Hydrazine itself could also be directly used as a fuel, as shown by Reaction 3.25.

$$N_2H_4 \ (l) + O_2 = N_2 + 2H_2O \qquad \Delta G^\circ = -632.9 \ kJ \ mol^{-1} \qquad (R. \ 3.25)$$

The thermodynamic voltage is 1.64 V [(632.9 × 1000)/(4 × 96485)], about 400 mV higher than that of H_2/O_2 fuel cells.

However, hydrazine is extremely toxic and dangerously unstable. Due to these serious problems, it is not recommended that hydrazine be used as an H_2 source or directly oxidized at the fuel cell anode for the general public to use.

3.3 Hydrogen Storage

3.3.1 Compressed Gas

Although H_2 has the highest specific energy density it has the lowest volumetric energy density due to its extremely low density in the gas form (0.084 mg cm^{-3}). Even in the liquid form its density is only 71 mg cm^{-3}. Liquefaction of H_2 can only occur when the temperature is lower than the critical temperature, which is –240°C (and at a suitable pressure). Achieving such a low temperature is not a trivial task, and it consumes a lot of energy (theoretically about 1/10, but in practice about 1/3 of the lower heating value of H_2). Only for large-scale industrial use will H_2 be liquefied. For fuel cells, the cheapest way of storing H_2 at the moment is using the popularly available steel bottles. Each 40-liter bottle weighs about 54 kg and can store about 0.37 g H_2 at a pressure of around 125 bars, with an H_2 mass percentage of 0.69%. For automobile applications, due to the space and weight limitations, people use lightweight high-pressure bottles made of Al, or a certain polymer inner liner wrapped and reinforced with carbon fiber and fiberglass. The pressures of these bottles are typically at either 350 bars or 700 bars, and the pressurization process will consume about 12 and 15% of the LHV of H_2, respectively. For example, a 30-liter, 350-bar bottle manufactured by Beijing Jonton Hydrogen Tech. Company weighs 16.8 kg and stores about 0.70 kg of H_2, achieving an H_2 mass percentage of 4.2%. The amazing lightness of these bottles can be seen in Figure 3.7, a picture of a 350-bar bottle taken during the 2011 Tokyo Fuel Cell Expo. Figure 3.8 shows a picture of the Al inner liner of a 350-bar bottle. Figure 3.9 shows a 350-bar H_2 supply system consisting of 15 12-liter bottles.

3.3.2 Metal Hydride

Certain metals, such as Pd, Ti, Mn, Ni, Mg, and La-series metals or their alloys, can store several percentage points of H_2 by weight at ambient temperatures and pressures around 10–40 bars. The major types of metal hydrides include AB_5 (such as $LaNi_5$), AB_2 (such as $TiMn_2$), AB (such as TiFe), and A_2B (such as Mg_2Ni). H_2 enters into the crystal structures of these metals via a chemical–physical process. These metals or alloys behave like a sponge for H_2. It is believed that H_2 dissociates into H atoms and stays in the empty space within the metal or metal alloy crystal structures. This is an exothermic process, and heat is released. In order to keep the temperature low (lower temperatures favor H_2 storage in metal hydrides), heat is best removed through circulation of liquid water. When hydrogen is released from the metal hydride, it is an endothermic process, and the tank temperature will decrease. In order to aid in the release of H_2, hot water is normally circulated around the tank.

FIGURE 3.7
Picture of a 350-bar H_2 bottle taken during the 2011 Tokyo Fuel Cell Expo.

FIGURE 3.8
Picture of Al inner liner.

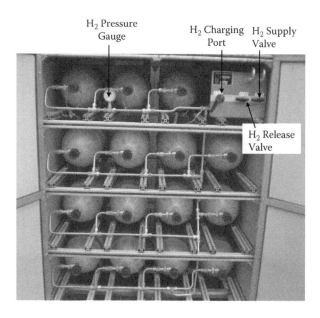

FIGURE 3.9
Picture of 350-bar H$_2$ supply system consisting of 15 12-liter bottles. Courtesy of Tongji University.

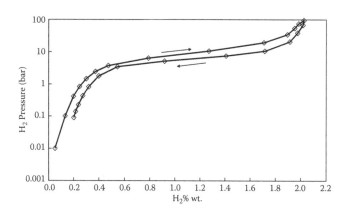

FIGURE 3.10
H$_2$ charging and discharging processes with Ti-series AB$_2$-type metal hydride. Courtesy of Beijing General Research Institute of Nonferrous Metals.

Figure 3.10 shows the H$_2$ charging and discharging processes with a Ti-series AB$_2$-type metal hydride. In the beginning of the charging process, the H$_2$ pressure increases rapidly to 2.4 bars and stores about 0.4% weight of H$_2$. Afterward, the pressure increase slows down and achieves 1.9% weight of H$_2$ when the pressure reaches 34.4 bars. This second stage can be regarded as a plateau region. Further increasing the pressure to 94.5 bars only adds about 0.1% weight more H$_2$ to reach 2.0% weight. In this last stage of pressure increase, most of the added H_2 is believed to stay as compressed gas instead of getting into the lattice of the metal hydride. The H_2 storage capacity for this metal hydride material can be considered nearly full when charging H_2 to 30–40 bars. During the discharging process, the pressure versus H_2 content curve follows the same trend as that during the charging process. The discharging has a hysteresis, and it depends on how easily the hydrogen can get out of the lattice of the metal hydride material and how fast the discharging rate is. The hysteresis will be smaller if the hydrogen can leave the lattice more easily and the discharging rate is slower.

It can be seen that before the metal hydride reaches its hydrogen storage limit, the hydrogen pressure within the tank increases quite slowly. When the metal hydride gets close to its hydrogen storage limit, little more hydrogen can be stored. After that point, the hydrogen pressure will increase rapidly. We can use this point to determine when the storage tank has reached its full capacity. Similarly, during the discharging process, when the H_2 pressure drop starts to accelerate, the tank gets close to empty, and the discharging process should be stopped. Although there is still about 0.4% weight H_2 remaining, it cannot be released because its pressure gets close to the 1-bar ambient pressure. Taking this fact into consideration, this metal hydride can store about 1.5% weight of releasable H_2.

When used to provide H_2 for a fuel cell, the liquid circulation loop for the metal hydride H_2 storage device is best connected with the liquid circulation loop for the fuel cell stack. During the operation of the fuel cell, the heat generated will help the metal hydride maintain the temperature. At the same time, the heat taken by the metal hydride helps control the temperature of the stack. With such a configuration, the heat exchanger incorporated in the fuel cell system can be reduced, which in turn lowers the parasitic power consumed by the fan mounted on the heat exchanger.

As shown by Figure 3.11, metal hydride can store hydrogen near its full capacity in a few minutes if the heat released in the process can be removed fast enough. In practice, since the heat cannot be removed as quickly as desired, the charging process may take a few hours, especially when the metal hydride cylinder is large.

Metal hydride can be charged and discharged with hydrogen a few thousand times without significantly losing its hydrogen storage capacity. Figure 3.12 shows that less than 5% hydrogen storage capacity was lost when the metal hydride was charged and discharged 2000 times. For a metal hydride hydrogen storage system, it is very important to make sure that there is no water getting into the tank with the H_2 gas; otherwise, the metal hydride will lose its hydrogen storage capacity quickly because water will disintegrate the metal hydride.

If the wrong gas is fed to a metal hydride tank, the gas pressure will increase rapidly without showing a pressure plateau. Since this wrong gas cannot get into the crystal structure of the metal hydride, it can only take the

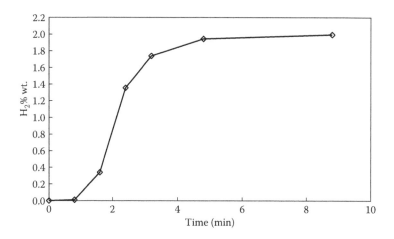

FIGURE 3.11
Hydrogen charging rate of Ti-series AB$_2$-type metal hydride when heat generated is released fast enough. Courtesy of Beijing General Research Institute of Nonferrous Metals.

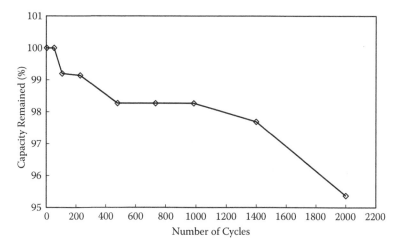

FIGURE 3.12
Hydrogen storage capacity loss in 2000 charging and discharging cycles for Ti-series AB$_2$-type metal hydride. Courtesy of Beijing General Research Institute of Nonferrous Metals.

very limited available empty space within the tank. The pressure of the tank will follow the gas law:

$$P = nRT/V \qquad \text{(Eq. 3.14)}$$

Storing H$_2$ in metal hydride is one of the best methods to achieve high volumetric hydrogen storage capacity. It can store as much as or even more hydrogen than liquid H$_2$ on a volume basis. Therefore, when volume is a critical factor, metal hydride becomes one of the preferred choices. However, metal hydride has poor specific hydrogen storage capacity. Other drawbacks of metal hydride are that it is quite expensive and the stored H_2 must be water-free.

3.4 Hydrogen Transportation and Refueling

Either in the liquid or compressed gas form, H_2 is currently transported by trucks. The truck needs an issued permit to transport dangerous or combustible materials. The driver is also trained accordingly before being allowed to drive the truck. Due to the use of special trucks and drivers, half or more of the H_2 cost to end users arises from H_2 transportation. The gas delivery company often charges the user the same fees for the truck and the driver regardless of whether one bottle or a full truckload of H_2 is delivered. The high cost of H_2 transportation seriously discourages people from using H_2. Even using large trailer trucks to transport H_2, the energy consumed by the trucks may account for about 10% of the energy contained in the H_2.

It is expected that in the future most of the H_2 will be delivered to customers via pipelines that are similar to those used today for delivering natural gas, but are able to handle the hydrogen embrittlement problems. The cost of H_2 delivery will be reduced significantly by that time. This will also free customers from storing H_2 onsite. If H_2 is used for stationary and backup power generation, H_2 only needs to be slightly pressurized, even at H_2 production sites, thus avoiding the energy losses associated with liquefaction or pressurization.

For automobile applications, H_2 must be in either liquid or highly pressurized form. Due to the additional cost of liquefaction or pressurization, H_2 costs for automobile applications may be about 20% higher than that for stationary use. Refueling of a car or a bus at an H_2 station can be done in 3 or 10 minutes because the H_2 supplied by the station is already in either liquid or highly pressurized form.

The refueling process looks similar to refueling today's ICE-cars with gasoline in appearance. Through flexible pipelines and a nozzle and receptacle combination, H_2 is sent from the storage tank to the receiving tank. When the receiving tank is full, the nozzle and the receptacle combination will shut down automatically. The nozzle and receptacle combination involves many technologies to handle the high pressure of H_2. The entire refueling system is quite smart and various safety protection mechanisms have been designed. For example, if somehow the temperature or the pressure of the receiving tank exceeds the design specifications during the refueling process, the system will shut down the refueling process automatically. For example, current regulation limits the maximum temperature to be 85°C for cylinders used for transportation applications. Figure 3.13 shows two refueling stands displayed during the 2011 Tokyo Fuel Cell Expo.

(a)

(b)

FIGURE 3.13
Refueling stands displayed during the 2011 Tokyo Fuel Cell Expo.

FIGURE 3.14
Shanghai Anting H_2 refueling station.

Since weight is not a concern at either H_2 production sites or refueling stations, high-pressure steel storage tanks can be used. However, the tanks carried by automobiles should be the high-pressure lightweight ones. Figure 3.14 shows the Shanghai Anting H_2 refueling station for refueling fuel cell vehicles during the 2010 Shanghai World Expo. There are 15 steel cylinders with a length and diameter of 10 m and 406 mm, respectively. They store H_2 at 450 bars with a storage capacity of 4768 Nm^3 H_2.

3.5 Summary

H_2 is the lightest gas and it can rise at a rate of around 20 m s^{-1}. Unless in a completely gas-tight environment, it is unlikely that H_2 will accumulate in the environment, making it relatively safer than most of the common gaseous and liquid fuels. When H_2 combusts or explodes, the process finishes instantly, making rescue efforts easier.

Although H_2 does not exist naturally on earth, it can be generated by a variety of methods such as electrolysis of water, thermal splitting of water, reforming of hydrocarbons, cracking of ammonia, reaction of certain chemicals with water, and biological processes using enzymes and bacteria.

Through steam reforming of CH_4, C_2H_6, C_3H_8, CH_3OH, and C_2H_5OH, the theoretical H_2% volume ranges from 75% to 80%, while the actual H_2% volume will be a few percentage points lower due to the incomplete conversion of the raw fuel, burning of some fuels for the heat needs of the reformer, and the H_2 consumption by O_2 in the preferential oxidizer. The suggested reforming temperatures for the above fuels are 1200, 900, 800, 500, and 600 K, respectively for achieving high conversion efficiency and fast reaction kinetics. Due to the high CO content after the reforming step, water–gas shift reactions will follow to reduce the CO% volume to a few percentage points, which is then further reduced to ppm levels by a preferential oxidizer. Cracking NH_3 at 700 K can convert 98.8% NH_3 into H_2.

Electrolysis of water is able to produce H_2 that is completely CO-free, which is best for a PEMFC. The theoretical efficiency of electrolyzing liquid water is 121%. A thermoneutral efficiency of 100% is achieved when the heat generated by the electrolysis process is fully utilized by the electrolyzer. By using a 175-μm-thick PEM, the pressure of H_2 and O_2 produced can be pressurized up to about 100 bars inside the cell without causing a combustible gas mixture in the anode or the cathode chamber at an electrolysis current density of 400 mA cm^{-2}. Producing H_2 by water electrolysis can use otherwise wasted or low-quality electrical power, and this will stimulate larger-scale deployment of other renewable energy such as solar and wind energy.

Producing H_2 by reacting certain metals, such as Be, Li, Al, and Mg, with water can achieve H_2 weight percentages of 7.41%, 6.29%, 5.56%, and 4.73%, respectively, when all the water produced by the fuel cell is recovered and used for the chemical reactions. When reacting some metal hydrides such as $LiBH_4$, $NaBH_4$, $Mg(BH_4)_2$, and $Ca(BH_4)_2$ with water that is recovered from the fuel cell reaction, the H_2 weight percentage can reach 36.9%, 21.2%, 29.7%, and 23.0%, respectively.

Today, H_2 is most commonly supplied in steel cylinders at a pressure of less than 150 bars. For vehicle applications, high-strength, lightweight cylinders with H_2 pressure at 350 bars or 700 bars have been developed and used in trials. Today, a significant portion of the H_2 cost arises from its delivery to end users. Due to the lack of an H_2 infrastructure, it is not convenient for the public to use H_2 at the moment, which seriously impedes the commercialization of fuel cells.

4

Evaluation

From the basic components to parts, modules, and the final system, many types of evaluations are involved in a proton exchange membrane fuel cell (PEMFC). Some evaluations can be done in minutes or hours, while some other evaluations may take weeks, months, or even years. This chapter briefly describes the major evaluations.

4.1 Catalyst

The catalyst is typically evaluated for its composition, size, size distribution, shape, activity, durability, poisoning tolerance, and mass loss.

4.1.1 Composition

The surface composition can be determined by using the x-ray photoelectron spectroscopy (XPS) technique. In order to be more CO tolerant, the anode for a direct methanol fuel cell (DMFC) using methanol as the fuel or a PEMFC using H_2 containing trace amounts of CO as the fuel normally utilizes a Pt alloy such as $PtRu_x$, $PtMo_x$, or $PtSn_x$ as the catalyst. Pt alloys such as $PtCo_x$ and $PtNi_x$, are sometimes used as the cathode catalysts to achieve higher catalytic activities for the oxygen reduction reaction (ORR). Those alloys are synthesized with a target composition in mind by using the metal sources at the corresponding molar ratio accordingly. The bulk composition of the resulting alloys is often very close to the targeted ratio. During the usage of the alloys, some element tends to be enriched at the surface of the catalyst particles, and some transition metals may leach out. The surface composition can be measured by the XPS.

At the cathode, the metallic elements are more or less oxidized to certain extents, depending on the alloy itself and the operating conditions (e.g., voltage, relative humidity, and temperature). For example, Pt will coexist with PtO_x at a cathode voltage higher than about 0.8 V. The oxidation states of platinum in PtO_x (e.g., Pt^{II} and Pt^{IV}), along with the relative percentages of Pt^0, Pt^{II}, and Pt^{IV} can be determined by the XPS. PtO_x is less active than Pt for the ORR. Even for freshly made Pt catalysts, different oxidation states may

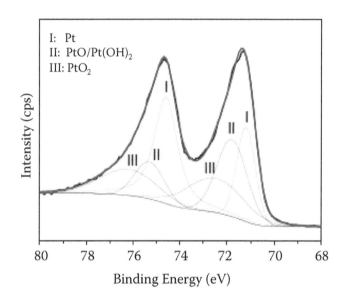

FIGURE 4.1
XPS of Pt catalyst particles. Courtesy of East China University of Science and Technology.

exist, as shown by Figure 4.1. High-resolution XPS indicates that Pt 4f has three pairs of peaks, with the strongest pair at 71.2 eV $(4f_{7/2})$/74.6 eV $(4f_{5/2})$ from Pt^0, followed by a pair at 71.8/75.3 eV from Pt^{II}, and the weakest pair at 72.5/76.2 eV from Pt^{IV}.

4.1.2 Size and Size Distribution

In most synthetic techniques, the sizes of the catalyst particles typically range between 1 and 10 nm, even though the targeted size is around 3 nm. The size often follows a Gaussian-type distribution. For supported catalysts such as Pt/C, the average size of the Pt particles will depend on the Pt content, the C type, and the synthetic methods. Higher Pt contents and lower C surface areas normally lead to the formation of larger Pt particles.

Pt particle size and its distribution can be estimated by using transmission electron microscopy (TEM). Under TEM, representative regions are first located. Then the diameters of enough Pt particles (e.g., 100) are measured, one by one, to get the average Pt particle size and the size distribution.

Through use in fuel cells the Pt size increases by two mechanisms. First, particles migrate on the surface of the support to become larger agglomerates. Second, Pt dissolves from smaller particles and deposit on larger ones to make the latter even larger.

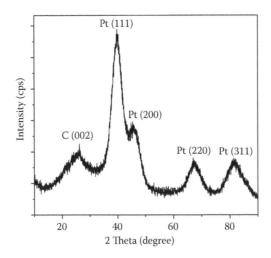

FIGURE 4.2
XRD of Pt/C. Courtesy of East China University of Science and Technology.

4.1.3 Crystallite Size

People often use the x-ray diffraction (XRD) technique to estimate the crystal size based on the crystallite facets. For Pt, the prominent crystalline facets are the (111), (200), and (220) and (311), respectively. Based on the broadening of the x-ray diffraction peaks, the crystallite sizes are calculated according to the Debye–Scherrer equation,

$$\text{Crystallite Size} = (0.9\lambda)/(\text{FWHM}\cos\theta) \qquad \text{(Eq. 4.1)}$$

where λ, FWHM, and θ are the wavelength of the X-ray, the full width at the half peak height, and the incidence angle of the X-ray. Please note that the crystallite size estimated using Equation 4.1 depends on and corresponds to the crystallite facets.

Figure 4.2 shows that the XRD of Pt/C exhibits 5 predominant peaks at 2θ of 26°, 39.8°, 46.2°, 67.5°, and 81.3°, respectively. The peak at 26° arises from the (002) crystallite facet of the graphitized carbon support, and the peaks at 39.8°, 46.2°, 67.5°, 81.3°, and 85.7° (not shown) correspond to the (111), (200), (220), (311), and (222) crystallite facets of Pt, respectively. Based on the Pt(220) peak that is least affected by the support, the Pt crystallite size is 3.4 nm according to Equation 4.1.

4.1.4 Surface Area

For spherical catalyst particles, the surface area per unit mass can be calculated. If the radius of the particle is r and the density of the metal is d, then the volume (v), the surface area (s), and the mass (m) of each particle are, respectively,

$$v = (4/3)\pi r^3 \qquad \text{(Eq. 4.2)}$$

$$s = 4\pi r^2 \qquad \text{(Eq. 4.3)}$$

$$m = (4/3)d\pi r^3 \qquad \text{(Eq. 4.4)}$$

The number of Pt particles within 1 g Pt is

$$N = 3/(4d\pi r^3) \qquad \text{(Eq. 4.5)}$$

Then the total Pt surface area is

$$S = Ns = 3/(rd) \qquad \text{(Eq. 4.6)}$$

where the units of S, r, and d are $cm^2\,g^{-1}$, cm, and $g\,cm^{-3}$, respectively. Table 4.1 lists the Pt surface areas of different-sized Pt particles ($d = 21.45\,g\,cm^{-3}$). It is clear that the unit mass surface area decreases steeply with an increase in the Pt particle size, especially when its diameter is less than 5 nm.

Since Pt particles less than 1 nm or larger than 10 nm are rarely used in fuel cells, the unit mass surface area is plotted for a size range between 1.5 and 10 nm, as shown in Figure 4.3. To achieve a high surface area without suffering

TABLE 4.1

Unit Mass Surface Areas of Different-Sized Pt Particles

Particle Diameter (nm)	Surface Area $(cm^2\,g^{-1})$	Surface Area $(m^2\,g^{-1})$
0.5	5594406	559
1	2797203	280
1.5	1864802	186
2	1398601	140
2.5	1118881	112
3	932401	93
3.5	799201	80
4	699301	70
5	559441	56
6	466200	47
7	399600	40
8	349650	35
9	310800	31
10	279720	28
20	139860	14
30	93240	9
40	69930	7
50	55944	6

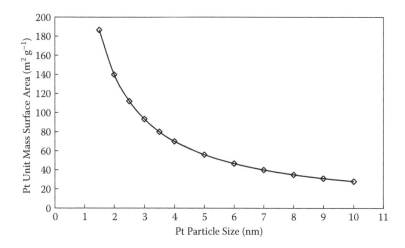

FIGURE 4.3
Pt unit mass surface area versus particle diameter.

from lower activity (particles less than about 2 nm are not as active as larger ones), the best Pt particle size should be controlled at about 3 nm. The specific lower activity of particles smaller than 2 nm are probably because they are more readily to be oxidized, and therefore, a higher percentage of Pt surface is covered by an oxide layer at the cathode potential during the fuel cell operation.

In order to aid the formation of catalyst nanoparticles, the support particles normally need to have a very high surface area (e.g., >100 $m^2 g^{-1}$). The tiny primary support particles are often connected together to form much larger secondary particles. If some catalyst particles form within the small pores of the support materials, and the ionomer used for making the catalyst ink formulation cannot penetrate into those small pores, those Pt particles may not be able to participate in the electrochemical reaction due to the lack of proton conductance nearby, although the presence of liquid water may be able to aid proton transport. Therefore, it is generally believed that Pt particles should not be produced within those small pores of the support materials. Similarly, when a porous material is used as the catalyst support, the pore size should not be too small. It is a misunderstanding that the smaller the pore size of a support, the higher the catalyst utilization is.

4.1.5 Activity

The activity of a catalyst toward the reduction of O_2 can be evaluated by using the rotating disc electrode (RDE). The electrode rotates from several to 10,000 revolutions per minute (f). The thickness of the diffusion layer at the electrode surface is determined by the rotation rate of the electrode.

In the diffusion-controlled region where the electrode overpotential is high enough, the current is proportional to the square root of the angular velocity of the electrode according to the Levich equation:

$$i_d = 0.62nFAD^{2/3}\omega^{1/2}v^{-1/6}C \qquad\qquad \text{(Eq. 4.7)}$$

where i_d is the limiting current, n is the number of electrons involved in the reaction per reactant molecule, F is the Faraday constant, A is the electrode area, D and C are the diffusion coefficient and the bulk concentration of the reactant, ω is the angular velocity (= $2\pi f$), v is the kinematic viscosity of the electrolyte solution (= η/d, where η and d are the viscosity and density of the electrolyte solution, respectively). The i_d versus $\omega^{1/2}$ is a straight line.

In the activation region, where the overpotential is low, the reaction rate k_f is controlled by the reaction kinetics, and the current i_k is linearly related to k_f:

$$i_k = nFAk_fC \qquad\qquad \text{(Eq. 4.8)}$$

where k_f is a function of the overpotential, and a higher overpotential leads to a larger k_f.

For investigating the ORR properties of catalyst particles such as Pt particles, the particles are first mixed with an ionomer solution, a tiny amount of which is then applied to the surface of the RDE. The electrode itself should not catalyze the ORR, and glassy carbon is a good choice. After the catalyst ink is dried, it forms a thin layer on the glassy carbon electrode. From the percentage of Pt in the ink and the ink amount applied to the electrode surface, the amount of Pt can be estimated. In order for the Pt particles to stay on the glassy carbon electrode strongly during the following RDE test, a drop of ionomer solution can be applied on top of the dried catalyst layer. After drying, an ionomer thin layer forms on top of the catalyst layer. The ionemer layer should be thin enough (e.g., less than about 0.2 μm) so that its impedance to the diffusion of O_2 toward the underneath catalyst layer is not significant at the current density used in the test.

During the RDE test, at each electrode rotating rate, the potential applied on the electrode is often scanned from a low overpotential to a high overpotential (e.g., starting from the open-circuit voltage [OCV]). At the OCV there is no net current; as the overpotential increases, the current first increases slowly, and when the overpotential is higher than about 200 mV (for ORR), the current increase starts to accelerate and a very steep increase follows. Finally, the current reaches a plateau. Figure 4.4 shows an RDE test of 20% Pt/C in 0.5 M H_2SO_4 at 25°C and 1600 rpm.

Several parameters can be used to judge the catalytic activity of the Pt particles toward the ORR. First is the onset potential. There are different understandings on onset potential. Some people take the voltage when the current is at 0 as the onset potential, while others use the potential when the current increase starts to accelerate as the onset potential, and therefore, many data

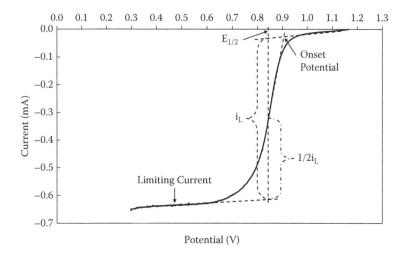

FIGURE 4.4
RDE test of 20% Pt/C in 0.5 M H_2SO_4 at 25°C and 1600 rpm. Courtesy of Dalian Institute of Chemical Physics, Chinese Academy of Sciences.

from different studies lost base for comparison. In order to unify the method and considering the fact that experimental setup and errors can significantly impact the voltage at 0 current, it is suggested that we use the crossing point of the two tangent lines shown in Figure 4.4 as the onset potential. A better catalyst will show a higher onset potential (i.e., lower overpotential). Second is the current density at 0.90 V. A better catalyst generates higher current density than a poor one. People often use the current density at 0.90 V to estimate the mass and surface activities of the catalyst. Better catalysts possess higher mass and surface activities than poor ones. Third is the half-wave potential, $E_{1/2}$, the potential at the half height of the limiting current, as shown in Figure 4.4. A better catalyst has a higher $E_{1/2}$ than a poor one. A difference of 20 mV in $E_{1/2}$ may seem insignificant, but the activity of the catalysts differs significantly (about 40% higher as shown in Table 2.4).

When Figure 4.4 is rotated counter clockwise for 90° it becomes the familiar V-I curve as shown in Figure 2.2. It is not surprising because they basically represent the same thing but carried out under different conditions.

4.1.6 Cyclic Voltammetry

Cyclic voltammetry (CV) is a very popular technique for studying electrochemically active species that scans the potential and observes the corresponding current. It can quickly identify at what potential(s) the species will be electrochemically active and how many redox processes the species will present. The potential scanning rate can range from 1 to 1000 mV s^{-1}, although for most studies the scanning rate is lower than 50 mV s^{-1}.

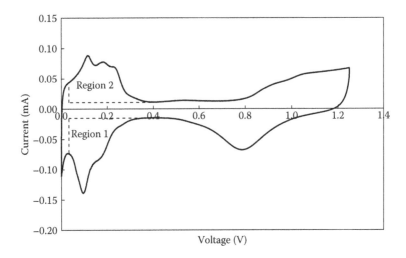

FIGURE 4.5
Cyclic voltammetry of Pt/C in 0.5 M H_2SO_4 at 25°C and 50 mV s^{-1} voltage scan rate. Courtesy of Dalian Institute of Chemical Physics, Chinese Academy of Sciences.

In fuel cells, the CV technique is widely used to estimate the electrochemically active surface area (ECSA) of a catalyst such as Pt. In the potential region between 0.0 and 0.4 V, acidic electrolyte will normally show two pairs of peaks on Pt due to the underpotential hydrogen adsorption and desorption processes, as shown in Figure 4.5 and Reactions 4.1 and 4.2, respectively.

$$\text{Adsorption:} \quad Pt + H^+ + e^- = Pt\text{-}H_{ad} \quad\quad (R.\ 4.1)$$

$$\text{Desorption:} \quad Pt\text{-}H_{ad} = Pt + H^+ + e^- \quad\quad (R.\ 4.2)$$

One H atom adsorbs on the surface of one Pt atom. For polycrystalline Pt surfaces, the adsorption (or desorption) charge is 0.21 mC cm^{-2} Pt. Then, by dividing the integrated charge under the hydrogen adsorption (Region 1 in Figure 4.5) or desorption (Region 2 in Figure 4.5) peak area with 0.21 mC cm^{-2} Pt, the ECSA of the total Pt particles is calculated. Dividing the ECSA by the amount of Pt, the ECSA per-unit mass of Pt is obtained; a higher value means that the Pt atoms are more efficiently used. The Pt surface area can also be determined by various physical methods (e.g., the BET method or calculated based on the particle size), and this area is often regarded as the total surface area. The ratio of ECSA to the total surface area is taken as the catalyst utilization within the electrode.

In the higher potential region (>0.75 V) the redox processes are due to the oxidation of Pt to PtO_x and the corresponding reduction of PtO_x to Pt. PtO_x has lower catalytic activity than Pt toward the ORR. During the operation of a fuel cell, the cathode is typically at a potential of around 0.65–0.70 V. It can be seen from Figure 4.5 that the peak potential for the reduction of PtO_x to

Pt is around 0.78 V (in this particular case). Since 0.65–0.70 V is considerably lower than 0.78 V, most of the cathode catalyst surface should be in the Pt form during the operation of a fuel cell. But at OCV, the cathode catalyst will be dominant in the PtO_x form. Between 0.78 and OCV, the cathode catalyst will be in both the Pt and the PtO_x forms.

Studies have shown that as the Pt particle size decrease the peak position for the reduction of PtO_x shifts to lower potentials, indicating that PtO_x forming on smaller particles is more difficult to be reduced back to Pt (Maillard 2009 and Hayden 2009). This will result in a relatively higher Pt surface not to be free of oxide layer at the fuel cell cathode operating voltage, leading to lower activity towards to ORR and thus the cathode performance.

The H adsorption process occurs before the H_2 evolution starts during a cathodic potential scan (i.e., scanning the potential from high to low). In other words, H_2 evolution begins after the completion of the second hydrogen adsorption peak (i.e., the one at the lower potential). If the second hydrogen adsorption peak develops well, as shown in Figure 4.5, it is easier to determine when the H_2 evolution starts, and therefore the hydrogen adsorption region and thus the charge can be identified with less error. For many electrodes with Pt/C as the catalyst, the second hydrogen adsorption may not form a peak before the H_2 evolution starts, making it difficult to identify at exactly what potential the H_2 evolution starts. Then the integrated charge from the hydrogen adsorption may be either under- or overestimated. In such a situation, it is better to use the oxidative stripping peak area of CO to estimate the ECSA. CO can quickly adsorb on the surface of Pt to form a monolayer, and this monolayer will be oxidized to CO_2 according to Reaction 4.3 at potentials between about 0.65 and 0.95 V to form a sharp peak, as shown in Figure 4.6.

$$CO + H_2O = CO_2 + 2H^+ + 2e^- \quad E° = -0.106 \text{ V} \qquad (\text{R. 4.3})$$

Holding the potential of the electrode at around 0.0 V and letting the electrode be exposed to a gas mixture containing a small amount of CO will dominantly result in linear adsorption of CO on Pt, that is, one CO occupies one Pt atom. As can be seen from Figure 4.6, there are no H adsorption and desorption currents or peaks during the potential holding at around 0.0 V. The reason is that since the Pt surface is completely occupied by CO, underpotential deposition of H^+ (i.e., H adsorption) becomes impossible. After the CO adsorption completes, the CO-containing gas is purged out by an inert gas such as N_2 so that only those CO molecules that adsorb on the Pt surface remain. Then the potential is scanned from low to high until the oxidation of CO is complete. Accompanying the CO oxidation, the cathode Pt is gradually changed to the PtO_x form. During the following cathodic scan from high to low potentials, there is no current or peak corresponding to the CO oxidation, because its product, CO_2, does not adsorb on the surface of Pt. The result is the reduction of PtO_x to Pt in the higher potential region and the H adsorption in the lower potential region. The following scan from low to high potentials will first show

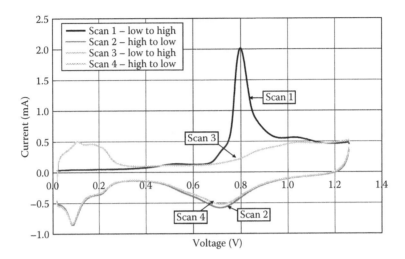

FIGURE 4.6
CO oxidative stripping. Courtesy of Dalian Institute of Chemical Physics, Chinese Academy
of Sciences.

the H desorption peaks in the lower potential region and the oxidation of Pt to
PtO_x in the higher potential region. There is no CO oxidation current or peak
anymore because CO has been completely removed from the catalyst surface
during the first anodic scan. Returning the scan from high to low potentials
will again show the reduction of PtO_x to Pt and H adsorption processes.

Although the standard reduction potential for R. 4.3 is as low as −0.106 V,
CO oxidation does not occur until ca. 0.65 V or higher indicating the reaction encounters extremely high overpotential. It is interesting to see from
Figure 4.6 that the onset potentials for CO oxidation and for Pt oxidation
coincide, indicating that the oxide formation at Pt surface is necessary for
oxidizing CO. In other words, only after the potential is high enough to lead
formation of Pt-OH from activating water will the CO oxidation reaction
start. This also explains the observation that the CO oxidation potential on
PtRu catalyst is about 0.2 V lower than on Pt catalyst because Ru can form
Ru-OH ca. 0.2 V lower than on Pt.

ECSA decreases with time during fuel cell operation due to Pt particle size
increase and decrease in three-phase regions. Some Pt can even be found
in the PEM because some positively charged Pt ions that form at the cathode migrate into the PEM, where they are reduced to Pt particles by H_2 that
crosses through the PEM.

Since each CO gives out two electrons in the oxidation process, as shown by
Reaction 4.3, and each CO occupies one Pt atom when the adsorption process is
held at near 0 V, the unit Pt surface area charge is 0.42 mC cm^{-2} Pt. Dividing the
integrated charge under the CO oxidation peak by 0.42 mC cm^{-2} Pt will result

in the ECSA. The CO oxidation peak shifts to higher potentials as the temperature decreases. Also, studies have shown that the CO oxidation peak location shifts to higher potentials as the Pt particle size decrease probably because the stronger adsorption of both CO and –OH on smaller particles (Li 2008).

CV can use the electrode discussed in the RDE section. Very often, RDE and CV are performed in succession.

4.2 Membrane

4.2.1 Proton Conductivity

As described in Chapter 2, the incapability of a PEMFC to achieve a current density higher than 2 A cm^{-2} at a good voltage is primarily because the resistance is not low enough. The resistance includes the ionic resistance from the PEM and the catalyst layers (CLs), the contact resistance between the gas diffusion media (GDMs) and both the neighboring plates and the neighboring CLs, and the electronic resistance of the plates, the GDMs, and the CLs. The majority of the resistance arises from the ionic resistance, followed by the resistance of the plates and then the resistance of the GDMs.

In order to lower the ionic resistance, the conductivity of the ionomer should be as high as possible. Making the ionomer with a higher –SO$_3$H concentration, that is, a smaller equivalent weight (EW), will reduce the proton transport resistance. For the Nafion series ionomer with an EW of 1100 g per mole of –SO$_3$H groups, the molar concentration of –SO$_3$H within dry Nafion 117, whose density is about 2.0 g cm^{-3}, is

$$C_{-SO_3H} = (2 \text{ g cm}^{-3}/1100 \text{ g mol}^{-1}) \times 1000 \text{ cm}^3 \text{ L}^{-1} = 1.8 \text{ M} \quad \text{(Eq. 4.9)}$$

Table 4.2 shows the –SO$_3$H (i.e., H$^+$) concentrations for Nafion-like ionomers (d = 2.0 g cm^{-3}) with different EWs. Smaller EWs result in higher proton concentrations. For known perfluorosulfonic acids (PFSAs), the EW typically ranges between 800 and 1100. When the EW is lowered further, the mechanical properties of the membrane are not adequate for long-time operations. The results shown in Table 4.2 are plotted in Figure 4.7.

The bulk proton conductivity of a piece of PEM can be measured by the impedance technique. The membrane is placed between two electrodes, one working electrode and one counter electrode. An alternating current (AC) with a magnitude typically smaller than 10 mV is applied to the electrodes at a frequency of 1000 Hz or higher. The real impedance Z' is the resistance of the PEM. If the AC voltage is applied for a frequency range (e.g., 1000 Hz to 1 Hz), the real impedance at the highest frequency is the resistance of the

TABLE 4.2

H⁺ Concentrations of Ionomers with Different EWs

Equivalent Weight (g mol⁻¹)	−SO₃H Molarity (M)
1500	1.3
1400	1.4
1300	1.5
1200	1.7
1100	1.8
1000	2.0
900	2.2
800	2.5
700	2.9
600	3.3
500	4.0

Note: $d = 2.0$ g cm⁻³.

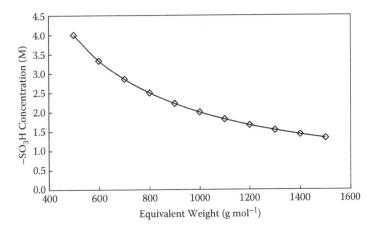

FIGURE 4.7
−SO₃H group concentration versus ionomer equivalent weight.

PEM. If the thickness and the area (under the electrode) of the PEM are L and S, respectively, then the conductivity of the PEM is

$$\sigma = L/(Z'\,S) \qquad \text{(Eq. 4.10)}$$

PEM soaked in liquid water has higher conductivities than PEM equilibrated with water vapor. Also, the conductivities increase with the RH.

Figure 4.8 shows the proton conductivity measurement result at 25°C and 100% RH using the impedance technique. The difference from the previously described arrangement was that the two electrodes were placed on

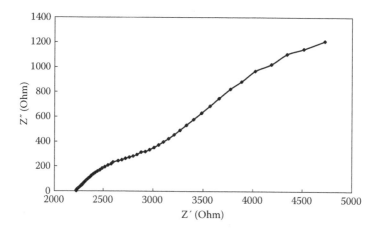

FIGURE 4.8
Proton conductivity measurement of PEM by the two-probe impedance technique at 25°C and 100% RH. Courtesy of Dalian Institute of Chemical Physics, Chinese Academy of Sciences.

the two opposite edges of the membrane, and thus the conductivity measured was the in-plane conductivity of the PEM. The PEM had a thickness of 51.5 μm, a length of 2 cm between the two electrodes, and a width of 2 cm. The real impedance Z' at the x-axis is 2225 Ω, and therefore, the in-plane proton conductivity is

$$\sigma = L/(Z'\,S) = 2\ cm/(2225\ \Omega \times 2\ cm \times 51.5\ \mu m \times 10^{-4}\ cm\ \mu m^{-1}) = 0.087\ S\ cm^{-1}$$

The voltage drop caused by such a PEM is then around 60 mV at a current density of 1 A cm^{-2} [(1 A cm^{-2} × 51.5 μm × 10^{-4} cm μm^{-1})/(0.087 S cm^{-1} × 1 cm^2)].

4.2.2 Water Content

Water content affects the proton conductivity and the mechanical strength of the PEM significantly. A PEM with a smaller EW can take in more water. The water content λ, which is defined as the number of water molecules held by each –SO$_3$H group, can be determined by the weight difference between dry and wet PEM:

$$\lambda = 1100(m_{wet} - m_{dry})/(18m_{dry}) \qquad \text{(Eq. 4.11)}$$

where 1100 is the EW of the PEM, 18 is the molar mass of H$_2$O, and m_{wet} and m_{dry} are the mass of the PEM at the wet and dry states, respectively. For example, if a piece of PEM gains 30% mass after being hydrated, then $\lambda = 18.3$, that is, each –SO$_3$H group shares, on average, 18.3 H$_2$O molecules. λ has been found to be about 22 when Nafion (EW = 1100) is fully hydrated in liquid water, indicating that the membrane can gain as much as 36% weight of water. Clearly, a PEM can absorb a significant amount of water in the fully hydrated state. More water will be taken in if the EW is smaller, but this could lower the mechanical properties of the PEM significantly. That is why a durable PEM cannot have an EW smaller than 800. For example, if such a PEM contains 22 water molecules per $-SO_3H$ group, the water gain will be 49.5% ($22 \times 18/800$), which is obviously too high to have good enough mechanical strength.

When Nafion membrane is hydrated by exposing it to water vapor, its water content per $-SO_3H$ group (λ) was correlated with the water activity (a) by the following empirical equation (Springer 1991):

$$\lambda = 0.043 + 17.18a - 39.85a^2 + 36a^3 \qquad \text{(Eq. 4.12)}$$

a is equivalent to the relative humidity (RH) in this case. For example, at 100% RH, λ will be 13.4, which is much lower than 22, the λ when Nafion is hydrated in boiling liquid water.

4.2.3 Durability

During the operation of a fuel cell, some hydroxide radicals, $\bullet OH$, will form. They can attack the PEM and cause its gradual degradation. It has been found (Curtin et al. 2004) that the degradation is mainly through the main chain unzipping for a PEM where the main chains' ends are $-COOH$:

$$HOOCCF_2-CF_2-R + \bullet OH \rightarrow \bullet CF_2-CF_2-R + CO_2 + H_2O \qquad \text{(R. 4.4)}$$

$$\bullet CF_2-CF_2-R + \bullet OH \rightarrow HOCF_2-CF_2-R \rightarrow OCF-CF_2-R + HF \qquad \text{(R. 4.5)}$$

$$OCF-CF_2-R + H_2O \rightarrow HOOC-CF_2-R + HF \qquad \text{(R. 4.6)}$$

After Reaction 4.4 to 4.6, the main chain of the PEM is one carbon atom shorter, and the process will continue repeatedly because $\bullet OH$ radicals are generated continuously and gradually by the fuel cell.

In order to quickly evaluate the ability of a PEM against radical attack, ex situ tests can be performed using Fenton's reagent such as 3% H_2O_2 + 10 ppm Fe^2 solution. Fe^{2+} reacts with H_2O_2 to form $\bullet OH$ radicals according to Reaction 4.7:

$$Fe^{2+} + H_2O_2 = \bullet OH + OH^- + Fe^{3+} \qquad \text{(R. 4.7)}$$

After immersing a piece of PEM in the Fenton reagent solution for a period of time (hours to days) to allow the $\bullet OH$ radicals to attack the PEM, the solution is analyzed for F^- concentration (such as using an F^- selective electrode).

Higher F⁻ concentration means that the PEM can be attacked more easily. The PEM is weighed to determine how much mass loss occurs. Also, the morphology change of the PEM surface can be examined by the scanning electron microscope (SEM).

After finding the major degradation mechanism of the PEM mentioned previously, manufacturers have modified the ionomers by replacing the end –COOH groups with –CF$_3$ groups. The durability of the resulting ionomer and PEM has shown significant improvement. The new ionomers are often called *chemically stabilized ionomers*.

Pendent side chain cleavage at the C-O bond between the main chain and the side chain also occurs, but at much lower extent. This kind of cleavage was found to be less severe for an ionomer with shorter side chains.

Since hydroxide radicals form in the catalyst layers, the ionomer within the catalyst layers have a higher chance to be attacked than the membrane.

4.2.4 Dimensional Changes

During the lifetime of a PEMFC, the PEM will encounter numerous dimensional changes accompanying its hydration states and temperatures. When the fuel cell is under operation, the PEM becomes hotter and more hydrated, leading to an increase in its dimension; when the fuel cell is not in operation, the PEM becomes cooler and gradually gets drier, resulting in a decrease in its dimension. Such swelling and shrinking processes impose mechanical stress on the PEM, particularly in the length and width directions. Repeated actions like those will weaken the PEM, especially the portions near the sealing regions and near the edges of the GDMs. Controlling the dimensional change amplitude of the PEM is therefore useful.

In order to get good proton conductivity, the PEM should have a low EW, but such a PEM will contain more water, leading to a higher amplitude of dimensional change. Some approaches are used to achieve a reasonable balance between the two opposing factors. The most popular approach is to reinforce the PEM by incorporating an expanded polytetrafluoroethylene (ePTFE) in the middle of the PEM. Since the dimension of the ePTFE changes little with the temperature and the humidity level, the resulting composite PEM improves dramatically with regard to dimensional stability. Another approach to reinforce the PEM is to crosslink the ionomer molecules. A third approach is to incorporate fillers, especially those with fibrous structures, into the PEM. These reinforcement methods will lower the proton conductivity of the resulting PEM, so it is typically made thinner than a regular PEM.

The planar dimensional change of a PEM with its hydration level can be easily measured using a ruler.

4.3 GDM

4.3.1 Electronic Conductivity

Measuring the in-plane and the through-plane electronic conductivities of a GDM is slightly different. Higher electronic conductivity of a GDM will result in smaller voltage losses due to resistance. For the through-plane measurement, a piece of GDM is placed between two copper plates (or gold-coated copper plates to minimize the contact resistance, but the measured result still includes the contact resistance between the GDM and the two plates on either side), a small current passes through the plates and the GDM, the corresponding voltage is measured, and the resistance is calculated according to

$$R_P = V/I \qquad\qquad \text{(Eq. 4.13)}$$

The resistance changes with the pressure applied to the GDM through the two copper plates; higher pressure results in smaller resistance, as shown in Figure 4.9. The thickness of the GDM decreases when the pressure increases, so its conductivity increases because the GDM becomes denser. The increase in conductivity slows down after ca. 0.4 MPa, and the change from 1.0 to 1.8 MPa is insignificant. In a stack, the GDM may be compressed up to 30%, so the conductivity measured at around 30% thickness reduction represents the actual conductivity of the GDM in a fuel cell. From the resistance R_P and the thickness L_P under different pressures, the conductivities of the GDM in the thickness direction can be calculated using Equation 4.14 (S is the area).

$$\sigma_P = L_P/(R_P S) \qquad\qquad \text{(Eq. 4.14)}$$

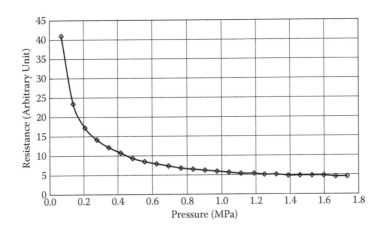

FIGURE 4.9
Conductivities of a GDM under different pressures. Courtesy of Dalian Institute of Chemical Physics, Chinese Academy of Sciences.

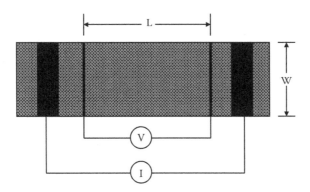

FIGURE 4.10
Illustration of in-plane conductivity measurement using the four-point probe technique.

The four-point probe method can be used to measure the in-plane electronic conductivity of the GDM, as shown in Figure 4.10. Two copper strips are placed near the ends of the GDM, and two copper wires separated at a distance of L are placed between the two strips. The strips and the wires are pressed against the GDM to form good contacts. A current passes from one strip through the GDM along the in-plane direction to the other strip, and the voltage drop between the two wires is measured. Then the in-plane conductivity is

$$\sigma = L/(dRW) \qquad\qquad (Eq.\ 4.15)$$

where d and W are the thickness and the width of the GDM, and R is the resistance ($= V/I$).

For a GDM made of carbon fiber (e.g., carbon paper), the in-plane conductivity is normally much higher than the through-plane conductivity because most of the carbon fibers stay in the planar direction. Fortunately, since the thickness of the GDM is around 0.2 mm, the voltage loss caused by the through-plane resistance is not significant. For example, if the through-plane conductivity is 20 S cm^{-1}, then the voltage drop caused by one membrane electrode assembly (MEA) (with two pieces of GDM) will be 2 mV at a current density of 1 A cm^{-2}.

4.3.2 Hydrophobicity

In order to prevent the pores of the GDM from being fully filled by liquid water during the operation of a fuel cell, the GDM is treated to achieve higher hydrophobicity. PTFE is typically used for the treatment. The GDM is immersed in a PTFE aqueous suspension with a suitable concentration to take in some PTFE. It is then dried at different temperatures in a few steps. The final temperature is around 340°C. Initially, PTFE particles are spherical, and after the temperature treatment, PTFE becomes more or less

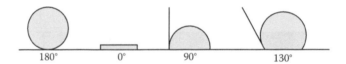

FIGURE 4.11
Schematic illustration of contact angles.

fibrous. From the weight difference of the GDM before and after the PTFE treatment, the PTFE content (%) can be obtained. Adding PTFE to the GDM will reduce its electronic conductivity slightly, especially the conductance between the GDM and the flow-field plates. The porosity of the GDM will also be reduced. For the conductivity, the PTFE content should not be too high, but for assisting the transport of gaseous reactants, a higher PTFE content is more effective. Generally, the PTFE is between 15% and 30% to achieve a balance between the two factors. The optimal PTFE content depends on the fuel cell operating conditions (e.g., current density, T, P, and RH).

The PTFE content in the GDM is controlled by the concentration of the PTFE suspension during the soaking process. A higher concentration will lead to a higher PTFE %. Sometimes, the GDM is soaked in a PTFE suspension more than once in order to get the desired amount of PTFE into the GDM. It is important to get an even distribution of PTFE throughout the GDM. During the soaking process, PTFE can evenly distribute into the GDM, but during the initial drying process, when liquid water is not fully evaporated, liquid PTFE will flow downward, potentially causing the bottom side of the GDM to have more PTFE.

The surface contact angle is a parameter for gauging the hydrophobicity of the GDM. A drop of liquid water is applied to the surface of the GDM, and the contact angle is measured with the help of an optical scope. Figure 4.11 schematically illustrates the shapes of water droplets at contact angles of 180°, 0°, 90°, and 130°. The angle between the tangent line and the horizontal line is the contact angle. After the teflonization step, the surface contact angle of a carbon paper–type GDM can increase about 20° to 30°.

Contact angle measurement appears to be simple, but an accurate value may not be easy to get, because the water droplet is often not in the ideal shape on the rough surface of the GDM.

4.3.3 Porosity

The initial porosity of a carbon paper–type GDM is often higher than 75% (Toray paper and SGL paper as high as 80 and 90%, respectively). After teflonization, it may decrease to around 70%, which is still very high. The porosity of the GDM is measured by porosimetry, a technique ideal for measuring pores ranging from a few nm to a few hundred μm. The porosimeter first creates a vacuum condition to remove gases within the pores of the GDM, then mercury is pushed into the pores by gradually increasing the pressure from about 4 kPa to 400 MPa. The total amount of mercury intruding into the GDM is its total pore volume. Mercury gets into the large pores at lower pressures and into smaller pores at higher pressures. The net amounts of mercury intruding into the GDM at different pressures correlate to the different pore sizes. The pressure and the pore size follow the Washburn equation shown in Equation 4.16:

$$d = 4\sigma\cos\theta/(P_L - P_G) \qquad\qquad \text{(Eq. 4.16)}$$

where d is the diameter of the pores, σ and θ are the surface tension and the contact angle between mercury and the GDM, respectively, P_L is the pressure applied to mercury (including the pressure created by the mercury column itself), and P_G is the gas pressure (if there is any gas in the sample chamber; otherwise, it is 0). The σ and θ of mercury on most surfaces are quite similar and generally taken at 0.485 N m^{-1} and 140°, respectively.

Since mercury cannot intrude into closed pores, those pores are not included in the result from the porosimetry measurement. Closed pores are useless for mass transport, and therefore their exclusion from the porosimetry measurement is a good thing. Blind pores (i.e., dead-ended pores) are of no use for mass transport either, but they are included in the porosimetry measurement. It is expected that a GDM made of carbon fiber contains few closed pores and blind pores.

Within a single cell or a stack the porosity of the GDM will be reduced due to the clamping pressure, especially at the portion in contact with the lands of the flow-field plates. The reduction is nearly 15% (e.g., from 70 to 60%).

4.4 Plate

4.4.1 Conductivity

One important characteristic is the electronic conductivity of the plate. Before the corrosion problem is satisfactorily solved, the plate will be predominantly made of graphite. Due to its brittleness, a graphite plate cannot be made too thin, typically >1 mm. If the conductivity of the plate is 20 s cm^{-1} (i.e., resistivity is 0.05 ohm cm), the voltage loss per cell (having two plates, each 1 mm thick) is 10 mV at a current density of 1 A cm^{-2}.

Since the plate (metal or graphite) is not compressible, it cannot be in direct contact with the two copper plates of a two-probe conductivity measurement device. Often, a piece of GDM is put between a blank plate (i.e., without flow-filed channels) and each of the copper plates. The resistance of the GDM/plate/GDM assembly is measured the same way as described in the GDM section. Subtracting the resistance of the two pieces of GDM measured under the same condition results in the resistance of the plate. Then the conductivity σ is calculated with Equation 4.17:

$$\sigma = L/(RS) \qquad\qquad (Eq.\ 4.17)$$

where L and S are the thickness and area of the plate, respectively.

The needle-type four-point probe method may be more convenient to use to measure the conductivity of a plate because the solid plate will not be penetrated by the needles and the needles can achieve good contact with the plate.

4.4.2 Density and Permeability

Graphite plates are normally made by compressing a mixture of graphite powders (typically less than 0.1 mm in size) and a binding resin at a high pressure (e.g., up to 5000 psi) and a suitable temperature. The size and shape of the graphite power affect the electrical conductivity and other properties of the resulting plate as well. The temperature causes the resin to soften or even melt to effectively bind the graphite particles. The resulting plates may then go through a high-temperature treatment to convert the resin to carbon or graphite in order to increase its electronic conductivity. It will require a higher temperature to change the resin to graphite than to amorphous carbon. After the high-temperature treatment, the plate is likely to become more porous, and therefore carbon infiltration is often needed. It is important to mix the graphite particles well with the resin before compressing; otherwise, an unacceptable percentage of the graphite particles will not be bonded by the resin, and the unbound graphite particles may get loose or dust off from the plates during usage, which could then lead to blockage of flow channels and leakage of fluids among the air chamber, the fuel chamber, and the coolant chamber. Graphite itself has a density of around 2.15 g cm^{-3}. If the density of the plate exceeds 1.8 g cm^{-3}, it is acceptable for fuel cell applications (higher density means lower porosity and thus a lower fluid crossover rate). The porosity of the plate should be less than ca. 2%.

When expandable graphite or exfoliated graphite foil is used as the starting material, the resulting plate will possess higher electrical conductivities due to the presence of three-dimensional network in such material. Adding carbon fibers into the graphite powder can also increase the electrical conductivity of the resulting plate.

The flow-field channels can be made either during the plate-making process or machined on blank plates afterward. Machining is a slow process, leading to higher plate cost due to labor. Molding is faster and makes the plate cost less, but requires skill to control the dimensions of the plate and the features of the flow fields (such as the channels, the lands, and the sealing areas). The surface of the molded plates often have excess amounts of resin due to the movement of the resin toward the surface under the high-pressure compression process, raising the surface and the contact resistances. If this happens, the surface of the plates should be sanded or polished. Improperly molded plates may warp during the following high-temperature treatment step.

Determining the permeability of the plates with flow-field channels is more useful than determining the permeability of the blank plates, because the channel-making process may create some problems, or some regions with loose graphite may be exposed. A leak-rate test can be performed by sealing the plate and a piece of nonporous flat solid together. After the flow-field channels are filled with a gas such as helium under a certain pressure, the pressure change for a time period, t, is measured. From the rate of the pressure decrease, the leak rate can be calculated according to Equation 4.18 based on the ideal gas law $PV = nRT$.

$$\Delta n/t = (n_1 - n_2)/t = n_1(1 - P_2/P_1)/t \qquad \text{(Eq. 4.18)}$$

where Δn is the change of helium (mole) during time period t, n_1 and n_2 are the amounts of helium, and P_1 and P_2 are the pressures of helium before and after the pressure holding time period, respectively. P_1 should be slightly higher than the pressure of the gaseous reactants intended for the fuel cell operation. The US Department of Energy (DOE) target for plate permeability is less than 2×10^{-6} cm^3 s^{-1} cm^{-2}.

The flexibility and strength of the plates should be good enough to guard against brittleness and deformation. The US DOE targets for the flexural strength and the tensile strength are 59 and 41 MPa, respectively.

4.4.3 Corrosion Current

If there were no corrosion problems, metals would be the ideal materials for making plates because of their high electronic conductivity, high mechanical strength, and ease of manufacturability. Metal sheets can be as thin as 0.1 mm and the flow-field channels can be stamped, which dramatically reduces the cost, volume, and weight of the metallic plates. A stack using metallic plates will be several times lighter than one using graphite plates. The challenge is that metals corrode in the fuel cell environment, which is humid, hot, and has potentials higher than the standard reduction potentials of metals. Corrosion releases metal ions that will replace the protons in the ionomer to lower its proton conductivity. Corrosion also leads to the formation of an insulating metal oxide layer, which significantly increases the contact resistance of the metallic plates. Stainless steel, titanium, nickel, and aluminum are the most widely used materials for making metallic plates. In order to overcome the corrosion problem, metallic plates need to be coated with a protective layer such as noble metals (e.g., Au), metal nitrides, metal carbides, or carbon. The coating process increases the manufacturing cost, and noble metals such as gold are too expensive. Due to the presence of coating defects, such as microcracks and inadequate or weak coating at edges and corners, metal corrosion and dissolution can start at those defective regions. Even without any coating defects, the difference of the thermal expansion between the plate metal and the coating materials may result in localized lifting or detachment of the coating materials. The corrosion current density should be lower than 16 μA cm^{-2}.

Corrosion current measurement is a first step to gauge the corrosion resistance of a metallic plate. To more or less simulate the PEMFC environment, a solution containing ppm levels of F$^-$ ion such as 0.1 M H$_2$SO$_4$ + 2 ppm NaF bubbled with either H$_2$ or O$_2$ is normally used in the test, and the plate (or a coupon of the plate materials) is used as the working electrode. Using the CV technique, the current at different voltages can be easily obtained. The corrosion current density should be less than 16 μA cm^{-2} for the coated metal to pass the initial test. Figure 4.12 shows the corrosion test result of a molded graphite plate. The current density at the cross point of the two V versus logj Tafel slopes is the corrosion current density, which is 0.22 μA cm^{-2} in this case.

After the corrosion test or after a metallic plate has been used in a fuel cell for some time, the contact resistance should be evaluated using the method described earlier in this section. When the metal is corroded and forms a metal oxide layer, it can prevent further corrosion of the metal below it, but the high resistance of the metal oxide is detrimental as it significantly lowers the fuel cell voltage due to the iR drop.

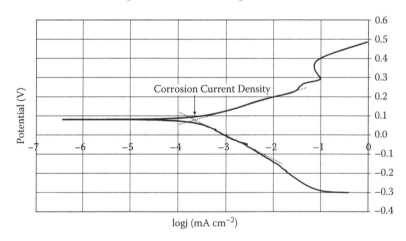

FIGURE 4.12
Corrosion test of molded graphite plate. Courtesy of Dalian Institute of Chemical Physics, Chinese Academy of Sciences.

4.5 MEA

4.5.1 Catalyst Loading

Catalyst loading (Pt or Pt alloy) on the anode and the cathode should be estimated. It can be determined in two ways. First, if the catalyst layer is applied to the GDM to form a gas diffusion electrode (GDE) (also called a catalyst-coated backing, or CCB), the total mass of the catalyst layer is obtained by weighing the dry mass of the GDM before and after the catalyst layer is applied. Then, using the Pt weight percentage in the catalyst ink formulation (the portion of solid material excluding the mass of the liquid solvents) and the electrode area, the Pt loading is calculated. Today, the cathode Pt loading is between 0.2 to 4 mg cm^{-2}. If the catalyst is a Pt alloy, PtM$_x$, the loading of Pt, M, and PtM$_x$ can be further calculated based on the composition of the PtM$_x$. If the catalyst layer is applied to a PEM or a decal and then transferred to a PEM, the catalyst loading can be estimated similarly. For a GDE or CCM received from others, the catalyst layer can be scraped off the GDM or the PEM, and the amount of the solid materials within the catalyst layer is roughly estimated by weight difference. By using various chemical methods, the Pt amount within the solid materials can be determined. A simple chemical method is to burn the solid catalyst mixture in a furnace to convert all the nonmetal materials to gases. From the remaining solid material, and assuming that it is PtO$_2$, the amount of Pt is estimated.

4.5.2 Electrochemically Active Surface Area

Since only the catalysts at the three-phase boundaries can participate in the fuel cell reactions, the ECSA of the electrode should be determined. The MEA is assembled in a single-cell test fixture, and humidified H$_2$ and N$_2$ are fed into the anode (or cathode) and cathode (or anode), respectively. The side with H$_2$ will function as the counter electrode, and the side with N$_2$ functions as the working electrode. Then, we use the CV technique described in the catalyst section to measure the ECSA of the working electrode. The ECSA is often measured on an MEA before, during, and after being used. The difference reveals how much and how fast the ECSA decreases. Good MEAs should have a low ECSA loss rate. It is necessary to humidify H$_2$ and N$_2$ during the test; otherwise, the MEA will become too dry to allow a good measurement.

4.5.3 H$_2$ Crossover

H$_2$ crossover shows how much H$_2$ crosses through the PEM to the cathode side. The US DOE target for the H$_2$ crossover current density is less than 2 mA cm^{-2}. For a pinhole-free PEM the H$_2$ crossover current density is around 0.82, 0.66, 0.33, 0.19, 0.13 and 0.09 mA cm^{-2} for Nafion membranes with a thickness of 20, 25, 50, 87.5, 125, and 175 μm, respectively, calculated according to Equation 2.28 (H_2 permeability through Nafion is ca. 8.5×10^{-12} mol cm^{-2} s^{-1}). As the membrane ages during usage, the H_2 crossover rate may increase gradually. When the rate shows a steep increase, the MEA is near the end of its lifetime. For a multicell stack, the OCV of the cell with an unacceptably high H_2 crossover rate will drop much faster than that of the other cells. This can be used to quickly identify the bad cells. Bad cells also show lower performance, but lower performance in cells does not necessarily mean that they have higher H_2 crossover rates. (Internal shortening between the anode and the cathode also lowers the cell performance.)

The H_2 crossover rate can be determined similarly with the setup discussed in the ECSA section. One side is fed with humidified H_2 and the other side with humidified N_2. The N_2 side is the working electrode and the H_2 side is the counter electrode. The voltage of the N_2 side can be set at around 0.5 V or scanned between 0.3 and 0.5 V. The H_2 that crosses through the PEM to the N_2 side will be quickly oxidized. The observed current density is the crossover current density. The amount of H_2 can be calculated by Equation 4.19.

$$n = i_{crossover}/(2FS) \text{ (mmol cm}^{-2} \text{ s}^{-1}) \qquad \text{(Eq. 4.19)}$$

where $i_{crossover}$ is in units of mA, and S is the electrode geometric area (cm^2).

Figure 4.13 shows the H_2 crossover current density through an ePTFE-reinforced PEM with a thickness of around 18 μm. The H_2 crossover current density is affected by its hydration level, and higher hydration levels lead to

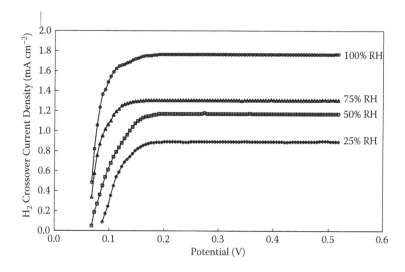

FIGURE 4.13

H_2 crossover current density through an ePTFE-reinforced PEM. Courtesy of Wuhan University of Technology.

higher H_2 crossover currents. Although the PEM is pretty thin, its H_2 crossover current density at 100% RH is less than the $2\ \text{mA cm}^{-2}$ target set by the US DOE.

4.6 Single Cell

4.6.1 V-I Curve

The most important characteristic of a single cell is its V-I polarization curve. It shows the cell voltage at different currents (or current densities). In the beginning of the test, the cell needs to be conditioned (or activated). The conditioning is normally carried out by running the cell between high and low current densities until the cell does not show any further increase in performance. This normal conditioning process will take several hours at least, and may not be able to fully activate the MEA. In order to facilitate the conditioning process, some special methods can be used. One simple method is to first run the cell using pressured reactants (up to 3 barg) and at higher temperatures (up to 75°C). It can condition the cell within about 1 hour. Another method is to pump hydrogen from one electrode to the other electrode by applying a small voltage (e.g., 0.05 V). The half reactions at the positive electrode, the negative electrode, and the overall reaction are shown in Reactions 4.8, 4.9, and 4.10, respectively:

$$\text{Positive electrode:} \quad H_2 = 2H^+ + 2e^- \quad\quad \text{(R. 4.8)}$$

$$\text{Negative electrode:} \quad 2H^+ + 2e^- = H_2 \quad\quad \text{(R. 4.9)}$$

$$\text{Overall reaction:} \quad H_2 \text{ (positive electrode)} \rightarrow H_2 \text{ (negative electrode)} \quad \text{(R. 4.10)}$$

At the positive electrode, the applied voltage causes H_2 to dissociate into protons and electrons. The protons and electrons transport to the negative electrode through the PEM and the external circuit, respectively, and combine to form H_2. Humidified H_2 is fed to the positive electrode, and nothing or humidified N_2 is fed to the negative electrode. H_2 pumping can condition a fuel cell in about 1 hour. During this hour, fuel cell mode and pumping mode are normally alternated, such as alternating 5 minutes pumping mode and 10 minutes fuel cell mode. The electrode that is conditioned is the negative electrode where H_2 evaluation occurs. Figure 4.14 shows the H_2 pumping process. If H_2 contains trace amounts of CO, the positive electrode will be poisoned, and the required external voltage will increase as the extent of CO poisoning proceeds.

After a fuel cell is conditioned, it will be evaluated under various conditions, such as different temperatures, reactant stoichiometric ratios, reactant

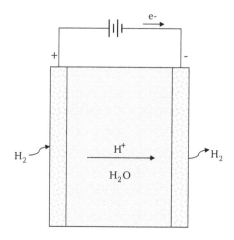

FIGURE 4.14
Schematic of the H_2 pumping process.

pressures, RH, and CO concentrations. These tests will enable an operator to form general opinions about the performance of the MEA for intended usage.

4.6.2 Durability Assessment

Besides performance, the durability of an MEA is important. Durability involves the performance decay rate and the lifetime of the MEA. The test is normally carried out at a selected current (or current density), and the cell voltage change with time is monitored. After a period of testing, the decay rate is calculated by dividing the voltage change (μV) by the testing hours. The decay typically proceeds faster in the beginning of the test (e.g., the first 100 hours), and then it slows down. When the performance of the cell starts to decline quickly (like a sudden death) after certain hours of operation, the end of the cell's lifetime is reached. Sometimes, such a late-stage quick voltage decline does not appear, but the voltage drops below the set target. This is also regarded as the end of the cell's lifetime, because the cell cannot offer the needed voltage, or power output, or electrical efficiency anymore. During the durability test, the OCV, ECSA, and H_2 crossover rate are measured periodically. The results from those tests illustrate the overall healthy status of the cell. Durability tests can take thousands of hours, and maintaining the external conditions, especially the lab power, becomes the greatest challenge. For example, an unexpected power shortage could damage the cell.

Accelerated tests are often performed to shorten the time needed for durability tests. Accelerated tests are carried out in a condition that is not good for the MEA, such as high temperature and low RH. An MEA decays much faster during an accelerated test, and could shorten the durability test from thousands of hours to hundreds of hours. If a regular durability test and an accelerated test are performed on a particular type of MEA, the accelerated test result can be correlated to the regular lifetime of similar MEAs in the future. However, applying the accelerated test result of one type of MEA to another type of MEA could be misleading.

Setting at OCV also accelerates the decay of the cell due to the higher cathode voltage and the reactant crossover rates. Besides H_2, O_2 can also diffuse through the PEM forming peroxide radicals at the anode/PEM interface. The radical attack to the ionomer causes its gradual degradation. Since O_2 crossover rate is higher through higher hydrated PEM, the decay may be faster when the hydration level of the PEM is higher.

4.6.3 Load Cycling

A fuel cell often needs to generate power according to the need of the load. The fuel cell is then operated in a load-following mode. The load (i.e., the current) could change significantly in a short time. For example, the load during the acceleration of a car can be several times higher than the load when the car cruises. In order to gauge the lifetime in real situations, a cell is often tested in the lab by simulating real situations. During the test, the number of loads, the amplitude of each load, the duration of each load, and the time interval between loads are programmed, and the H_2 and air flow rates are set at constant stoichiometric ratios (i.e., the flow rates change with the loads). A downward voltage overshoot with a 1-s time scale is commonly observed when the load is dramatically increased. It may be possible that the H_2 flow rate cannot follow a sudden and significant load surge instantly; the anode may then experience a short period of fuel starvation at certain locations, causing it to decay faster than at a constant load mode. The Pt oxidation state change between the reduced and the oxidized forms also accelerates the decay rate. Adding some pseudocapacitive materials such as $RuO_2 \cdot xH_2O$ that can give out electrons can help minimize damage to the anode when there is a very short fuel starvation.

4.6.4 Temperature Cycling

When a PEMFC is in operation, its temperature can be around 75°C, while when it is idling, its temperature is the ambient temperature, which can be as low as −40°C when there is no external heating means. Components of the stack will change dimensions with temperature. Repeated dimensional change of the MEA will weaken its mechanical strength, or even cause its rupture. In addition, if the stack drops to subzero temperatures in the winter, most of the water in the stack will freeze, even though an antifreeze-type coolant does not freeze. Formation of ice in the GDM and the CL will cause their volume to expand, which in turn can lead to physical damage to those components and neighboring components such as the PEM. In order to minimize the damage caused by ice formation, water should be blown out after the fuel cell is shut down when the ambient temperature is lower than 0°C. When external heating is available, keeping the stack temperature above the water freezing point can be a good choice.

The purpose of the temperature cycling test is to evaluate the impact of temperature change on the performance and the durability of the cell. The temperature at the higher end is the fuel cell temperature under operation, while the temperature at the lower end is the ambient temperature, which can be above or below 0°C. The fuel cell does not have to be running during the test. The impact of the temperature cycling can be estimated by periodically measuring the H_2 crossover rate and the fuel cell performance.

4.6.5 Poisoning Tolerance

H_2 obtained through fossil fuel reforming and processing always contains ppm levels of CO that can strongly adsorb on the surface of catalysts to lower the fuel cell performance (i.e., increasing the anode overpotential). Various methods such as using CO-tolerance catalysts (e.g., PtRu), adding CO removal catalyst (e.g., Au or Au-Fe_2O_3) within the MPL or the GDM, and bleeding a few percentage points of air into the fuel stream have been used to reduce the impact of CO. A test is typically carried out by switching between CO-free and CO-containing H_2 to evaluate the actual impact of CO on performance and thus the effectiveness of those methods under the chosen test condition (e.g., CO concentration, current density, and cell temperature). The current density should not be too low because the impact of CO is not significant at such current densities and thus it is difficult to discern the difference in effectiveness. Air bleeding technique is typically not used when the CO concentration is lower than ca. 10 ppm because other methods are adequate. Besides not easy to control, air bleeding consumes some H_2 to lower the fuel utilization, and leads to the formation of some H_2O_2 and thus peroxide radicals that can attack the ionomer.

CO_2 is generally considered as a diluent to H_2 in the reformate with PtRu as the anode catalyst. But if the fuel cell condition leads to the reversed water gas shift (RWGS) to occur (please refer to Section 5.2.1), trace amounts of CO may be able to form according to Reactions 4.11 and 4.12:

$$2Pt + H_2 = 2Pt\text{-}H_{ad} \tag{R. 4.11}$$

$$2Pt\text{-}H_{ad} + CO_2 = Pt\text{-}CO_{ad} + H_2O + Pt \tag{R. 4.12}$$

A study reported that the presence of 40% CO_2 lowered the fuel cell performance by 10% (Rajalakshmi 2004). Thermodynamically, there is no doubt that some CO can form. Assuming a gas mixture from steam reforming CH_4 containing 75% H_2 and 25% CO_2 vol is fully saturated with water at 27°C, the mixture will contain 3.5% H_2O, 72% H_2, and 24% CO_2 vol. Using the reaction equilibrium constant K_2 shown in Table 3.3 (K_2 = 93944 at 27°C), the equilibrium CO concentration will be 52.6 ppm, which would severely poison the anode. At a typical fuel cell operating temperature that is around 60~70°C, since K_2 will become significantly smaller (K_2 = 1549 at 127°C), the CO concentration will be dramatically higher, although the increase in the saturated vapor concentration and the corresponding decrease in the concentrations of CO_2 and H_2 favors lower CO concentration. It is also possible that some CO will form through the electrochemical reduction of CO_2. Kinetics plays a major role here. Even though PtMo is an excellent CO-tolerant catalyst, since it can catalyze the water-gas shift reaction and the reduction of CO_2, using PtMo as the anode catalyst is likely to encounter CO_2 problem.

Voltage loss caused by CO poisoning is fully recoverable when CO-free H_2 is subsequently used.

H_2S is a worse poison than CO to the anode, and it can severely poison the anode even at ppb levels. It forms a S layer on the surface of the catalyst according to Reaction 4.13:

$$Pt + H_2S = Pt\text{-}S_{ad} + H_2 \qquad \text{(R. 4.13)}$$

And thus the poisoning is not recoverable when switching to S-free H_2. Fortunately, H_2S and all the S-containing compounds should have been removed prior to the fuel processing steps and thus the fuel stream is not expected to contain any of it before entering the fuel cell.

The most popular poisons to the cathode are NO_x and SO_2 emitted out from the ICE vehicles and some chemical plants. They can poison the cathode at ca. 1 ppb level. The voltage loss caused by NO_x poisoning is fully recoverable, but that caused by SO_2 poisoning is not due to the formation of a S layer according to Reactions 4.14 and 4.15:

$$Pt + SO_2 + 2H^+ + 2e^- = Pt\text{-}SO_{ad} + H_2O \qquad \text{(R. 4.14)}$$

$$Pt\text{-}SO_{ad} + 2H^+ + 2e^- = Pt\text{-}S_{ad} + H_2O \qquad \text{(R. 4.15)}$$

Ammonia gas (NH_3) can low the performance of both the anode and the cathode by combining with H^+ to form NH_4^+ according to Reaction 4.16:

$$NH_3 + H^+ = NH_4^+ \qquad \text{(R. 4.16)}$$

The reaction reduces the proton concentration and thus the proton conductivity of the ionomer in both the PEM and the catalyst layers. Metal ions will do similar harm, and the impact of highly charged cations such as Fe^{3+} is most harmful. It not only replaces H^+ but may also restructure the hydrophilic domains of the ionomer (e.g., crosslinking the side chains by ionically interacting with the $-SO_3^-$ groups). Obviously, the impact caused by cations is cumulative.

4.7 Stack

4.7.1 Performance and Durability

Most of the evaluations on a stack are similar to those on a single cell. The V-I polarization curve is used to estimate the stack's performance from lower to middle and to higher current densities. Parameters such as the humidification temperature, temperatures of the coolant at the stack inlet and the outlet, the coolant flow rate, the reactant stoichiometric ratio, and the RH should be optimized through experiments. The compression force on the stack assembly should seal the stack but not cause damage to the GDM.

The performance and the lifetime of the stack often depend on the weakest cell. It is better to have a stack that has even performance and durability among all the cells, even though the average performance is lower than a stack with a higher average performance but uneven performance and durability among cells. So, one of the most important things to do is to achieve evenness among cells. Durability is defined as the number of hours the stack can provide power for the load. If a stack reaches the end of its lifetime, it is likely to show a quick decline in performance (like a "sudden death"). The sudden death is often associated with serious gas crossover through the PEM due to the aging of the PEM. Aging weakens the mechanical strength of the PEM, makes it become thinner, and leads to the formation of larger pinholes on the PEM or even the breaking up of the PEM. It is also possible that the stack will not experience a sudden death and can still provide power after operating for many hours, but the power output is not able to meet the needs of the load; this also signals that the stack has reached the end of its lifetime for this use (the stack may be used for powering a smaller load in another use). This is often because the catalyst has lost enough of the ECSA due to aging. The ECSA loss can be due to agglomeration of the catalyst, migration of catalyst materials (into the PEM or even to the other electrode), detachment of the catalyst particles from the support and a local lack of either electronic or protonic conductivity, and a catalyst surface composition change due to enrichment of a certain element or deposition of a foreign element. In general, the end of a stack's lifetime due to sudden death is caused by damage to the PEM, while that due to inadequate power output arises from the loss of the ECSA of the catalyst. After a stack reaches the end of its lifetime, the stack cannot be fixed, because either the PEM or the catalysts or both have reached the end of their lifetime.

If a newly built stack shows problems, they are often related to reactant leakage, either internally or externally, due to improper seal or misalignment of some components. After the problems are identified, the stack can be fixed. It is worth the effort to fix the stack because the key components, such as the PEM and the catalysts, are new and have the potential to operate for thousands of hours.

TABLE 4.3

Lifetime Targets and Maximum Allowable Decay Rates of Fuel
Cells for Different Applications[a]

Application	Targeted Lifetime (h)	Max. Decay Rate ($\mu V\ h^{-1}$)
Backup	1,500	100
Cars	5,000	30
Buses	25,000	6
Primary stationary	60,000	2.5
Portable	5,000	30

[a] Assuming unit cell voltages are 0.70 and 0.55 V at BOL and EOL,
respectively.

Durability is typically gauged by the performance decay rate. The decay
rate is faster in the beginning of the test (e.g., the first 100 hours), typically
because the smaller catalyst particles are not stable and aggregate into larger
ones that are more stable. Afterward, the decay rate slows down. In this time
period, the decay rate is between 5 and 30 $\mu V\ h^{-1}$ at the current stage of technology development. For example, if the decay rate is 10 $\mu V\ h^{-1}$, the beginning of life (BOL) cell voltage is 0.70 V, and the designed end of life (EOL) cell
voltage is 0.55 V, then the lifetime of the cell is 15,000 hours (150 mV × 1000 μV
mV^{-1}/10 $\mu V\ h^{-1}$). The decay rate starts to accelerate near the end of life for the
sudden death–type failure; so, when an accelerated decay rate is observed, it
signals that the stack's end of life is near.

For different applications, there are different lifetime requirements. If
the BOL cell voltage is 0.70 V, and the designed EOL cell voltage is 0.55 V, the
required decay rate limits are summarized in Table 4.3.

It can be seen from Table 4.3 that the current technology is more than
enough to meet the backup power requirements, and adequate for cars, and
reachable for buses and portable applications (using DMFCs), but difficult for
stationary applications. In August of 2011, a bus developed by UTC reached
10,000 hours of cumulated operation time without replacement of the stack.

How to start up and shut down a stack impact its lifetime significantly.
The most important thing is to minimize or better prevent the formation
of H_2/air boundary at the anode, followed by shortening the time at OCV.
Several methods can achieve those goals. One method invented by the
Author only resulted in 25 mV voltage loss (at 0.6 A cm^{-2}) after 20,000 times
of startup and shutdown cycles.

4.7.2 Startup Rate and Subzero Temperature Challenge

A fuel cell should be able to send enough power out when it is needed,
although there are different startup-time requirements for different applications. For portable, backup, and vehicle applications, the fuel cell system
must send enough power out instantly once it is started, while for primary

stationary applications, where the system is intended to keep running for most of the time, the fuel cell system is not required to be able to start instantly but should be able to adapt instantly to the load change. All fuel cell systems have batteries (or supercapacitors, or both) to start the systems and to offer additional power before the fuel cell stacks can handle all the power needs. If the stack can offer all the power needs in a shorter time, the duration for the batteries to provide power will be shorter. Therefore, the startup time of the stack itself is a parameter to gauge its capability. From ambient temperatures, an air-cooled stack can reach the nominal power output in about 1 minute, while a liquid-cooled stack may take 10 minutes or longer to reach the nominal output power. People often measure the time to reach 80% and 100% of the nominal power, respectively, to assess the startup rates of a stack.

It is a challenge to start a fuel cell system in subzero temperatures because the freezing of water forms ice that hinders the transport of the reactants. For an air-cooled stack using graphite plates, it can start from about −10°C, while for a liquid-cooled stack using graphite plates, it is difficult to start the stack below 0°C because the liquid coolant absorbs a lot of heat during the starting up process. For backup applications, since there is grid power available while the fuel cell system is in the idling state, the stack can be kept at temperatures higher than 0°C by using a heating mechanism. For example, for a liquid-cooled stack, the coolant can be electrically heated up when the environment temperature drops to near 0°C, and once the coolant reaches about 20°C, the heater will be shut down. Starting up and shutting down the heater are programmed automatically. Keeping the MEAs above 0°C using an external heater also reduces their performance losses caused otherwise by the freeze–thaw cycles. For vehicle applications, it is not feasible to keep the stack above subzero temperatures by using external power, and efforts must be put on the stack itself. If thin metallic plates are used, due to the smaller heat capacity of the entire stack, it is possible to start the stack at around −30°C. The major problems that need to be solved are the corrosion problems of the metallic plates and the accelerated decay rate of the MEAs due to the freeze–thaw cycles. For primary stationary applications, since the fuel cell system is in operation for most of the time (except during the first startup stage), startup is not a concern. The concern is whether the system can follow the load surge quickly enough; the stack has no problems, but the fuel processing system may not be able to generate hydrogen quickly enough to meet the load surge needs. So it is critical to make the capacity of the fuel processing subsystem adequate enough.

Some ice will form when starting up a stack from subzero temperatures, and the formation of ice may cause damage to unit cell components such as the GDM, CL, and even the PEM. The preference of ice formation is likely to decrease in the following order: the interface between the GDM and the plate, the interface between the CL and the GDM, within the GDM, within the CL, and the interface between the CL and the PEM, due to the differences in microstructures and temperature distribution. For example the smaller pores and the presence of ionomer should cause water more difficult to freeze in the CL than in the GDM. A study showed that water generated by a fuel cell at $-10°C$ did not freeze within the catalyst layer; instead, it moved as super-cooled liquid to the surface of the catalyst layer and then frozen (Ishikawa 2007). The temperature is expected to decrease from the CL, to the GDM, and to the plate. The formation of ice at the GDM/CL and the CL/PEM interfaces may cause delamination of the CL from either the GDM or the PEM; the formation of ice within the GDM may cause damage to its structure and the segregation of PTFE from the carbon fibers; and the formation of ice may create large pinholes through the PEM although ice is not expected to form within the PEM. In order to minimize the negative impact of ice and to successfully start up a stack from subzero temperatures, the water within the stack should be removed after it is shutdown from a previous operation through gas purging for a few minutes. Higher gas flow rate and temperature are more effective.

4.8 Balance of Plant

Except for the stack, the remaining portion of a fuel cell system is referred to as the balance of plant (BOP). Several major items will be discussed here.

4.8.1 DC–DC Converter

The output voltage of a stack changes with the current, and it ranges from the OCV to about 1/3–2/3 of the OCV. The DC device powered by the stack often has a nominal input voltage (within a certain range), so the output voltage of the stack must be regulated to the input voltage range of the device. This task is done by a DC–DC converter. The DC–DC converter will either just decrease, or just increase, or both decrease and increase the stack output voltage. A DC–DC converter with a voltage step-down (i.e., decrease) is called a buck converter, with a voltage step-up (i.e., increase) a boost converter, and with a bidirectional (step-up and step-down) a boost–buck converter.

A DC–DC converter consists of electrical circuits with inductors, capacitors, and resistors as the major components. The input energy from the stack is stored in inductors and capacitors and then released to the load at a desired voltage. The energy storage and release are carried out at very high frequencies by controlling the duty cycle, that is, the input voltage on/off time. Due to the high on/off frequency electronic noise, and electromagnetic disturbance may be generated. They must be controlled carefully when the fuel cell is for a telecommunications backup power application; otherwise, these

noises will make communications unpleasant. The electromagnetic disturbance may also interfere with the proper operations of other components
within the fuel cell system. A well-designed DC–DC converter can reach an
energy conversion efficiency higher than 90%. The energy loss is caused
by resistance in the electrical circuits. The conversion efficiency is shown in
Equation 4.20:

$$\text{Conversion Efficiency} = (I_{out} \times V_{out})/(I_{in} \times V_{in}) \qquad \text{(Eq. 4.20)}$$

Although there are many DC–DC converters on the market for conventional applications, there are almost no DC–DC converters for fuel cell applications at kW levels. A major reason is that the output voltage of a fuel cell
stack is too wide, and the width is different in different stacks. For example,
a 10-cell stack is likely to have an output voltage range between 4 V and
10 V, while a 200-cell stack may have an output voltage range between 80 V
and 200 V. Without more or less uniform stack output, voltage range, and
large enough demand, a DC–DC converter manufacturer is not interested in
manufacturing DC–DC converters for fuel cell applications.

4.8.2 Air Supply Device

The devices sending air to the stack can be fans, blowers, or compressors.
They must meet two requirements: (1) being able to send enough air to the
stack for reaction (typically 2.5 to 3.0 times stoichiometric ratio), (2) being
able to overcome the pressure drop caused by the stack. Fans can only be
used for air-cooled stacks because their pressure boost normally lowers than
kPa. The advantages of fans are that they have the lowest power consumption and noise levels. Blowers and compressors are used for liquid-cooled
stacks. Compressors have the highest power consumption and noise levels.
Whenever possible, blowers should be used over compressors. A blower can
achieve a pressure boost up to about 25 kPa. When the pressure need is more
than a blower can provide, a compressor must be used.

The air flow rate generated by a blower or a compressor is inversely related
to the pressure boost at the same energy consumption. Without any air flowing out, the pressure boost will be highest. The air flow rate and the pressure boost should be evaluated before a blower or a compressor is used.
Figure 4.15 shows the relationship between the power consumption versus
the air flow rate and pressure of a blower. With an increase in the air flow
rate, the pressure boost decreases at the same power consumption. With
such data in hand, you can determine whether or not a blower can meet the
flow and pressure requirements of a particular stack.

It is often worth the effort to build a stack with a lower air pressure drop
so that a blower can be used. A lower pressure drop can be achieved by
making the flow channels straighter, shorter, wider, and deeper. The stack
may offer a slightly lower performance due to the lower air partial pressure,

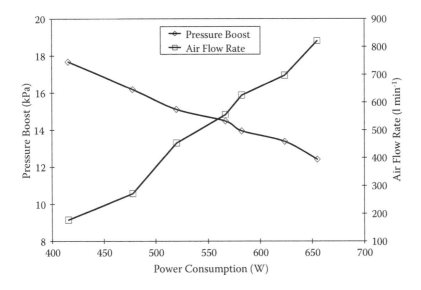

FIGURE 4.15
Relationship between power consumption versus air flow rate and pressure boost.

but the smaller power consumption by a blower than a compressor (maybe several times) will more than compensate for the slight performance loss of the stack. In addition, a blower will be significantly smaller and lighter than a compressor, making the integrated fuel cell system more attractive.

When choosing a blower, some excess air flow capacity should be available. If a blower always works at the maximum air flow rate, it not only consumes more power but also has a risk of overheating. When overheating happens, the blower should be able to shut itself down automatically to prevent it from being burned. Although the pressure drop is mainly caused by the stack, pressure drops caused by other components, such as the humidifier, the air flow meter, and the connection piping, should also be considered. If a blower meets the pressure requirement but not the air flow rate requirement, consider using two blowers together in parallel.

4.8.3 Liquid Pump

For a liquid-cooled stack, a pump is needed to circulate the liquid coolant through the stack and the heat exchanger, which are the major sources that resist the coolant flow. If a flow meter is incorporated in the coolant flow loop, it can also generate significant resistance to the coolant flow. A pump must be able to meet the coolant flow rate and overcome the resistance in the entire flow loop.

For the proper operation of a stack, the humidification temperature and the coolant temperatures at the stack inlet and outlet and their difference (typically 5° to 10°C) should be controlled. The difference between the

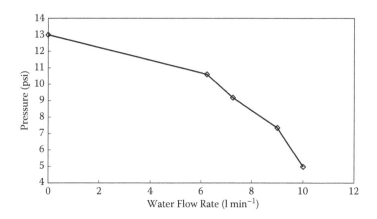

FIGURE 4.16
Pressure versus water flow rate of a liquid pump.

humidification temperature and the stack outlet temperature determines how much water will be taken out by the air exhaust as vapor. A higher difference will enable more water vapor to be taken out. The coolant flow rate will affect the coolant temperature difference between the stack inlet and outlet, with a higher flow rate leading to a smaller temperature difference. The coolant temperature at the stack inlet is mainly affected by the heat exchanger. A higher heat exchange capacity will result in a lower inlet temperature.

The flow rate versus pressure of a pump should be evaluated before it is used. Figure 4.16 shows that the pressure decreases with the flow rate. The maximum pressure of the pump is 14 psig at 0 flow rate. The power consumption by a liquid pump is quite small, so it is better to choose a large enough pump to be able to handle unexpected coolant flow needs.

4.8.4 Heat Exchanger

A stack may generate as much heat as the electrical power. The heat must be dissipated in order to prevent the stack from overheating. The stack of a PEMFC is typically operated at a coolant outlet temperature of less than 75°C to achieve a longer lifetime. The excess heat is released by a heat exchanger.

A heat exchanger is composed of heat radiation fins and a fan. The coolant flowing in the heat exchanger radiates heat out through the fins, and the radiated heat is sent out of the fuel cell system by the fan. The radiation fins are made with thin metal sheets with large surface areas in order to achieve high heat radiation efficiency. Without the fan, the air temperature around the fins will quickly get close to the coolant temperature after the fuel cell is running for a short time because the heat accumulates inside the fuel cell system's enclosure. Then, the heat that is able to radiate out of the fins will become smaller and smaller due to the decrease in the temperature difference between the temperature of the fins and the temperature of the surrounding air. Therefore, the fan is a must-have device to effectively draw the hot air out of the enclosure and allow cooler air into the enclosure. At higher fan spin rates, faster air exchange will occur, sending more heat from the coolant out of the fuel cell enclosure.

The heat exchange capacity of a heat exchanger needs to be evaluated to determine whether it can meet the heat removal requirement. The test can be carried out with a simple setup. Using an electrical power source to simulate the heat power from the stack first raises the temperature of the coolant in a reservoir and the entire circulation loop to a value near the stack coolant outlet temperature. While keeping the electrical power on, the temperature of the coolant entering and leaving the heat exchanger is measured at different coolant circulation rates and different fan spin rates. From the difference in the two temperatures and the heat capacity and the flow rate of the coolant, the amount of heat removed by the heat exchanger can be calculated.

Let us have a look at an example. An electrical heater that generates 5000 W of heat is used as the heat source to simulate the stack. Water is the liquid coolant and its flow rate is 10 L min^{-1}. The density and the heat capacity of water are 0.98 g cm^{-3} and $C_p = 4.18$ J g^{-1} K^{-1}, respectively, at around 60°C. The water flow rate is

$$v = 10 \text{ L min}^{-1} = 0.98 \text{ g cm}^{-3} \times 10 \text{ L} \times 1000 \text{ cm}^3 \text{ L}^{-1} \times 60 \text{ min h}^{-1} = 589920 \text{ g h}^{-1}$$

The amount of heat dissipated by the heat exchanger is given in Equation 4.21:

$$\Delta H = (T_1 - T_2) \times C_p \times v \qquad \text{(Eq. 4.21)}$$

where T_1 and T_2 are the temperatures of water at the points of entering and leaving the heat exchanger, respectively. Table 4.4 shows some results with hypothetical temperature differences.

TABLE 4.4

Removal by Heat Exchanger

			Heat Removed by Heat Exchanger	
T_1(° C)	T_2 (°C)	T_1-T_2 (°C)	kJ h^{-1}	W
60	60	0	0	0
60	58	2	4937	1371
60	55	5	12342	3429
60	53	7	17279	4800
60	50	10	24684	6857

4.8.5 Solenoid Valve

There are at least two solenoid valves in a fuel cell system, the first one for H_2 entering the stack and the second one for H_2 leaving the stack. When the fuel cell is in the idling state, the second solenoid valve is always closed to prevent air from getting into the anode chamber (without some special arrangements, air can always get into the anode chamber from the cathode chamber by diffusing through the PEM). During the operation of the fuel cell, the first solenoid valve is always kept open while the second opens only at purging times. The opening or closing of the solenoid valves is triggered by electrical signals. Some solenoid valves are kept open without electrical or magnetic power, while others are kept closed without electrical or magnetic power. The latter type is most suitable for fuel cells because they do not need power to keep closed. If the former type is used, the fuel cell system has to provide power to keep the solenoid valve closed when the fuel cell is in the idling state.

Assuming that the solenoid valve can be properly opened and closed, its closeness and resistance to H_2 flow should be evaluated. If the solenoid valve is not completely closed when it is in the closed state, H_2 will seep into the stack from the first solenoid valve and air will seep into the stack from the second one. Then there will form an H_2–air boundary at the anode side after the stack is in the idling state, which creates a voltage well over the OCV at a portion of the cathode (this portion corresponds to the air portion at the anode), leading to its quick decay. The closed state of a solenoid can be tested by many methods. One simple setup is shown in Figure 4.17. Connected to one side of the solenoid valve is an air-tight, sealed enclosure (e.g., a tank, or a cylinder, or a piece of pipe) with a pressure sensor to monitor the H_2 pressure inside. Connected to the other side of the solenoid is a higher-pressure H_2 source through a pressure regulator. First, fill the sealed enclosure with H_2 at pressure P_1 by opening the solenoid valve. Then close the solenoid valve and set the pressure regulator at pressure P_2 that is higher than P_1. P_1 should be close to the ambient pressure, while P_2 should be around the H_2 inlet pressure in

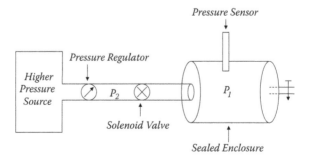

FIGURE 4.17
A schematic setup to evaluate the closeness of a solenoid valve.

front of the solenoid valve in the actual fuel cell system. If P_1 does not change with time, then the solenoid valve does not leak in the closed state.

If the opening of the solenoid valve is not large enough, it will resist the flow of H_2. When a larger amount of H_2 is needed during purging, the H_2 flow rate may not meet the requirement, and the effectiveness of purging will be affected. It is better to choose a solenoid valve whose opening is as large as the internal diameter of the pipes connected to it.

The H_2 flow rate during purging can be monitored by a flow meter. Alternatively, the amount of H_2 from each purging can be estimated by collecting the H_2 in a water-filled graduated cylinder that stands inverted in a larger water bucket. H_2 displaces some water out of the graduated cylinder, and its top "empty" portion is the volume of H_2. The volume of H_2 can also be estimated by collecting it in something like a balloon.

4.8.6 Control Board

Many devices need to be controlled in a fuel cell system. Controls are critical tasks for a fuel cell system to work. Without good controls, the stack can be damaged in seconds. Typically, there is a mother board for carrying out the system-level control algorithms, and there are several subboards for controlling individual devices such as the blower, the fan(s), the pump, the solenoid valves, the DC–DC converter, and the switches. A central computing unit (CPU) is typically integrated in the mother board. The mother board coordinates the functionalities of the subboards to achieve a unified working fuel cell system.

The control algorithms are written in a language such as C or C++ to form the software, which is then embedded in the control boards. The software needs to be tested repeatedly, either through simulation or actual hardware to make sure that it fully reflects the control algorithms. The subboards will communicate with the mother board based on a mutual communications agreement.

The controls are carried out by the control boards. These boards are first made as printed circuit boards (PCBs), and then various components are assembled and soldered onto the PCB. Since wires and some components will generate heat when the boards are functioning, adequate heat dissipation must be considered to prevent the overheating of the boards during use. The entire board needs to be evaluated to determine whether it will be able to carry out the designed functionality and whether or not it will overheat.

After all the control boards pass individual tests, they will be tested together. Problems such as miscommunication and bugs can be fixed in this process.

4.8.7 Flow Meter and Sensor

If a fuel cell system uses flow meters to measure the flow rates of H_2, air, or coolant, the flow meters should be evaluated for whether or not they offer too much resistance to the flow of the corresponding fluids. A fuel cell system needs a variety of sensors, such as the H_2 concentration sensor, H_2 pressure sensor, coolant-level sensor (to monitor the liquid coolant level within the coolant tank), current sensor, voltage sensor, flooding sensor, motion sensor, door-open sensor, and so on. Their functionalities should be evaluated before being installed in the fuel cell system.

4.8.8 Internal Power Supply

The largest internal power is often the battery (or supercapacitor, or both) that is used to start the fuel cell system and provide additional power for the load when needed. Its voltage and capacity should be determined according to the fuel cell system. For example, if the fuel cell is for generating 48 DC power for telecommunications backup power applications, the voltage of the battery should be around 48 V; otherwise, it needs a DC–DC converter to convert to 48 V, which increases the system's complexity, cost, volume, and weight. The capacity of the battery must be able to power the full load during the startup of the fuel cell system because the stack cannot send out either any power or enough power for this time period. Since the startup time is very short, a battery with smaller capacity but higher discharging rate is preferred in order to reduce the cost and volume of the battery. Li-ion batteries, nickel metal hydride batteries, and Ni-Zn batteries are most suitable for this purpose because of their higher energy densities and higher discharging rates. The voltage of the battery decreases with an increase in the discharging rate, so its output voltage must also be suitable for the load (besides power output) at the discharging rate. The capacity of a battery is often tested at a discharging rate of 0.05 or 0.10 C, but if the battery is used at a discharging rate of 2 C, its capacity will be much lower. Also, the voltage and the capacity of a battery are largely affected by the environmental temperature. For example, a Li-ion battery capacity at subzero temperature is much lower than at 20°C. Therefore, the output voltage, power, and capacity of a battery at the targeted discharging rate must be evaluated thoroughly under the worst condition that a fuel cell is designed to withstand. For example, if a fuel cell is designed to work at temperatures between –10°C and 50°C, the battery must be tested at –10°C and 50°C.

Figure 4.18 shows the voltage and the capacity when a 53V/40 Ah Li-battery discharges at 20°C and 1.5 C discharging rate (which means the discharging current is 60 A). It gave out 35 Ah instead of 40 Ah charge when its voltage dropped to 43 V, the cutoff voltage for being used to supply power to a load whose minimum input voltage cannot be lower than 43 V. If the voltage loss through wires is taken into consideration, the battery will reach the cutoff voltage before sending 35 Ah charge out. The cumulative energy output of this battery is 1580 Wh when the voltage drops to 43 V. If the power need of the load is 2580 W (43 V × 60 A), this battery can last for about 43 min (1580 Wh × 60 min h^{-1}/2580 W).

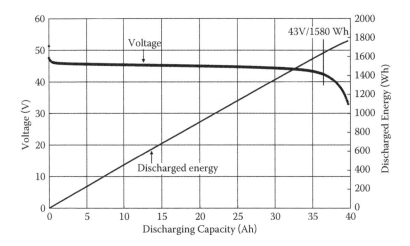

FIGURE 4.18
Voltage and capacity of a 40-Ah Li battery discharged at 20°C and 1.5 C.

However, when the same battery was discharged at −10°C and 1.5 C, its voltage dropped to below 43 V instantly, and therefore it was not adequate enough to provide power for a load that required a minimum input voltage of 43 V.

Since a battery typically shows poor performance at subzero temperatures, external heating may be needed to keep its temperature above 0°C.

Components such as the control boards, solenoid valve, blower, fans, pump, switches, and sensors also need power from the battery during the startup process. These components may need 48, 24, 12, or 5 V, and DC–DC converters are needed to convert the voltage from the internal power supply module to those values. When the fuel cell system is in the idling state, the mother board and various sensors are also powered by the internal power supply module.

Once the stack starts to provide power, it shares the load with the battery in the beginning, and all of the load will be quickly powered by the stack. The power needs of the components mentioned above can then come entirely from the system's main DC–DC converter. The battery could also be charged by the system's main DC–DC converter.

If the internal battery is charged by the stack while the stack is in operation, the stack must be able to meet the power needs of both the load and the battery. If the battery has a large capacity and is quite empty, its power needs can be significant. It is better to limit the charging current to the battery to prevent it from withdrawing too much power. The combined power need from the load and the battery should not exceed the stack's designed capacity.

4.9 System

All the efforts mentioned previously are necessary to build a reliable and durable fuel cell system. The system will be first constructed in a bread-board fashion with all the needed components and modules. It is evaluated for functionality, performance, decay rate, and so on. The best orientation of the stack and how to effectively connect different components will be tested. Various problems will be encountered in this stage, and they must be resolved.

Afterward, the design of the entire system layout can be started. It may start with a pro-engineering (ProE) design. An engineer plots out the entire system structure in three dimensions using ProE software. Every component will be drawn to reflect its actual size and location. After the ProE design passes a critical evaluation by a group of scientists and engineers, it will be converted to an engineering plot for parts manufacturing. The parts are mainly the system enclosure and those that are used to host the various components and modules discussed in this chapter.

Finally, the entire fuel cell system can be built inside the enclosure, followed by various tests.

The current versus voltage and power results will elucidate the fundamental functionality of the fuel cell system. Parameters such as the coolant flow rate, the spin rate of the fan on the heat exchanger, the RH, and the stoichiometric ratios of the reactants will be adjusted to achieve the best power output and the lowest decay rate. The power output must meet or exceed the total load (internal and external); otherwise, the fuel cell system is underrated and cannot be used for the load.

The enclosure must have enough openings for ventilation and heat dissipation purposes. It is not uncommon to find that that the heat exchanger dissipates enough heat on a bread-board test, but not in the enclosed system. So, it is wise to use a heat exchanger with enough capability to dissipate excess heat at the highest ambient temperature required during the bread-board test. The openings on the enclosure should also be made in such a way that water will not be able to get into the system during rainy times.

In addition to performance and durability, the fuel cell system needs to be tested for other characteristics such as audible noise level, electronic noise level, overloading capacity, startup time (especially at the lowest designated temperature), and various warning and protective functions.

Some limitations should be recognized when dealing with a fuel cell system. For example, the output voltage of the stack cannot become too low due to the input voltage requirement of the main DC–DC converter. When the output voltage from the stack drops below the lower input voltage limit of the DC–DC converter, the system will shut itself down to prevent the converter from being damaged. Such a protection mechanism is written into the software and the system will perform it automatically. Due to this limitation, the method to activate or reactivate a stack by dropping its voltage to near 0 V (using either low air supply or high current density) cannot be performed, unless the stack is disconnected from the DC-DC converter.

4.10 Summary

Various evaluations, ranging from the basic components to the entire system, are needed in the development of a fuel cell system. The evaluation may include physical, mechanical, chemical, electrochemical, electrical, and software systems. Performance evaluation is short while durability evaluation may take hours to years. Accelerated evaluation can be utilized once a baseline is established, but establishing the baseline itself may take years.

5

Stationary Power

A fuel cell can be used for any applications that require electricity (and heat). It was first used in manned spacecraft in the United States. Currently, the major markets for fuel cells are stationary, transportation, and portable power. This chapter will discuss its application for stationary power. Stationary power includes backup power and primary power.

5.1 Backup Power

Backup power is becoming more and more important in modern society. It plays a critical role in computing centers, information centers, hospitals, banks, and telecommunications stations. Lead-acid batteries are the most commonly used backup power nowadays, complemented by diesel generators. Lead-acid batteries are low in cost and safe, but they are bulky, heavy, and of limited backup time. Stations consume a lot of electricity yearly to maintain these batteries at around 25°C. Lead pollution happens if the manufacturing process and the after-use disposal are not strictly controlled. Lead pollution is a serious problem in developing countries because people do not take it seriously and the punishment for polluters is not severe enough. Diesel generators are cheap, but they are inefficient, noisy, cause pollution, and require frequent maintenance or repair.

Proton exchange membrane fuel cells (PEMFCs) do not have the problems associated with batteries and diesel generators, and are believed to be the ideal backup power. The following discussion will focus on telecommunications backup power applications.

5.1.1 Telecommunications Station

Wireless communications are becoming dominant worldwide, including in developed countries where a lined communication infrastructure is well established. In many developing countries, many families do not install lined phones at all, and people use cell phones exclusively for communication.

Wireless communications are made available by telecommunications stations (i.e., transceiver stations or cell towers). There are about 5 million such stations worldwide. By the end of 2011, China itself had about one

FIGURE 5.1
Cell tower of a telecommunications station.

million telecommunications stations. Figure 5.1 shows the signal receiving–transmitting tower of a telecommunications station.

The signal is processed by devices within the telecommunications station, and the transceiver device uses DC power. The input voltage of the device ranges from −43 V to −58 V, with a nominal voltage being set at around −53 V. The DC power is obtained from the grid AC power through an AC–DC converting process. Lighting and air conditioning in the stations are powered by the AC power. Forty-eight-volt lead-acid battery banks are currently used as the backup power for the DC power equipment. During a power shortage, the batteries power the DC device at a discharging rate of 0.1 C.

As backup power, the DC output of a PEMFC is connected to the DC buss of the station, as shown in Figure 5.2. The solid arrows represent the power flows in the system. The dashed arrows show potential power flows, and they depend on the system design. The dashed arrow between the DC–DC

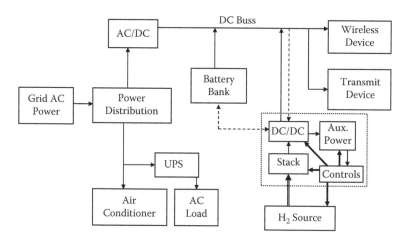

FIGURE 5.2
Power distribution within a telecommunications station.

converter of the fuel cell system and the battery bank means that the battery may be used to provide power for the fuel cell system when the latter is in the idling state through the fuel cell system's DC–DC converter. It also means that the fuel cell may be used to charge the battery bank when the fuel cell is in operation. The battery bank can be either the existing battery bank in the telecom stations or a smaller battery bank incorporated into the fuel cell system. The dashed arrow between the DC buss and the fuel cell system means that the DC buss power may be used to provide power for the fuel cell system when the latter is in the idling state. Please note that the power flows from either the battery bank or the DC buss to the fuel cell system send electricity to the auxiliary power unit of the fuel cell system, and does not send any electricity to the fuel cell stack; otherwise, the stack would be damaged. The thicker solid arrows indicate the flows of the control signals. They control the stack, the DC–DC converter, the auxiliary power, the H_2 source, and so on. The power needed comes from the auxiliary power unit.

When the grid power is off, the fuel cell system will start automatically to power the DC load; when the grid power is back, the fuel cell system will go into standby mode automatically. There are several ways for the fuel cell system to detect the presence or absence of the grid power. One method is to monitor the AC current getting into the station: when the disappearance of the AC current is sensed, the fuel cell system starts, and when the reappearance of the AC current is sensed, the fuel cell system shuts down. Another method is to monitor the buss voltage for the load. Table 5.1 gives examples of the preset voltages when using DC voltage to control the startup or standby mode of the fuel cell system. The buss voltage is typically set at around −53.5 V. When the grid power is off, the buss voltage will drop; when the buss voltage dropping to a value such as −49.5 V is detected, the fuel cell system starts, and the buss voltage will be the fuel cell system output voltage

TABLE 5.1

Example Voltages Used to Start and Stop a Fuel Cell System

Parameter	Preset Voltage (V)	Comments
DC buss voltage	−53.5	This is a common preset DC buss voltage when there is grid power.
FC startup voltage	−49.5	When the DC buss voltage drops to this preset value, the fuel cell system starts.
FC output voltage	−51.5	This will be the DC buss voltage when the fuel cell is in operation.
FC entering standby mode voltage	−52.5	When the DC buss voltage increases to this preset value, the fuel cell system goes into standby mode.

set at a value such as −51.5 V. When the grid power comes back, the buss voltage will rise (to −53.5 V in about a minute); after the buss voltage becomes higher than −51.5 V, the fuel cell system will stop sending power out to the load, and when the buss voltage reaches a preset value such as −52.5 V, the fuel cell system automatically enters standby mode.

5.1.2 Fuel Cell System with Air-Cooled Stack

5.1.2.1 System Architecture

The DC load of telecommunications stations typically ranges from less than 1 kW to more than 5 kW. In China, most of the stations have a DC load of around 2 kW. For these kinds of loads, fuel cell systems with air-cooled stacks are the preferred choice.

Systems with air-cooled stacks are simpler and more cost effective than those with liquid-cooled stacks. Figure 5.3 shows the major modules within

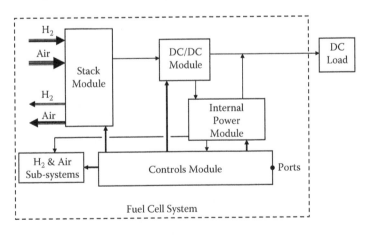

FIGURE 5.3

Major modules in a PEMFC system using an air-cooled stack.

such a system. The system includes a stack, a DC–DC converter, air supply, H_2 supply, internal power, and the controls. The thinner and thicker solid arrows represent the flow of electrical power and the control signals, respectively. The fan(s), either mounted on the stack or separated from the stack, provide air for the cathode reaction and for stack cooling. The air stoichiometric ratios are determined by the stack cooling need (not by the cathode reaction need), and they can be 20 times the stoichiometric ratio or more depending on the environment temperature. A PEMFC system built using a Ballard FCgen-1020ACS series stack will be first presented as an example in this chapter.

5.1.2.2 Fuel Cell System with Ballard FCgen-1020ACS Stack

5.1.2.2.1 Ballard FCgen-1020ACS Stack and Fuel Cell System

Ballard Power Systems (Canada) is the world leader in stack development and manufacturing. Its stacks are reliable and offer very uniform cell-to-cell and stack-to-stack performance. Figure 5.4 is a picture of a Ballard FCgen-1020ACS series air-cooled stack with 80 unit cells. The honeycomb-like

FIGURE 5.4
Picture of Ballard 1020ACS Series air-cooled stack. Courtesy of Ballard Power Systems.

TABLE 5.2

Specifications of Ballard FCgen-1020ACS Stacks

Parameter	Unit	Value
Rated power at ambient condition (20°C, 30% RH)	W cell^{-1}	45
Current at rated power	A	65
Voltage at rated current	V cell^{-1}	0.683
Suggested H$_2$ purity	%	≥99.95
Suggested H$_2$ supply pressure	barg	0.16–0.36
Suggested H$_2$ flow rate at rated power	slpm cell^{-1}	~0.5
Coolant	—	Air
Suggested air flow rate at rated power	slpm cell^{-1}	~50
Stack operating env. temperature	°C	−40–52
Stack startup env. temperature	°C	−10–52
Startup time to 80% rated power	s	~20
Dimension of 56-cell stack (L/W/H)	mm	363/103/351
Mass of 56-cell stack	kg	11.0
Certification	—	CAN/CSA-C22.2 No. 62282-2 fuel cell modules

Source: Adapted from Ballard Power Systems website, FCgen-1020ACS Spec Sheet,
http://www.ballard.com/fuel-cell-products/fcgen-1020ACS.aspx.

openings are the channels for air to pass through for both the cathode reaction and stack cooling. Spring caps are used to make sure that H$_2$ will not leak when the stack goes through volume changes between operation and idling states. Table 5.2 shows the basic parameters of FCgen-1029ACS stacks.

Figure 5.5 is a picture of a 2.5 kW fuel cell system using a Ballard FCgen-1020ACS stack consisting of 80 unit cells. Since the stack in this case had no fans mounted on it, two separate fans were used. The fan was first evaluated to see whether it could meet the stack requirements on air flow rates and air pressure. Since the air flow is determined by the stack cooling need, and higher ambient temperatures require larger air flow rates due to the smaller temperature difference between the stack temperature and the surrounding temperature, we can estimate how much cooling air the stack needs at 40°C environment temperature when the stack is running at 65 A/0.68V (the current/voltage at the rated power).

The heat generation rate of the 80-cell stack is

$$q_{stack} = (V_{LHV} - V)\,I\,n = (1.25 - 0.68) \times 65 \times 80 = 2964\ \text{W} = 177.7\ \text{kJ min}^{-1} \quad \text{(Eq. 5.1)}$$

$(1\ \text{kJ} = 0.278\ \text{Wh} = 16.68\ \text{W min})$

This amount of heat must be removed by the cooling air,

$$q_{stack} = q_{heat\ removal} = (T_{stack} - T_{env})mC_p \quad\quad \text{(Eq. 5.2)}$$

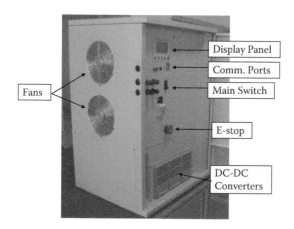

FIGURE 5.5

Picture of 2.5 kW fuel cell system using a Ballard 80-cell FCgen-1020ACS stack. Courtesy of Wuhan Intepower Fuel Cells.

where T_{stack}, T_{env}, m, and C_p are the stack temperature, environment temperature, air flow rate through the stack, and air heat capacity, respectively. $T_{stack} = 61°C$ (the optimal stack temperature at 65 A), $C_p = 1.007$ kJ kg^{-1} K^{-1}. Rearranging Equation 5.2 leads to

$$m = q_{stack}/[(T_{stack} - T_{env})C_p] \qquad \text{(Eq. 5.3)}$$

$$= 177.7 \text{ kJ min}^{-1}/[(334-313 \text{ K}) \times 1.007 \text{ kJ kg}^{-1} \text{ K}^{-1}]$$

$$= 8.4 \text{ (kg min}^{-1})$$

$$= 7434 \text{ (l min}^{-1})$$

The air stoichiometric ratio is ca. 86 {[0.21 × 7434 L min^{-1}/(60 s min^{-1} × 22.4 L mol^{-1})]/[(80 × 65 A)/(4 × 96485)]}. (Dry air density is about 1.13 kg m^{-3} at 40°C and 1 bar.)

The fan must also meet pressure needs. Tests have shown that an 80-cell stack causes an air pressure drop of around 90 Pa. So, the fan must be able to generate 7434 L min^{-1} air flow rate at 90 Pa pressure. Considering potentially higher ambient temperature (>40°C) and additional air pressure losses along its path, the fan should be able to generate an air flow rate and a pressure boost slightly higher than those two values.

Figure 5.6 shows the characteristics of a fan. At 100% PMW (pulse-width modulation) it can achieve an air flow rate of 28384 slpm and an air pressure of 380 Pa, well over the needs of a Ballard 80-cell FCgen-1020ACS stack. Actually, at a PMW of 32%, the air flow rate and the pressure are 8207 slpm and 94.5 Pa, respectively, which exceed the stack's needs. The power consumption by the fan at this point is 50 W. The fan's power consumption

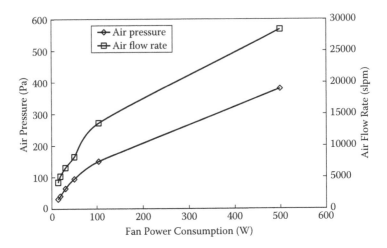

FIGURE 5.6
Characteristics of a fan.

increases more rapidly after about 50% PWM (corresponding to about 100 W power consumption), so it is better to keep the fan working at a PWM smaller than 50%, which also lowers the noise level generated by the fan. Both the fan's air flow rate and the pressure increase with the fan's power consumption. In order to pull air evenly through every cell of the stack in Figure 5.5, two fans were used and they were kept at a suitable distance from the stack for the fuel cell system. Basically empty space was left between the fans and the stack in order to not obstruct the air flow.

The section below the stack (and the fans) was used to host the electronic control boards, the DC–DC converters, and the H_2 supply structure. Small fans were used to keep the control boards from overheating, and the hotter air was sent out of the system from the openings on the enclosure, as shown in Figure 5.7. The H_2 supply structure included a pressure sensor, an inlet solenoid valve, an outlet solenoid valve, and metal piping. Table 5.3 shows the key parameters of the system.

Two Li-ion battery banks, 10 Ah and 40 Ah, respectively, with the OCV around 55 V were incorporated into the fuel cell system as the internal power. The 40 Ah battery bank was directly connected to the DC buss of the load (i.e., it is connected in parallel between the fuel cell system's DC–DC converter and the load). During the startup of the fuel cell system, this battery bank supplied DC power to the load directly without DC–DC conversion because it was connected to the DC buss directly. This battery bank itself can supply 2500 W power output for more than 40 minutes at a discharge rate of 1.25 C in case the fuel cell stack fails to start. The system set a 5-second delay for starting the stack, because if the grid power was back within 5 seconds, the stack would not start in order to reduce its start–stop numbers, and thus lengthen its lifetime (the performance decay caused by one start-stop cycle

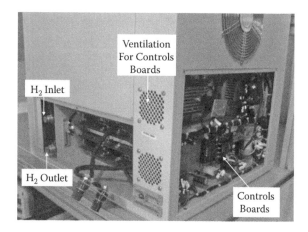

FIGURE 5.7

Picture of controls and H_2 supply subsystems of a 2.5 kW fuel cell system. Courtesy of Wuhan Intepower Fuel Cells.

TABLE 5.3

Key Parameters of a 2.5 kW Air-Cooled Fuel Cell System

Parameter	Unit	Value
Nominal system output power	kW	2.5
Overloading capacity	%	10
System startup time	ms	<10
Environment temperature	°C	−10–45
System electrical efficiency	%	43 (based on input H_2 LHV)
System output voltage range	Vdc	48–54
Communications port	—	RS232
Start–stop methods	—	Manual; remote; automatic
Remote communications	—	Control; messaging; measurement
Protective functions	—	Overload; H_2 inlet pressure high or low; system output voltage high or low; short circuiting; stack over-temperature; environment temperature high or low
Warning	—	System door opened; system flooded; H_2 leaked; H_2 source pressure low; H_2 inlet pressure too high or too low; stack temperature too high or too low; stack voltage too high or too low; stack current too high or too low; system output voltage too high or too low; system output power too high or too low; input voltage to the DC–DC converter too high or too low; input current to the DC–DC converter too high

Source: Courtesy of Wuhan Intepower Fuel Cells.

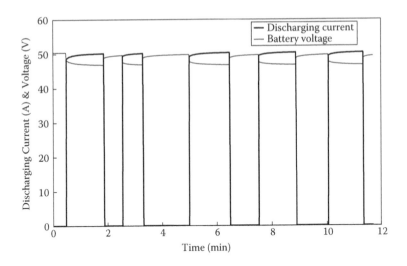

FIGURE 5.8
Discharging of a 40 Ah Li-ion battery bank. Courtesy of Wuhan Intepower Fuel Cells.

is equivalent to that caused by several hours of running time). Five seconds later, the two solenoid valves for H_2 opened to let H_2 into the stack. Then the inlet solenoid stayed open and the outlet solenoid closed. The 10 Ah battery bank and the stack DC power output were connected parallel to the fuel cell system's internal DC buss, which was located in front of the DC–DC converter; they will provide power for the load together after the start of the stack.

Figures 5.8 and 5.9 show the discharging processes of the 53 V/40 Ah and 53 V/10 Ah battery bank at 2350 and 2300 W constant power output with discharge rates at 1.25 and 5 C, respectively. The voltage of the 40 Ah battery bank was around 47 V, above the load's lower voltage limit of 43 V for telecommunications stations. However, the voltage of the 10 Ah battery bank dropped to 41.9 V in one minute, so this battery bank could not provide power by itself to the load with the lower input voltage limit of 43 V. Since the 10 Ah battery bank worked together with the stack to supply power to the load through the system's main DC–DC converter, tests showed that it was enough.

5.1.2.2.2 Tests

Figure 5.10 shows the power-sharing situation during the startup of the fuel cell system. The DC buss of the load was set at around 53.5 V. When the system started, after sensing that the grid AC current disappeared, battery 2 (i.e., the 40 Ah battery bank) instantly supplied 2500 W of power to the load, and its voltage was about 44.5 V, higher than the lower limit of 43 V. Five seconds later, the system started the stack, and the voltage of the stack jumped from 0 to 73 V. In another 5 seconds, battery 1 (i.e., the 10 Ah battery

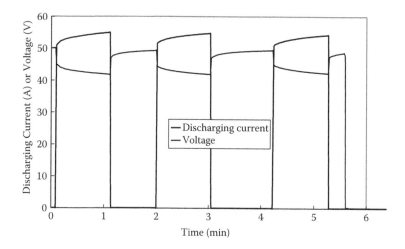

FIGURE 5.9
Discharging of a 10 Ah Li-ion battery bank. Courtesy of Wuhan Intepower Fuel Cells.

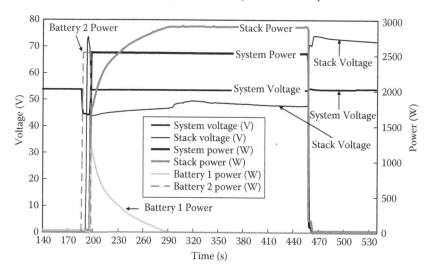

FIGURE 5.10
Power sharing situation during the startup of a 2.5 kW fuel cell system. Courtesy of Wuhan Intepower Fuel Cells.

bank) and the stack supplied DC power together. In the beginning, battery 1 and the stack offered 1205 and 1675 W of output power, respectively (the total power output was 2880 W). Then, the output of the stack increased and the output of battery 1 decreased in proportion like a mirror image, and 100 seconds later, all the load was powered by the stack itself. After the load was removed, the system entered standby mode, and the open-circuit voltage (OCV) of the stack jumped to 74 V.

It is clear that when battery 2 provided power to the load, the DC buss voltage of the load was determined by the battery voltage, and it dropped to around 44.5 V in the first 10 seconds. Once battery 1 and the stack supplied power to the load, the DC buss voltage of the load was determined by the output voltage of the system's main DC–DC converter, and therefore, it was very stable and close to the set point of 53.5 V.

For a 2530 W load, the stack provided about 2900 W of power, meaning that the sum of system parasitic power loss and the DC–DC converting loss was 370 W, 12.8% of the stack power output. The DC–DC conversion efficiency was around 90%, and therefore, the system parasitic power loss was 2.8% of the stack power, which was 81 W (2.8% × 2900 W). About half of it was consumed by the two fans for the stack, and the other half was consumed by other devices in the system such as the control boards, the small fans (for cooling the control boards and the contactors), sensors, solenoid valves, and wiring.

Figure 5.11 shows the voltage and power situation of the fuel cell system in a broad current range. The difference between the stack power output and the system power output increases nearly linearly with the stack current. The ratio of this difference to the stack power output in an expected power output range (e.g., 1000 to 2600 W) was less than 13% (the nominal power output of the system was 2.5 kW). At a system power output of 2526 W, the stack voltage was 46 V, and assuming that the H_2 utilization was 93%, then the system efficiency was 43% [(0.93 × 46)/(80 × 1.25)]. In the lower system power output region (e.g., less than 800 W), the percentage loss increased rapidly to about 34%, simply because some parasitic power consumption was either

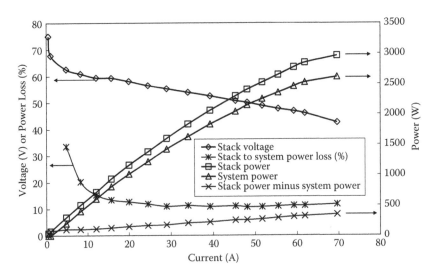

FIGURE 5.11
Voltage and power situation of a 2.5 kW fuel cell system. Courtesy of Wuhan Intepower Fuel Cells.

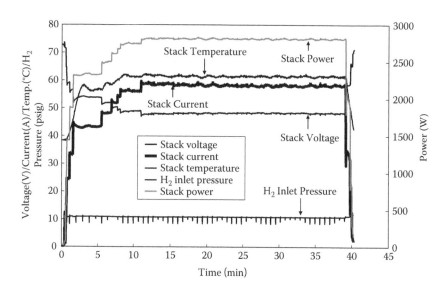

FIGURE 5.12

Stack current, voltage, power, temperature, and H_2 inlet pressure of a 2.5 kW fuel cell system. Courtesy of Wuhan Intepower Fuel Cells.

constant or decreased little (i.e., it did not decrease in proportion to the stack output power).

With the system output power and voltage set at 2520 W and 53.5 V, respectively, Figure 5.12 shows the stack current, voltage, power, temperature, and the H_2 inlet pressure. The stack temperature was controlled at around 61.5°C, and it was pretty steady. The H_2 inlet pressure was set at around 10.8 psig, and purging was carried out every 54 seconds with a purging duration of 500 ms. During each purge, the H_2 inlet pressure dropped to about 8.5 from 11 psig. The inlet pressure drop during purging did not affect the voltage stability of the stack.

The system was tested in a telecommunications station and the results are shown in Figure 5.13. Fluctuations of the current and power for both the system and the stack were synchronized. The fluctuation was because the load frequently changed due to the changing number of phone calls processed by this station. The system output voltage was basically a straight line and is not affected by the voltage of the stack, indicating that the output of the system's main DC–DC converter was very steady. The steadiness of the system output voltage is an important characteristic required by a telecommunications station in order to have high-quality communications.

Figure 5.14 illustrates the situation at the beginning of startup. The system was in an idling state, but the H_2 supply was turned on before the grid power was turned off. In such a standby mode, the parasitic power required by the system was supplied by the stack through the system's main DC–DC

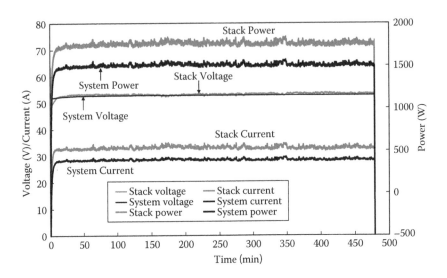

FIGURE 5.13
Eight-hour test of a 2.5 kW fuel cell system in a telecommunications station. Courtesy of Wuhan Intepower Fuel Cells.

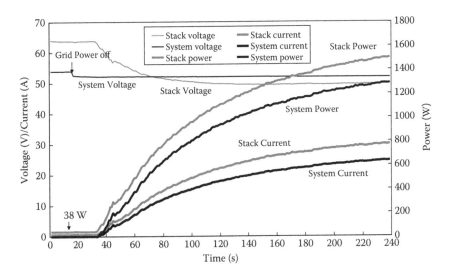

FIGURE 5.14
Features at beginning of startup of a 2.5 kW system. Courtesy of Wuhan Intepower Fuel Cells.

converter. Therefore, the voltage of the stack was not at the open circuit voltage. It can be seen that the stack provided 38 W of power during this time period. This was basically the parasitic power needs of the system. A variety of devices got their needed power from the output of the system's main DC–DC converter, and the voltage from this main DC–DC converter was converted to the desired input voltages through additional DC–DC converters. At the 16-second time mark, the grid power was turned off, and in 2 seconds, the DC buss voltage of the load dropped from 53.9 to 52.5 V and more or less stabilized near this voltage. The lead-acid battery banks used by the station were not disconnected from the DC buss during this test. So, after the grid power was off, it was the lead-acid battery bank that provided power for the station. Since this battery bank was connected directly to the DC buss of the load, its output voltage determined the DC buss voltage of the load. At the 34-second time mark, the stack started to send out power to the load, and its power output increased quickly. From the 18-second to the 34-second time mark, the stack did not send out any power to the load, not because it started slow, but because the battery bank's voltage was higher than the preset voltage output of the system's main DC–DC converter. Only after the battery bank's output voltage dropped to the preset output voltage of the fuel cell system's main DC–DC converter was the DC–DC converter able to send power out to the DC buss of the load. It is possible that if the battery bank has a very high capacity and its discharging voltage is higher than that of the output voltage of the fuel cell system's main DC–DC converter, it will take a long time for the fuel cell system to be able to send out power to the load. So, when selecting the battery bank for a fuel cell system, its capacity and discharging voltage must be carefully considered.

5.1.2.3 ReliOn Fuel Cell Systems

ReliOn Inc. is a world leader in developing and marketing self-humidified air-cooled PEMFC systems for backup and emergency applications with modular and hot-swappable substack cartridge configurations. The company has sold and delivered more than 4000 kW of fuel cells, which is equivalent to more than 2000 2-kW systems, one of the largest sales worldwide. The product portfolio includes the E-200, E-1000x, E-1100, E-1100v, E-2200x, E-2500, T-1000, and T-2000, with the rated output power at 200, 1000, 1100, 2000, 2200, and 2500 W, respectively. When higher output power is needed, several systems can be connected in parallel.

The uniqueness of ReliOn's systems involves the use of modular substacks and modular electronics cards that are hot-swappable as the basic building blocks. ReliOn's T-2000 fuel cell system was made commercial in 2006. Each substack had a rated output power of 200 W, and the T-2000 system consisted of 10 such substacks. A substack could be pulled out and replaced quickly without affecting the functionality of the remaining substacks during the replacement process. This is an important feature when the reliability of the stacks is not adequate. Also, by using one or two extra substacks, the entire system can achieve N + 1 or N + 2 redundancies. If one substack malfunctions, it can be pulled out and the remaining system still offers the rated power to the load even if the substack could not be replaced immediately. Therefore, the system can achieve extremely high reliability.

TABLE 5.4

Specifications of ReliOn T-2000 and E-2500 Systems

Parameter	Unit	Value	
		T-2000	E-2500
Rated power	kW	2.0	2.5
Nominal voltage	Vdc	24 or 48	
Rated current	A	80 @ 24 V	105 @ 24 V
		40 @ 48 V	52.5 @ 48 V
H_2 purity	%	>99.95 (standard industrial grade)	
H_2 inlet pressure	barg	0.24–0.41	0.55–0.83
H_2 consumption rate	slpm	~30 @ rated power	
Operational env. temperature	°C	2–46	−5–50
Operational env. RH	%	0–95 (non-condensing)	
Operational altitude	m	−60–4206	
Installation	—	Indoors or hardened outdoor cabinet	
Maximum noise level	dB	53	65
Remote operation	—	System configuration and status; historical and operational data	
Communications	—	RJ45/DB9/dry contact	Standard: USB/ethernet/dry contact/front panel display/ SNMP/Web Interface
			Optional: wireless modem-CDMA/ GSM/ethernet switch
Mounting	—	23-inch rack mount	
Scalability	—	To 12 kW	To 20 kW
Certification	—	UL/CSA/CE	CSA; CE; NEBS

Source: Adapted from ReliOn website, specifications in English: E-2500 Hydrogen Fuel Cell, http://www.relion-inc.com/pdf/ReliOn_E-2500spec_2012.pdf; T-2000 Hydrogen Fuel Cell, http://www.relion-inc.com/pdf/T2000SpecSheet0508.pdf.

ReliOn has made great progress in recent years over the early T-series fuel cell systems, and the company currently focuses on the E-series products. Table 5.4 compares the major specifications of the early T-2000 system with the recent E-2500 system. It is clear that the E-series system is lighter and more compact, has higher system efficiency and H_2 utilization, a wider operational ambient temperature range, and offers a lot more communications means over the T-series system.

5.1.3 Fuel Cell System with Liquid-Cooled Stack

Liquid-cooled stacks have higher volumetric power density and they are preferred when the output power is around 5 kW or higher. The resulting fuel cell system, however, becomes much more complicated. Blowers or compressors are needed to supply air to the stack instead of fans, which increases the parasitic power consumption and the noise level. The air needs to be humidified, and a humidification subsystem is required. A coolant circulation subsystem is required, which needs a coolant storage tank, a heat exchanger, a pump, and the connection piping. An ion-exchange filter is needed to keep the ionic conductivity low by removing the ionic species that get into the coolant from the stack components (e.g., the plates) and the metallic piping, or the liquid coolant needs to be changed when its ionic conductivity rises to a preset value. Due to the presence of these additional components and subsystems, the controls become more complicated.

5.1.3.1 Features of a 5 kW System

Figure 5.15 shows the inside front of a 5 kW fuel cell system using a liquid-cooled stack. The display and control panel shows some key parameters and allows the user to operate the system onsite. The alarm makes a sound and emits light when it is triggered.

The inside back view of the system is shown in Figure 5.16. The stainless steel coolant tank is 2 liters, and there is a 1 kW electric heater tightly attached to its outer surface. The heater is used to keep the stack temperature above 0°C on cold days. When an antifreeze coolant is used, the coolant will not freeze until the temperature reaches about −40°C, but the liquid water

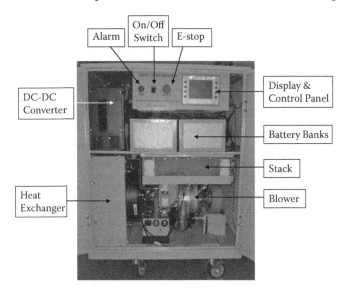

FIGURE 5.15

Inside front view of a 5 kW fuel cell system using a liquid-cooled stack. Courtesy of Wuhan Intepower Fuel Cells.

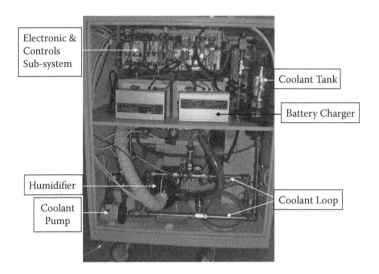

FIGURE 5.16
Inside back view of a 5 kW fuel cell system using a liquid-cooled stack. Courtesy of Wuhan Intepower Fuel Cells.

within the stack will freeze, which will shorten the life of the stack. Keeping the stack temperature above 0°C not only prevents the stack from suffering damage caused by subzero temperatures, but also enables the stack to start quickly on cold days. When the ambient temperature drops to a preset lower temperature limit (near 0°C), the heater turns on automatically and the coolant starts to flow slowly; when the coolant temperature increases to a preset higher temperature limit (e.g., 25°C), the heater turns off and the coolant stops flowing.

Part of the coolant loop has two separate paths: one path does not go through the heat exchanger and is used from the startup of the fuel cell system until the coolant temperature increases to a preset value (e.g., 60°C) to accelerate the rise of the coolant temperature; the other path goes through the heat exchanger and is used when the coolant temperature reaches the preset value. Based on the stack needs, the coolant flow rate is adjusted by the coolant pump, and the heat that is dissipated out of the system enclosure is controlled by the fan mounted on the heat exchanger.

Incoming air is humidified by a tubular humidifier made of many PEM tubules. It transfers some heat and moisture from the outgoing air from the cathode exhaust to the incoming air to make it hotter and humidified to a certain extent (e.g., 75%).

Each of the two battery banks has a charger. The battery banks are charged by the grid AC power. The parasitic power needs of the fuel cell system when it is in the idling state come from the battery banks. When the grid power is disrupted, the battery banks first provide power for the load and then start the fuel cell system after a few seconds of preset delay time.

TABLE 5.5

Key Characteristics of a 5 kW Liquid-Cooled Fuel Cell System

Parameter	Unit	Value
Nominal system output power	kW	5.0
Overloading capacity	%	10
System startup time	ms	<10
Environment temperature	°C	−10–50
System electrical efficiency	%	>39 (based on input H_2 LHV)
System output voltage range	Vdc	43–58
Communications port	—	RS232
Start-stop methods	—	Manual; remote; automatic
Remote communications	—	Control; messaging; measurement
Protective functions	—	Overload; H_2 inlet pressure high or low; system output voltage high or low; short circuiting; stack over-temperature; environment temperature high or low
Warning	—	System door opened; system flooded; coolant level low; H_2 leaked; H_2 source pressure low; H_2 inlet pressure too high or too low; stack temperature too high or too low; stack voltage too high or too low; stack current too high or too low; system output voltage too high or too low; system output power too high or too low; input voltage to the DC–DC converter too high or too low; input current to the DC–DC converter too high

Source: Courtesy of Wuhan Intepower Fuel Cells.

The electronic and controls subsystem includes a mother control board, several subcontrol boards, contactors, current sensors, voltage sensors, and switches. They control the operation of the entire fuel cell system.

Table 5.5 lists the key characteristics of the fuel cell system. The RS232 communication port is used because this is the type of port used by telecommunications stations. Some warning actions had two sets of values: one set for less-severe situations that triggers the warning alarm but does not shut down the fuel cell system; the other set is for more severe situations that trigger the warning alarm and shut down the fuel cell system. If the fuel cell system is not able to shut down when the severe values are reached, the fuel cell system or the load it powers can be damaged.

5.1.3.2 Stack from Sunrise Power

The stack in this case was evaluated on a test stand under various conditions. Figure 5.17 shows the stack V-I curves at air stoichiometric ratios of 2.0, 2.5, and 3.0, respectively. The stack performance increased with the air

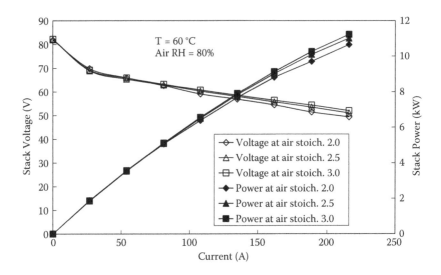

FIGURE 5.17
Stack V-I curves. Courtesy of Sunrise Power.

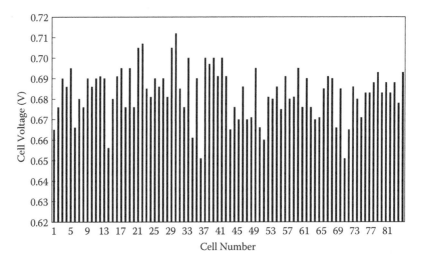

FIGURE 5.18
Voltage variation among cells. Courtesy of Sunrise Power.

flow rate, but the difference between 2.5 and 3.0 air stoichiometric ratios was negligible. Figure 5.18 shows the voltage variation among cells. The largest difference was 61 mV, 8.9% of the average cell voltage. Clearly, there was still room for improvement in performance evenness among cells.

The RH of air has significant impact on the performance of the stack, as shown in Figure 5.19. From 40% to 60% and to 80% RH, the performance

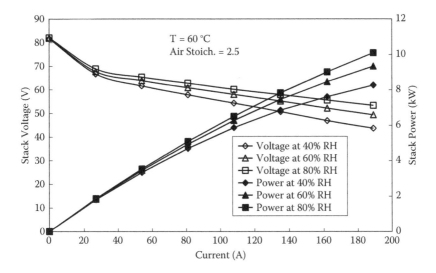

FIGURE 5.19
Impact of air RH on stack performance. Courtesy of Sunrise Power.

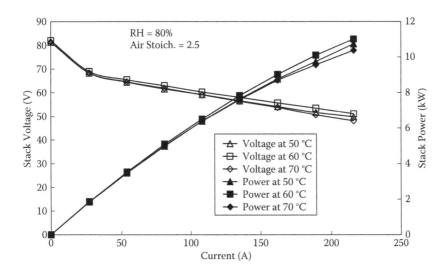

FIGURE 5.20
Effect of temperature on stack performance. Courtesy of Sunrise Power.

increased significantly. Testing on a 20-cell shorter stack found that from 80% to 100% RH had little impact on the stack performance.

The effect of temperature on the stack is shown in Figure 5.20. It can be seen that between 50°C and 70°C, the change in performance was not significant.

Figure 5.21 shows the air pressure drop of the stack using either dry or 80% RH air at 60°C. The pressure drop increases linearly with the air flow rate.

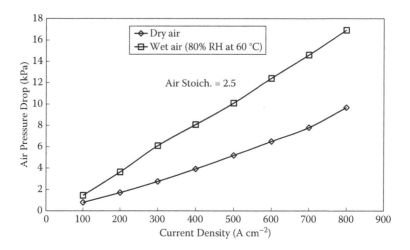

FIGURE 5.21
Air pressure drop through the stack. Courtesy of Sunrise Power.

The air pressure drop using 80% RH air was significantly higher (about 1 time more) than that using dry air. With 80% RH at 60°C, the total air volume entering the stack will be about 15% more than that of dry air, and this accounts for a smaller portion of the total air pressure drop. The larger portion is believed to be related to the partial blockage of the flow field channels by liquid water. Since the stack was intended to operate at a current density of less than 0.5 A cm^{-2}, the air supply device had to be able to overcome around a 10 kPa pressure drop.

Figure 5.22 shows the stack performance when air was supplied by a blower with a maximum pressure boost capability of 11.7 kPa. For a 5 kW output fuel cell system, the output from the stack needs to be slightly over 6 kW considering the system parasitic power loss and the DC–DC converting loss. It can be seen that with 7020 W stack output power, the current and voltage of the stack was 120 A and 58.6 V, respectively. The stack contained 84 unit cells, so the average cell voltage was 0.70 V, and the stack electric efficiency was 56% (0.70/1.25). The power consumption by the blower was 370 W, accounting for only 5.3% of the stack output power.

5.1.3.3 Breadboard System

A breadboard system is often a necessary step in evaluating the major components of a fuel cell system together, and it is easier to rearrange the parts during the evaluation process. Figure 5.23 shows a breadboard system with most of the major parts laid on a table, and some parts placed beside and underneath the table. Figure 5.24 shows the V-I and W-I curves of the stack obtained from this system. It should be noted that the performance of a stack tested in a system can be quite different from that tested on a test stand, because the system may not be able to offer the best conditions for the stack.

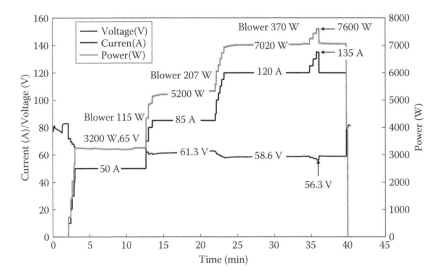

FIGURE 5.22
Stack performance and blower power consumption. Courtesy of Wuhan Intepower Fuel Cells.

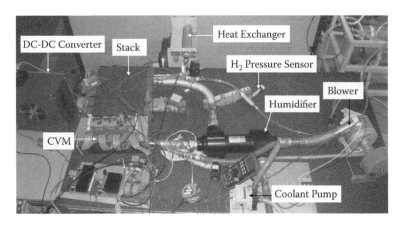

FIGURE 5.23
Picture of a 5 kW breadboard system layout. Courtesy of Wuhan Intepower Fuel Cells.

Figure 5.24 shows that the stack performed well. Figure 5.25 shows the stability of the DC–DC output voltage, which changed from 49.87 to 50.11 V, while the voltage setup point was at 50.00 V. Such a small voltage variation was acceptable.

The breadboard system was also tested by using metal hydride as the H_2 source (Figure 5.26). The H_2 supply system was composed of 4 metal hydride cylinders. Each cylinder contained 58 kg Ti-series AB_2-type metal hydride, and the total mass of each cylinder was 89 kg. Since the metal hydride could store 1.9% of H_2 by weight, each cylinder stored about 1.1 kg of H_2. The

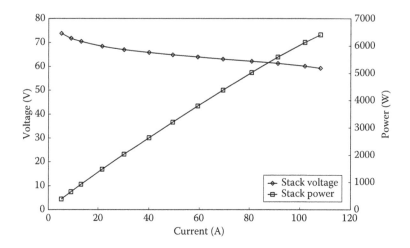

FIGURE 5.24
Stack performance in a 5 kW breadboard system. Courtesy of Wuhan Intepower Fuel Cells.

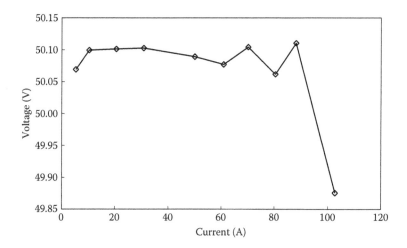

FIGURE 5.25
Output voltage variation of a 5 kW system. Courtesy of Wuhan Intepower Fuel Cells.

diameter and length of each cylinder are 14 and 135 cm, respectively, resulting in the volume of the cylinder being 20771 cm^3 (=3.14 × 49 cm^2 × 135 cm). Then the volumetric H_2 storage capacity is 0.053 g cm^{-3} (1100 g/20771 cm^3). Figure 5.27 shows the H_2 discharging process. At an H_2 discharging rate of 75 L min^{-1}, 33 Nm^3 of H_2 was discharged. The pressure of H_2 in the tank decreased gradually from 33 to 3 bars. Afterward, the H_2 discharging rate was reduced to 37.75 L min^{-1}, and finally to 8 L min^{-1}. It is important that the H_2 discharge rate be able to meet the stack needs; otherwise, H_2 starvation will happen and will cause quick damage to the stack. As a precaution, when

FIGURE 5.26
Metal hydride system. Courtesy of Beijing General Research Institute of Nonferrous Metals.

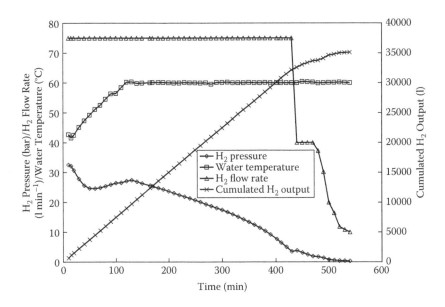

FIGURE 5.27
H_2 discharging process. Courtesy of Beijing General Research Institute of Nonferrous Metals.

the H_2 pressure drops to the lower end of the pressure plateau region (please refer to Figure 3.10), a metal hydride system should be stopped from further discharging H_2. To this point, it should be recharged by H_2.

The liquid water circulation loop through the H_2 cylinders was connected with the fuel cell system's coolant circulation loop. During the H_2 charging process to the cylinders, the heat exchanger in the fuel cell system was used

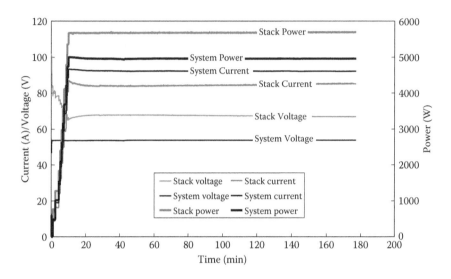

FIGURE 5.28
Fuel cell system performance at 5 kW system output using metal hydride. Courtesy of Wuhan Intepower Fuel Cells.

to dissipate the heat generated by this charging process. When the H_2 pressure reached about 40 bars, the charging process was complete. During the operation of the fuel cell, the temperature of the H_2 cylinder was maintained by the hot coolant from the fuel cell stack. The heat needed by the H_2 metal hydride cylinders took away some burden from the heat exchanger of the fuel cell system. Figure 5.28 shows the performance of the fuel cell system with output power of 5 kW. The system output voltage was steady at 53.7 V. The stack voltage showed very little decline.

5.1.3.4 Fuel Cell System

After evaluation and optimization of the breadboard system, the final fuel cell system shown in Figures 5.15 and 5.16 was built. Figure 5.29 shows the monitoring screen displayed on a computer. The section within the dashed rectangle indicates the system situation. It shows the stack, the H_2 supply subsystem, the air supply subsystem, the coolant subsystem, the pressure of the H_2 source, the liquid level of the coolant tank, the load, and so on. Critical parameters such as the H_2 inlet pressure, the PWM of the blower, the fan on the heat exchanger and the coolant pump, the coolant inlet and outlet temperatures, the stack voltage and current, and the minimal cell voltage are displayed. The Operation Mode section shows the mode in which the system is being operated currently. There are automatic, manual, remote, and maintenance modes. The System State section shows whether the system is on, off, or idling. The Stack and System Output section shows the voltage, current, and power of both the stack and the system, respectively.

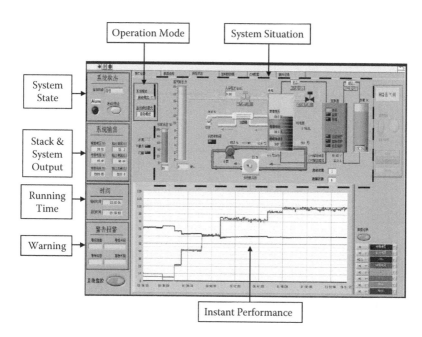

FIGURE 5.29
Monitoring screen displayed by computer. Courtesy of Wuhan Intepower Fuel Cells.

The Warning section shows the type of warning if there is one. The Instant Performance section shows the performance trend of selected parameters (such as current, voltage, power, temperature, PWM, H_2 inlet pressure, etc.).

A continuous test lasting more than 7 hours is shown in Figure 5.30. The fuel cell system output voltage was set at 53 V. The load was added gradually at 10 A every 2 minutes. Twenty minutes later the system output power reached 5 kW, and the system was tested continuously for more than 7 hours. The system output voltage was very stable, indicating that the DC–DC converter controlled its voltage output properly. But the stack voltage declined gradually from 60.3 to 56.6 V, which meant that the system condition was not optimal. Most of the voltage loss was recovered when the system was restarted the next day, indicating that most of the loss was recoverable.

As backup power, a fuel cell system must be able to support the full load when the grid power is disrupted. Figure 5.31 shows that the system sent out 5 kW of power in 10 seconds when at normal ambient temperatures. This rate is extremely fast for a liquid-cooled stack partly because the stack was oversized in power output capability.

The fuel cell system was trial tested in a telecommunications station located in Beijing. Figure 5.32 is a picture after the fuel cell and the H_2 systems were installed. The H_2 system contained four 35 MPa, 30-liter reinforced cylinders, shown as an inset in the figure. When two cylinders were empty, they could be replaced while the other two supplied H_2 to the fuel cell system. The H_2

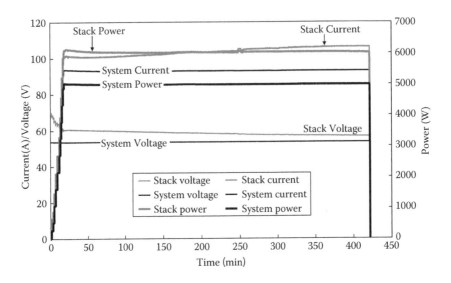

FIGURE 5.30
Continuous lab test lasting more than 7 hours. Courtesy of Wuhan Intepower Fuel Cells.

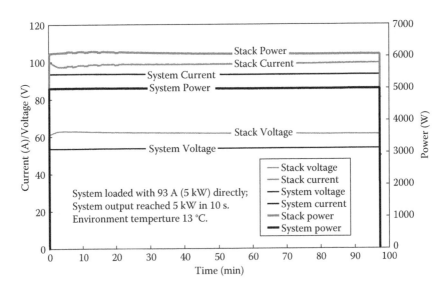

FIGURE 5.31
Fuel cell system loaded with 5 kW directly. Courtesy of Wuhan Intepower Fuel Cells.

piping and the electrical wires between the fuel cell system and the H$_2$ system were placed underneath the metallic board to prevent damage.

Figure 5.33 shows the test results for over 7 hours. The load was in constant change, and the power output from the stack and the fuel cell system changed accordingly. The output voltage of the fuel cell system was very

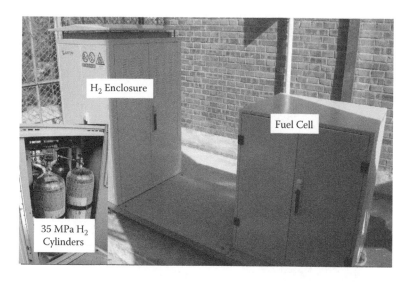

FIGURE 5.32
Picture of a 5 kW fuel cell system and H_2 supply unit. Courtesy of Wuhan Intepower Fuel Cells.

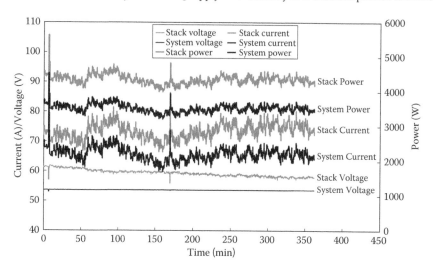

FIGURE 5.33
Trial test of a 5 kW system in a telecommunications station lasting more than 7 hours. Courtesy of Wuhan Intepower Fuel Cells.

stable. The average load was around 3500 W, while the stack average power output was around 4400 W. There were two unknown power surges during the tests, and the power outputs of the stack and the system went as high as 5631 and 4536 W, respectively.

Sunrise Power is a leading Chinese company in developing and manufacturing PEMFC components such as catalysts, membranes, MEAs, stacks, and

FIGURE 5.34
Picture of an FCP-5 backup power fuel cell system developed by Sunrise Power. Courtesy of Sunrise Power.

fuel cell systems. It has developed compact FCP-10 (nominal output power 10 kW, 220 VAC) and FCP-5 (nominal output power 5 kW, 48 VDC) fuel cell systems, both of which passed inspections by two relevant Chinese inspection agencies. Figure 5.34 shows a picture of its FCP-5 backup power fuel cell system. It can be operated manually, automatically, and remotely.

5.2 Primary Power

5.2.1 Fuel Processing

When a PEMFC is used as a primary power source to provide the power needs for a residence or a building, it will be in operational mode all the time, but its power output may change quite significantly according to variations in power needs. For example, during peak energy demand times, such as in the morning and in the evening, its power output will be higher, and if its power output cannot meet the demand, additional power sources such as grid power will be needed. In contrast, during the day, when residents go out to work, and during the night to early morning, when the residents are sleeping, its power output will be lower because there is less demand. If the fuel cell is still kept at the higher power output, the excess power generated can be sent to the grid. Whether keeping a fuel cell system at constant power output or varied power output (i.e., load following) depends on the users, the optimal fuel cell operation strategy, and the cost of fuel versus grid electricity. With constant power output, the fuel cell system will not go through frequent power transition times, which could help prolong its life. However, if the excess energy cannot be sent to the grid, or doing so is not economical, the fuel cell system will have to be in a variable power output mode.

Since a primary fuel cell system is in continuous operation, its fuel consumption is significant; this makes the use of H_2 infeasible because there is currently no H_2 infrastructure. But natural gas (or propane) pipelines have reached nearly all homes in developed countries and many homes in developing countries; therefore, it is very convenient to use those pipelined gases as the raw fuels for primary fuel cell systems.

Since natural gas (CH_4) does not react at the PEMFC anode, it needs to be reformed into an H_2-rich gas mixture called *reformate*, which is then sent to the fuel cell anode.

5.2.1.1 Desulfurization

If the pipeline pressure is adequate, the natural gas does not need to be further pressurized; otherwise, a pressurization step is needed. The fuel typically contains some odorant that is mainly composed of compounds containing a sulfur element such as ethanethiol (C_2H_5SH) or hydrogen sulfide (H_2S). These compounds can seriously poison the catalysts used in the fuel cell system (reformer, catalytic burner, water–gas shift reactor, preferential oxidizer, and the stack), so they must be removed first through a desulfurization (DS) process. One method to remove the S compounds is through adsorption by using materials such as active carbon. A study on a variety of adsorbents can be found in (Wild 2006). The capacity depends on the amount of the active materials, and they need to be replaced or rejuvenated periodically. Since the amount of S compounds that can be adsorbed per unit mass of active carbon is quite low, the volume of the DS unit can be quite large. Another method is called hydrodesulfurization (HDS), which allows the S compounds first to chemically react with H_2 in the presence of a catalyst such as NiMo at a temperature of around 300°C to convert the S compounds to H_2S according to Reaction 5.1:

$$RSH + H_2 = RH + H_2S \qquad \text{(R. 5.1)}$$

The resulting H_2S reacts with a suitable metal such as Zn according to Reaction 5.2:

$$H_2S + Zn = ZnS + H_2 \qquad \text{(R. 5.2)}$$

Clearly, the capacity of the HDS bed depends on how much Zn it contains. Since the chemical reaction can handle much more S compound than adsorption by active carbon, an HDS bed can be made much smaller. But in order for Reaction 5.1 to occur, H_2 needs to be added, at least in the beginning of the process, and the temperature should reach around 300°C, both of which make the entire system slightly more complicated.

5.2.1.2 Steam Reforming

After the DS step, the fuel is sent to a reformer. For steam reforming CH_4 at 1100 K (827°C), the reaction is endothermic:

$$CH_4 + H_2O = CO + 3H_2 \qquad \Delta H = 225.5 \text{ kJ mol}^{-1} \qquad \text{(R. 5.3)}$$

The ΔH without a superscript "o" refers to the value at the specified temperature, and the same is true for the following discussion. About 225.5 kJ of heat is needed to reform one mole of CH_4. Therefore, a catalytic burner is needed (Pt/Mo and Pt/W are active catalysts). During the start of the reformer, the fuel for the burner will be CH_4 itself, reacting with O_2 according to Reaction 5.4.

$$CH_4 + 2O_2 = CO_2 + 2H_2O \qquad \Delta H = -801.9 \text{ kJ mol}^{-1} \qquad \text{(R. 5.4)}$$

So, for reforming 1 mole of CH_4 at 1100 K (827°C), about 0.28 (225.5/801.9) mole of CH_4 needs to be burned to provide the needed heat.

After the fuel cell system starts to run, some unreacted H_2 will be purged from the anode, and it can be used to provide some heat for the burner. According to Reaction 5.5, 1 mole of H_2 provides 248.4 kJ of heat at 1100 K, and about 0.91 (225.2/248.4 kJ mol^{-1}) moles of H_2 will be needed to provide 225.5 kJ of heat.

$$H_2 + 0.5O_2 = H_2O \qquad \Delta H = -248.4 \text{ kJ mol}^{-1} \qquad \text{(R. 5.5)}$$

As presented in Chapter 3, if the reforming temperature is set at 1100 K, there will be 11.6% CH_4 that is unconverted (without considering the impact of the water–gas shift reaction concurring in the reformer on the equilibrium of the reforming reaction). This amount of CH_4 will be in the anode exhaust. If it can all be burned to provide heat for the burner, it can offer 93 kJ of heat (0.116 × 801.9 kJ mol^{-1}), and the remaining 132.5 kJ (225.5 − 93) of heat needed by the catalytic burner will come from 0.53 moles of H_2 (132.5/248.4 kJ mol^{-1}) from the anode exhaust.

In designing a reformer, people normally like to achieve a 100% conversion of the fuel to H_2 in order to maximize the fuel utilization. In practice, due to the heat needed for the reforming process, this is not necessary. Provided the reaction rate is fast enough to meet the fuel cell peak power need, the temperature for reforming CH_4 could be set at 1000 K (727°C), which will leave

TABLE 5.6

Molar Enthalpy of Formation of Some Chemicals

T (°C)	T (K)	$\Delta_f H°_{CO}$	$\Delta_f H°_{H_2O}$	$\Delta_f H°_{CH_4}$	$\Delta_f H°_{MeOH}$	$\Delta_f H°_{CO_2}$
25	298	−110.5	−241.8	−74.6	−201.0	−393.5
27	300	−110.5	−241.8	−74.7	−201.1	−393.5
127	400	−110.1	−242.8	−77.7	−204.6	−393.6
227	500	−110.0	−243.8	−80.5	−207.8	−393.7
327	600	−110.2	−244.8	−83.0	−210.4	−393.8
427	700	−110.5	−245.6	−85.0	−212.6	−393.9
527	800	−110.9	−246.4	−86.7	−214.4	−394.1
627	900	−111.4	−247.2	−88.0	−215.8	−394.3
727	1000	−112.0	−247.8	−89.0	−216.9	−394.6
827	1100	−112.6	−248.4	−89.7	−217.8	−394.8
927	1200	−113.2	−248.9	−90.1	−218.5	−395.0
1027	1300	−113.9	−249.4	−90.4	−218.9	−395.3
1127	1400	−114.6	−249.8	−90.4	−219.3	−395.5
1227	1500	−115.3	−250.1	−90.2	−219.5	−395.7

Source: L. V. Gurvich, V. S. Iorish, V. S. Yungman, and O. V. Dorofeeva, "Thermodynamic Properties as a Function of Temperature," in *CRC Handbook of Chemistry and Physics (1913–1995), 75th Edition (Special Student Edition)*, editor-in-chief, David R. Lide, 5-48–5-71 (Boca Raton: CRC Press, 1994).

27% CH_4 unconverted; burning of this amount of CH_4 (0.27 × 801.2 = 216.3 kJ) plus 0.034 ((224.8 − 216.3)/247.8) moles of unreacted H_2 from the anode exhaust can provide the needed heat. The numbers −801.2, 224.8, and −247.8 kJ mol⁻¹ are the ΔH for Reactions 5.4, 5.3, and 5.5, respectively, at 1000 K. The enthalpy values of related chemicals at different temperatures are listed in Table 5.6.

5.2.1.3 Water–Gas Shift Reaction

If all the CH_4 is converted and the water–gas shift (WGS) reaction concurring within the reformer is neglected, the product will contain 25% CO, which is too high for PEMFCs. Therefore, CO is converted to CO_2 in the following WGS reactions according to Reaction 5.6.

$$CO + H_2O = CO_2 + H_2 \qquad \Delta H° = -41.2 \text{ kJ mol}^{-1} \qquad \text{(R. 5.6)}$$

This is an exothermic reaction, and it releases −41.2 kJ mol⁻¹ of heat under standard conditions; therefore, it should be able to proceed without the addition of external heat. In order to shift this exothermic reaction to the right side to minimize the CO concentration, the WGS temperature should be kept as low as possible. But at lower temperatures, the reaction rate will be slower. So the optimal temperature is a compromise between the two factors. For reforming CH_4, there is normally a high-temperature shift (HTS) reactor operated at around 700 K and a low-temperature shift (LTS) reactor operated at around 500 K. After the HTS and LTS, the CO concentrations are around 8.3% and 2.4%, respectively (please refer to Table 3.5).

The overall reaction of reforming plus the WGS reaction is

$$CH_4 + 2H_2O = CO_2 + 4H_2 \qquad \Delta H^\circ = 164.7 \text{ kJ mol}^{-1} \qquad \text{(R. 5.7)}$$

The overall enthalpy change is 164.7 kJ mol^{-1} under the standard condition. It seems that these energy losses associated with converting 1 mole of CH_4 were wasted. In reality, the energy is not lost but stored in the 4 moles of H_2 produced. Under standard conditions, the enthalpy in 4 moles of H_2 is −967.3 kJ, the enthalpy of burning 1 mole of CH_4 is −802.6 kJ, and the difference is just −164.7 kJ. So, no energy is created or destroyed.

5.2.1.4 Preferential Oxidation

Since the 2.4% CO in the mixture after the LTS is still too high for a PEMFC, it is sent to a preferential oxidizer (Prox) for further removal of CO. In the presence of a suitable catalyst, CO is preferentially oxidized by O_2 to form CO_2 according to Reaction 5.8.

$$CO + 0.5O_2 = CO_2 \qquad \Delta H^\circ = -283 \text{ kJ mol}^{-1} \qquad \text{(R. 5.8)}$$

This is an exothermic reaction and can proceed by itself without adding heat. Even with the most selective catalyst, a percentage of H_2 will be oxidized by O_2. The selectivity is basically the ratio of reacted CO to the sum of reacted CO and H_2. Lowering the temperature favors selectivity, and it is normally set at around 120°C. If the selectivity is 50%, 2.4% of H_2 will be oxidized in the Prox as well, which means that about 5% (2.4% CO + 2.4% H_2) of the theoretical H_2 according to Reaction 5.7 is converted into heat in the Prox.

Due to the heat losses from various reactors and the heat exchange processes (there is a heat exchanger in each of the steps discussed above), the H_2 (plus CO) loss in the Prox, and the fuel loss during anode purging, achieving an overall CH_4 utilization of 90% for electrical power generation should be a good number for a fuel cell system incorporating fuel processing steps. In order to minimize the loss of CH_4, the heat collected from the WGS reactors and the Prox through heat exchangers should be effectively utilized, perhaps to generate steam for the reformer.

Good thermal insulation and integration is important in order to minimize heat losses in the reactors. It is best if the catalytic burner is made as part of the reformer (such as in a concentric reformer and burner configuration) in order to maximize the heat utilization produced by the burner and to achieve more even temperature distribution within the reformer. Since the reforming reaction is endothermic, the temperature will decrease along the gas flow path if there is no external adjustment. The capacity of the reformer, the WGS, and the Prox must be large enough to handle the highest fuel flow rate (i.e., the fuel flow rate at the peak power output) needed by the fuel cell system. Catalyst materials, surface area, and distribution within those reactors should be carefully considered. Varied catalyst loading along the gas flow path may be a better choice. Since the temperature of the burner can be as high as 850°C, it will unavoidably generate some NO_x (mainly NO) from the reaction between N_2 and O_2 in air, and the higher the temperature, the higher the NO_x concentration.

Clearly, reforming CH_4 is not a trivial task, and many materials, catalysts, and parameters must be properly selected and controlled. During the power transient process, a sudden change in the need for H_2 leads to corresponding changes in the flows of CH_4, H_2O, and O_2, which makes control more difficult. The transient response time of a reformer is normally on the order of 10 s or slightly longer.

5.2.1.5 Steam Reforming Methanol

Reforming liquid methanol is also very attractive, due to its higher volumetric energy density and simpler reforming processes, plus it is biologically achievable and degradable. Thermodynamically, methanol can be 100% converted at around 500 K (see Table 3.9), which is much lower than reforming CH_4. Reaction 5.9 shows the methanol steam reforming reaction at 500 K.

$$CH_3OH + H_2O = CO_2 + 3H_2 \quad \Delta H = 57.9 \text{ kJ mol}^{-1} \quad \text{(R. 5.9)}$$

Each mole of CH_3OH can generate three moles of H_2. Although the reaction is also endothermic, the enthalpy change is quite small. It needs to burn 0.086 (57.9/673.5 kJ mol^{-1}) moles of methanol or 0.24 (57.9/243.8 kJ mol^{-1}) moles of H_2 to achieve a heat balance at 500 K. The number −673.5 kJ mol^{-1} is the ΔH when 1 mole of methanol reacts with O_2, and the number −243.8 kJ mol^{-1} is the ΔH for Reaction 5.5 at 500 K. Since methanol will lower the anode performance of the stack, it should be completely converted. Therefore, the heat for maintaining the heat balance will come from H_2 in the anode exhaust once the fuel cell is in operation mode. Theoretically, at an anode stoichiometric ratio of 1.08, the excess 8% of H_2 that accounts for 0.24 moles of H_2 per 3 moles of H_2 produced from 1 mole of methanol will be enough to satisfy the reformer heat needs. So, if the anode stoichiometric ratio is set at 1.08 or slightly higher, the anode exhaust will not need to be recirculated into the anode inlet fuel stream. Instead, it is sent to the burner for the reformer.

Reaction 5.9 shows no CO formation in the reforming process. However, some CO will form through the reversed water–gas shift (RWGS) reaction according to Reaction 5.10.

$$CO_2 + H_2 = CO + H_2O \qquad \text{(R. 5.10)}$$

We can estimate how much CO will exist due to Reaction 5.10. Assuming 1 mole of methanol and 1 mole of water are fed to the reformer (normally the amount of water will be slightly more than that of methanol in order to make sure that the methanol is completed converted and to reduce the likelihood of coking) and they completely react, 1 mole of CO_2 and 3 moles of H_2 are produced according to Reaction 5.9. If n moles CO and H_2O are formed through Reaction 5.10, then the concentrations of CO_2 and H_2 will be $1 - n$, and $3 - n$, respectively. Then,

$$(1 - n)(3 - n)/n^2 = K_2 = 137 \qquad \text{(Eq. 5.4)}$$

where K_2 is the equilibrium constant at 500 K shown in Table 3.3. Solving Equation 5.4 results in $n = 0.135$. So, at the thermodynamic equilibrium, there is 3.4% vol CO (0.135/4) in the product mixture. If we assume that the amount of water being fed to the reformer is 10% more than that of methanol, the CO still accounts for 2.4%. Due to the presence of such a high CO concentration, the product mixture still needs to go through a Prox to lower the CO concentration to ppm levels.

From a cold start, the subsystem for processing CH_4 and methanol may take 30–60 and 10–30 minutes to reach the desired temperatures, respectively. During this time period, the reformate should not be sent to the stack because its low quality (i.e., high CO concentrations) can seriously poison the stack and shorten its life. If the fuel cell system carries an H_2 tank, it can use this H_2 to generate power for the load until the reformer is ready. Alternatively, if the fuel cell system has a high-capacity battery bank, it can use the battery to provide power. Fortunately, in contrast to backup power, once the fuel cell system is started as a primary power source, it will be kept in the running mode.

For operating a fuel cell system with varied loads, the fuel utilization for generating electric power will be lower during the low power output time periods because the percentage of fuel for maintaining the temperature of the reformer will be relatively higher.

5.2.2 IdaTech/Ballard Systems

IdaTech (acquired by Ballard Power Systems in 2012) is a world leader in developing fuel cell systems for backup and emergency applications. A picture of its ElectraGen™ ME system is shown in Figure 5.35. The system uses a methanol and water mixture as the fuel to provide higher specific energy density for the entire system and to avoid the difficulties in handling the heavy and bulky industrial H_2 cylinders. The fuel tank volume is 225 L, supplying energy equivalent to about 24 industrial H_2 bottles. One

FIGURE 5.35
IdaTech/Ballard ElectraGen™ ME System. Courtesy of IdaTech and Ballard Power Systems.

tank of fuel will allow 2.5 and 5.0 kW systems to run for 69 and 40 hours, respectively, at the rated power outputs. Methanol is reformed into H_2 to feed the fuel cell. Table 5.7 lists the major specifications of the system.

If we assume that the average cell voltage is 0.70 V, the overall fuel utilization efficiency is 90%, the parasitic power loss is 10%, and the DC–DC converting efficiency is 90%, then the electrical efficiency of the system will be 41% [(0.70/1.25) × 0.9 × 0.9 × 0.9]. Since the net output energy is around 200 kWh (173.1 and 204.5 kWh based on the fuel consumption rates of 1.3 and 1.1 L kWh^{-1} for the 2.5 and 5.0 kW systems, respectively), then the gross energy of the methanol in the 225 L tank will be 482 kWh (200/0.41). The lower heating value of pure methanol is 4.35 kWh l^{-1}, which implies that the 225 L tank contains around 110 L pure methanol (482/4.35), with the remaining 115 L being water. The numbers of moles of methanol and water are 2716 and 6389, respectively, and therefore, the amount of water is 2.3 times of that needed for reforming methanol.

It is not clear how long it takes for the methanol processing subsystem to reach the optimal operational point. Since the systems use valve-regulated

TABLE 5.7

Specifications of IdaTech/Ballard ElectraGen™ ME System

Parameter	Unit	Value	
Maximum power	kW	2.5	5
Voltage range	Vdc	−48 to −56	−48 to −56 and 24 to 28
Dimension (W/D/H)	m	1.35/1.15/1.76	
Environmental temperature	°C	−5–46	
Coolant	—	Air	
System weight	kg	281	295
Fuel	—	HydroPlus™ methanol–water mixture	
Internal fuel tank volume	L	225	
Fuel consumption rate	L kWh^{-1}	1.3	1.1
Run time	h	69	40
Bridging energy	—	VRLA batteries	
Installation location	—	Outdoor	
Communication interfaces	—	SNMPv2c; dry contact; LCD	
Certifications	—	CE; ANSI/CSA FC1:2004; Telcordia GR-63-CORE Seismic Zone 4; UL/CSA/EN61010-1	
Remote monitoring	—	Wireless (optional)	

Source: Adapted from Ballard Power Systems website, ElectraGen-ME Spec Sheet, http://www.ballard.com/fuel-cell-products/ElectraGenME.aspx.

lead-acid (VRLA) batteries as the bridging energy, it is likely that the batteries will be used to supply the power needs before high-quality reformate is generated by the fuel processor. This process is expected to take less than 30 minutes. If the fuel processing modules are kept at the operational temperatures when the fuel cell system is in the idling state by using external electrical energy, high quality H_2 can be generated at a time scale of ca. 1-minute.

Since the IdaTech/Ballard ElectraGen ME system incorporates fuel processing steps that are basically the same as for primary power applications, it is presented here as an example, although the system is mainly for backup power applications at the moment.

5.3 Summary

Stationary fuel cells include backup and primary power generation systems for various applications. For backup power applications, it is best to use pure H_2 as the fuel for instant startup, but the supply of H_2 is not convenient at the moment. For continuous primary power applications, it is best to use the existing fuel infrastructure for natural gas or propane, but the fuel cell system needs to process the fuel into an H_2-rich gas mixture, which increases the cost and the complexity of the fuel cell system. Liquid fuels such as methanol can also be used and they possess higher volumetric energy density than gaseous fuels.

About 100,000 stationary PEMFC systems have been deployed worldwide, and most of them are for primary power applications. The technology for backup power applications is mature enough for commercialization. The major resistance to commercialization appears to come from the higher initial investment cost due to the extremely low production volume (although the total operational cost of a fuel cell system for 10 years is not higher than that of using Pb-acid batteries) and the inconvenience in using H_2. It is believed that the situation will change dramatically once the H_2 supply becomes easier and the benefits of using fuel cells are realized by end users.

6

Motive Power

6.1 Fuel Cell Vehicles

Cars are the most popular transportation means in developed countries such as the United States and Canada, and their popularity is increasing rapidly in developing countries such as China. To some extent, cars make the routine commute more convenient and comfortable if roads are not heavily congested. In the United States, most of the crude oil is used by cars. What other fuel will power cars becomes an urgent issue as the world crude oil reserve decreases with time. In addition, cars are based on low-efficiency internal combustion engines (ICEs) with gasoline as the fuel, and emit harmful NO_x, SO_x, and a lot of CO_2 (a greenhouse gas, or GHG). Heavy use of cars has seriously deteriorated the air quality in many urban areas worldwide.

Gasoline can be represented by octane (C_8H_{18}). It combusts in the ICE according to Reaction 6.1.

$$C_8H_{18} + 12.5O_2 = 8CO_2 + 9H_2O \qquad \text{(R. 6.1)}$$

The combustion of each gram of C_8H_{18} will produce 3.09 grams of CO_2 ($8 \times 44/114$). If a car runs for 150,000 miles in its lifetime, and 1 gallon of gasoline allows a car to run for 25 miles, a car will consume 6,000 gallons (22,740 liters) of gasoline and generate 50,555 kg of CO_2 during its lifetime. It has been estimated that there are over 850 million cars worldwide today (excluding trucks and buses), translating to emissions of around 4.3 billion tons of CO_2 per year into the atmosphere (50.6 tons $\times$ 0.85 $\times$ 10^9 vehicles/ 10 years) when assuming the average car lifetime is 10 years.

The average atmospheric temperature between 1981 and 1990 increased by 0.48°C over the previous 100 years. If the temperature keeps increasing at the current trend, it has been estimated that the average atmospheric temperature will increase another 2–4°C by 2050, causing a significant portion of the ice in the Arctic and Antarctic to melt. If this happens, many coastal cities like New York, Tokyo, Sydney, and Shanghai may be submerged in water.

Global warming, deterioration of the air quality, and near depletion of the crude oil reserve all point to reducing or even abandoning the use of vehicles powered by conventional ICEs. Electric vehicles powered by proton exchange membrane fuel cells (PEMFCs) are believed to be the ideal replacement.

Many car companies have put significant effort into developing fuel cell vehicles in the past 1 to 2 decades. Figure 6.1 shows cars displayed by Toyota, Honda, and Nissan at the 2011 Tokyo Fuel Cell Expo. Visitors were welcome to test ride the cars, and some drivers purposely accelerated or decelerated the cars to show the performance of the cars during the test drive. Figure 6.2 shows fuel cell cars, tourist carts, and buses used for transporting visitors during the 2010 Shanghai World Expo. The fuel cell fleet consisted of

FIGURE 6.1
Fuel cell cars displayed by Toyota, Honda, and Nissan during the 2011 Tokyo Fuel Cell Expo.

FIGURE 6.2
Fuel cell cars, buses, and tourist carts used during the 2010 Shanghai World Expo. Courtesy of Shanghai Shenli High Tech. Company.

90 passenger cars, 6 buses, and 100 tourist carts, and was one of the largest fuel cell vehicle fleets in the world.

6.2 Vehicle Power Requirements

The power P (W) requirement of a vehicle is calculated by Equation 6.1.

$$P = P_{Aux} + (mgv\sin(\theta) + mav + mgC_Rv + 0.5\rho C_D A_F v^3)/\eta \quad \text{(Eq. 6.1)}$$

where P_{Aux} is the auxiliary power consumed by the vehicle during operation (W), m is the total mass of the vehicle (kg), a is the vehicle acceleration rate (m s^{-2}), v is the velocity of the vehicle (m s^{-1}), g is the acceleration of gravity (= 9.8 m s^{-2}), θ is the slope angle (degree), C_R is the rolling resistance between the tires and the road, ρ is the density of air, C_D is the aerodynamic drag coefficient, A_F is the frontal area of the vehicle, and η is the conversion efficiency from electrical to mechanical energy. Representative values of the general parameters, such as P_{Aux}, m, C_R, C_D, A_F, ρ, and η for cars, buses (mid-sized), and tourist carts are listed in Table 6.1.

TABLE 6.1

General Parameters of Three Types of Vehicles

Parameter	Symbol	Unit	Value	
Auxiliary power	P_{aux}	W	Car	500
			Bus	1000
			Tourist cart	200
Total mass of vehicle	m	kg	Car	1500
			Bus	10000
			Tourist cart	500
Frontal area of vehicle	A_F	m^2	Car	2
			Bus	7
			Tourist cart	1
Coefficient of rolling resistance	C_R	—	Car	0.008
			Bus	0.01
			Tourist cart	0.006
Aerodynamic drag coefficient	C_D	—	Car	0.25
			Bus	0.30
			Tourist cart	0.20
Efficiency of electrical to mechanical conversion	η	—		0.9
Density of air at 30°C	ρ	kg m^{-3}		1.16

TABLE 6.2

Power Needs during Constant Speed Cruising

Cruising Velocity		Power Needs (W)		
km h⁻¹	m s⁻¹	Car	Bus	Tourist Cart
10	2.8	870	4054	294
20	5.6	1281	7281	404
30	8.3	1775	10857	547
40	11.1	2394	14955	740
50	13.9	3178	19749	999
60	16.7	4170	25414	1341
70	19.4	5410	32122	1783
80	22.2	6940	40049	2340
90	25.0	8801	49368	3031
100	27.8	11036	**60254**	3870
110	30.6	13685	72879	4875
120	33.3	**16790**	87420	6063
150	41.7	29253	144268	10885
200	55.6	63010	293547	24115

The power needs of a particular vehicle depend on the velocity, the acceleration rate, the slope angle, and the road conditions. Unless otherwise stated, the following discussion is for situations on a flat surface, i.e., the slope angle is 0 degrees (or radian).

From Equation 6.1, and using the parameters in Table 6.1, we can estimate the power needs of a vehicle under several different scenarios. The first scenario is that the vehicle is cruising at a constant velocity, that is, the rate of acceleration is 0. Table 6.2 shows the results at different cruising velocities. The power needs increase with the cruising velocity. By assuming that the highest cruising velocities required are 120, 100, and 40 km h⁻¹ for cars, buses, and tourist carts, respectively, their power needs are around 16.8, 60.2, and 0.7 kW, respectively. So under a fixed cruising speed (e.g., without any acceleration), a normal passenger car only needs about 17 kW of power. Even at a 200 km h⁻¹ cruising speed, a passenger car only needs around 63 kW of power.

The power needs versus cruising velocity are plotted in Figure 6.3. Tourist carts represent specialty vehicles such as those that are used on golf courses and inside airports.

The second scenario is that the vehicle passes other vehicles during highway driving. If the velocity at the end of passing is 110 km h⁻¹, and the acceleration rate during the passing is 5 km h⁻¹ s⁻¹ (around 1.4 m s⁻²), then the passenger car needs about 85 kW of power.

The third scenario is that a passenger car accelerates from 0 to 100 km h⁻¹ in 10 seconds. If the acceleration rate is constant at 10 km h⁻¹ s⁻¹ (i.e., 2.8 m s⁻²) during the 10 seconds, then the power need at the end of acceleration (when the velocity reaches 100 km h⁻¹) will be around 141 kW, as shown in Table 6.3.

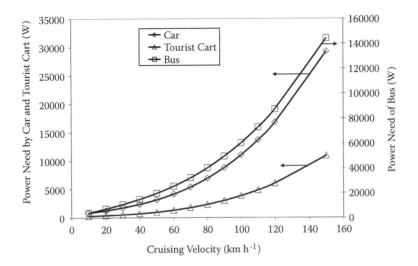

FIGURE 6.3
Power needs versus cruising velocity for cars, buses, and tourist carts.

TABLE 6.3

Power Need for a Passenger Car to Accelerate from 0 to 100 km h^{-1} in 10 Seconds at a Constant Acceleration Rate of 2.8 m s^{-2} and Acceleration Rates with 80 and 40 kW Constant Power Outputs

Velocity (km h^{-1})[a]	Power Needs (W)[b]	Acceleration Rate (m s^{-2})[c]		Time (s)[d]	
		80 kW	40 kW	80 kW	40 kW
10	13833	17.09	8.45	0.16	0.33
20	27207	8.50	4.18	0.33	0.66
30	40664	5.63	2.75	0.49	1.01
40	54246	4.19	2.03	0.66	1.37
50	67993	3.32	1.59	0.84	1.75
60	81947	2.73	1.29	1.02	2.15
70	96150	2.30	1.07	1.21	2.60
80	110643	1.97	0.89	1.41	3.11
90	125468	1.71	0.75	1.63	3.71
100	140666	1.49	0.63	1.86	4.44
	Total Time (s)			**9.6**	**21.1**

[a] Velocities at the end of every 10 km h^{-1} velocity increase period are used in the calculations for power needs and time.

[b] Power needs to accelerate from 0 to 100 km h^{-1} in 10 seconds at a constant acceleration rate of 2.8 m s^{-2}.

[c] Average acceleration rate in every 10 km h^{-1} increase period.

[d] Time for every 10 km h^{-1} velocity increase period.

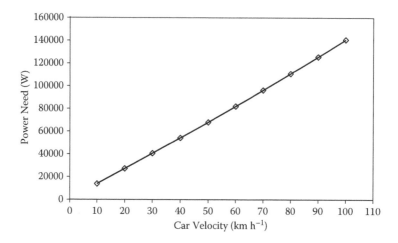

FIGURE 6.4

Power needs for a passenger car to accelerate from 0 to 100 km h^{-1} in 10 seconds at a constant acceleration rate of 2.8 m s^{-2}.

The speed at the end of each 10 km h^{-1} velocity increase segment is used in the calculation. With such a constant acceleration rate strategy, the power need about doubles when the velocity doubles, and in the early state of acceleration, the power needed is much less than the engine can provide. The results are plotted in Figure 6.4.

Obviously, this is not a good acceleration strategy. A better and actual strategy is to fully utilize the available power output during the entire acceleration process to achieve higher acceleration rates at lower velocities. Table 6.3 shows the acceleration rate and the velocity situation when the power is set at 80 and 40 kW constantly, where the passenger car reaches 100 km h^{-1} in 9.6 and 21.1 seconds, respectively. Again, the speed at the end of each 10 km h^{-1} velocity increase segment is used in the calculation. The acceleration rates are plotted in Figure 6.5. With this acceleration strategy, the passenger car accelerates from 0 to 100 km h^{-1} in less than 10 seconds with 80 kW of power, while the even acceleration strategy mentioned earlier needs 141 kW to achieve similar results.

A lighter car obviously needs less power than a heavier one. Figure 6.6 shows the power needs when cars of different weights are cruising at 110 km h^{-1} velocity. The power needs increase linearly with the mass of the car. Using a car with a total mass of 1000 kg as a comparison base, cars with 1500 and 2000 kg mass will need 11% and 22% more power. Higher power needs mean that more energy is consumed to travel the same distance. This is why a driver should not carry too much junk in the car in order to reduce the driving cost.

Considering the startup acceleration need (e.g., reaching 100 km h^{-1} in 10 seconds) and the highway driving passing need (1.4 m s^{-2}), a passenger

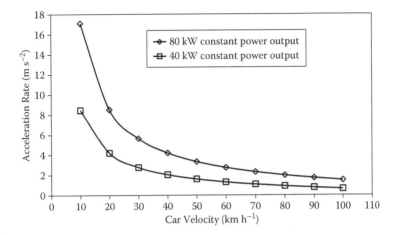

FIGURE 6.5
Acceleration rates of a car with 80 and 40 kW constant power outputs.

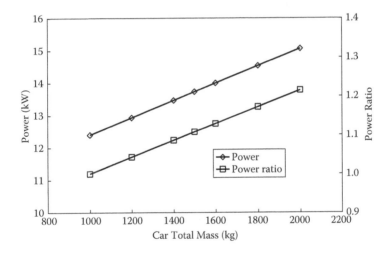

FIGURE 6.6
Power needs for cars with different total mass when cruising at 110 km h⁻¹.

car should be able to provide around 80 kW of power output. If the acceleration time and the highway passing time are both relaxed to double the above values (e.g., reaching 100 km h^{-1} in 20 seconds and the acceleration rate for highway passing at 0.7 m s^{-2}), a 50 kW power output is adequate for a passenger car (note: this may become a trend in the future).

During acceleration from 0 to 100 km h^{-1}, 80 kW of constant power in 9.6 seconds consumes 0.21 kWh of energy ($80 \times 9.6/3600$), while 40 kW constant power output consumes 0.23 kWh of energy ($40 \times 21.1/3600$). So it is interesting that lower acceleration rates do not lower energy consumption (provided that the electrical efficiency is similar).

6.3 Driving Distance with 6 kg H$_2$

Table 6.4 lists the potential driving distances of cars, buses, and tourist carts with 6 kg of H$_2$. The lower heating value (LHV) of H$_2$ is 241.8 kJ mol^{-1}, and thus the energy from 6 kg of H$_2$ is 201.5 kWh [(6000 × 241.8)/(2 × 3600)]. If the stack electrical efficiency is 53% (0.66 V/1.25 V), the H$_2$ utilization is 95%, the sum of the parasitic power loss and the DC–DC (or DC–AC) converting loss is 80%, and the efficiency of the electrical motor is 90%, then the net effective energy is 73 kWh (201.5 × 0.53 × 0.95 × 0.8 × 0.9). Using 73 kWh divides the power needs at different cruising velocity and multiplying the velocity, the potential cruising distance is obtained for cars, buses, and tourist carts; and the results are listed in Table 6.4. It may be surprising to see that 6 kg of H$_2$ enables a car, a bus, and a tourist cart to run for 1226, 200, and 3980 km, respectively, at a 30 km h^{-1} velocity. The results are plotted in Figure 6.7. For all three types of vehicles, the best cruising velocity is 30 km h^{-1}, although the difference is small, between 30 and 40 km h^{-1} cruising velocities for cars and tourist carts, and between 20 and 30 km h^{-1} for

TABLE 6.4

Potential Driving Distance with 6 kg of H$_2$ at Different Cruising Velocities

Velocity (km h^{-1})	Cruising Distance (km)			Ratio to Longest Distance		
	Car	Bus	Tourist Cart	Car	Bus	Tourist Cart
10	834	179	2472	0.68	0.89	0.62
20	1132	199	3595	0.92	0.99	0.90
30	**1226**	**200**	**3980**	1.00	1.00	1.00
40	1212	194	3922	0.99	0.97	0.99
50	1141	184	3631	0.93	0.92	0.91
60	1044	171	3245	0.85	0.85	0.82
70	939	158	2848	0.77	0.79	0.72
80	836	145	2480	0.68	0.72	0.62
90	742	132	2154	0.61	0.66	0.54
100	657	120	1874	0.54	0.60	0.47
110	583	109	1637	0.48	0.55	0.41
120	518	100	1436	0.42	0.50	0.36
150	372	75	1000	0.30	0.38	0.25
200	230	49	602	0.19	0.25	0.15

Note: the power needs shown in Table 6.2 are used in the calculations; these are the fuel cell engine power that is about 10% higher than the electrical motor output power, and thus the results listed here underestimate the driving distance for about 10%

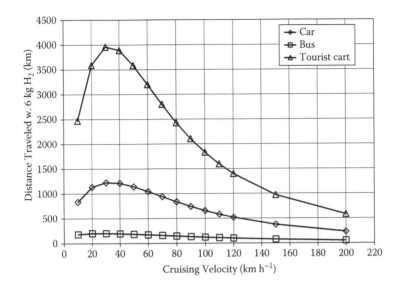

FIGURE 6.7
Potential cruising distance for cars, buses and tourist carts at different uninterrupted cruising velocities.

buses. Outside these cruising velocity regions, the cruising distance drops significantly. As shown in Figure 6.8, the cruising distance of a car, a bus, and a tourist cart at a cruising velocity of 108, 120, and 94 km h^{-1}, respectively, shortens to about half of that at 30 km h^{-1}.

The above results seem to contradict our daily driving experience, that highway driving (higher driving velocity) is more economical than local driving (lower driving velocity). The reason is that local driving involves too many stops, starts, decelerations, and accelerations, which are not considered in the above analysis where a constant cruising velocity is assumed. Local driving situations differ from community to community and from driver to driver, and they are impossible to standardize. Interested readers can refer to the U.S. federal guidelines for more information. Here, we can assume some local driving situations for a simple analysis. If the cruising velocity is 60 km h^{-1}, and the power is 80 kW, then the total energy consumed to accelerate a car from 0 to 60 km h^{-1} is 0.078 kWh (by multiplying 80 kW with the total time of 3.5 seconds up to 60 Km h^{-1} velocity shown in Table 6.3). The power need during 60 km h^{-1} cruising is 4.17 kW, as shown in the third column in Table 6.2. With different cruising times, the local travel distance can be estimated by Equation 6.2, and the results are shown in the third column in Table 6.5.

Local driving distance = ideal cruising distance ×
cruising energy consumption/(cruising energy consumption +
acceleration energy consumption)   (Eq. 6.2)

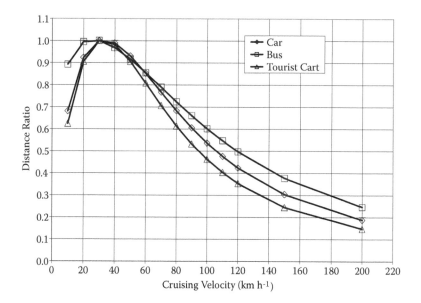

FIGURE 6.8

Percentage cruising distance at different uninterrupted cruising velocities with respect to that at optimal cruising velocity (i.e., 30 km h⁻¹) for cars, buses, and tourist carts.

TABLE 6.5

Driving Distance of a Car with 6 kg of H_2 for Different Cruising Durations at a Velocity of 60 km h⁻¹

	Without Regenerative Braking			With 40% Regenerative Braking		
Cruising Duration (min)	Energy Consumed during Cruising (kWh)	Cruising Distance (km)	% of Ideal Distance	Cruising Distance (km)	% of Ideal Distance	Efficiency Increase (%)
0.25	0.017	191	18	251	24	31.9
0.5	0.035	322	31	405	39	25.7
1	0.069	493	47	584	56	18.5
2	0.139	669	64	749	72	11.9
3	0.208	760	73	827	79	8.7
4	0.278	816	78	872	84	6.9
5	0.347	853	82	902	86	5.7
6	0.417	880	84	923	88	4.9
7	0.486	900	86	938	90	4.2
8	0.556	916	88	950	91	3.8
9	0.625	928	89	960	92	3.4
10	0.695	939	90	968	93	3.1
15	1.042	971	93	992	95	2.1
20	1.390	989	95	1004	96	1.6

Frequent stops significantly reduce travel distance. For example, at a cruising duration of 1 min (corresponding to about 1 km distance at 60 km h⁻¹), the car can travel for about 493 km (third column in Table 6.5), far less than the 1044 km that the car could potentially run without any interruption to the 60 km h⁻¹ cruising speed. One stop and following acceleration in every 500–1000 m are quite reasonable for local driving.

Typically, the fuel cell will be hybridized with an energy storage device, such as a battery or a supercapacitor, that can convert about 40% of the kinetic energy to electrical energy during each break through a regenerative breaking mechanism. The kinetic energy is expressed by Equation 6.3.

$$E_k = 0.5mv^2 \qquad \text{(Eq. 6.3)}$$

where m and v are the mass (kg) and the velocity (m s⁻¹) of the object (e.g., a car), and the unit of E_k is in Joules (J). Table 6.6 lists the kinetic energy that could be converted to electrical energy at different velocities for a 1500 kg car. For example, when a car stops from a velocity of 60 km h⁻¹, 0.023 kWh of electrical energy is stored in the electrical energy storage device.

With regenerative braking, the local travel distance is recalculated by Equation 6.4.

$$\text{Local driving distance} = \text{ideal cruising distance} \times \text{cruising}$$
$$\text{energy consumption/(cruising energy consumption +}$$
$$\text{acceleration energy consumption} -$$
$$\text{energy recovered via regenerative braking)} \qquad \text{(Eq. 6.4)}$$

The results are shown in the fifth column in Table 6.5. The percentage of the travel distance increase from 40% regenerative braking energy recovery is listed in the last column in Table 6.5. For example, if a stop happens every 1000 m of cruising, the travel distance increases from 493 to 584 km, an 18.5% increase. The travel distances at different cruising time durations with and without regenerative braking are shown in Figure 6.9.

TABLE 6.6

Kinetic Energy and 40% Regenerative Braking Energy Recovery at Different Velocities (Car Mass = 1500 kg)

Velocity before Stop		Kinetic Energy		Recovered Electrical Energy
(km h⁻¹)	(m s⁻¹)	(J)	(kWh)	during Each Stop (kWh)
20	5.6	23148	0.006	0.003
40	11.1	92593	0.026	0.010
60	16.7	208333	0.058	0.023
80	22.2	370370	0.103	0.041
100	27.8	578704	0.161	0.064
120	33.3	833333	0.231	0.093

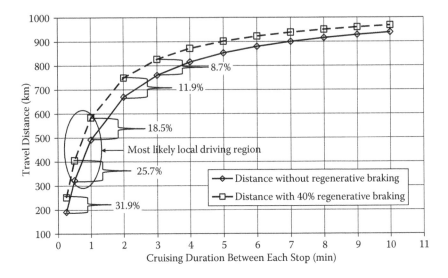

FIGURE 6.9
Travel distances of a car at different cruising time durations during local driving with and without regenerative braking.

With a similar analysis, a 10,000 kg bus that has 80 kW of electrical power accelerating from 0 to 100 km h^{-1} at a constant 80 kW of power requires 43.3 seconds. The local driving distance at 60 km h^{-1} cruising velocity with 6 kg of H$_2$ and one stop every 500 m will be 42 and 49 km without and with 40% regenerative breaking, respectively, which is only 14%–17% of the uninterrupted cruising distance. If the power is 40 kW, the bus can accelerate from 0 to 70 km h^{-1} in 162.8 seconds, and further acceleration to 80 km h^{-1} is not possible.

A tourist cart weighing 500 kg can accelerate from 0 to 100 km h^{-1} in 3.4 and 8.3 seconds with a constant power of 80 and 40 kW, respectively. The local driving distance at 30 km h^{-1} cruising velocity with 6 kg of H$_2$ and one stop every 100 m will be 446 and 583 km without and with 40% regenerative breaking, respectively, only 11%–15% of the uninterrupted cruising distance.

Clearly, acceleration after frequent stops and breaking significantly reduces the driving distance of a vehicle. Reducing the number of stops and breaking is crucial to increase driving distance.

Correspondingly, the detailed driving schedule must be given when describing the driving distance of a vehicle. Without such detailed information, it is meaningless to compare results reported to the public that are labeled as only urban (i.e., local) or highway driving. For example, for urban driving with different driving schedules, the driving distance can be a few times different, as shown by Figure 6.9.

According to Table 6.4, a car can drive 583 km at 110 km h^{-1} uninterrupted highway cruising, and according to Table 6.5, a car can drive 493 and

584 km without and with 40% regenerative braking when locally cruising at 60 km h^{-1} with one stop every 1 km. Making an estimated generalization (for both local and highway driving), a passenger car can drive for around 600 km per 6 kg of H$_2$, or per 201.5 kWh of H$_2$, translating to about 100 km per kg of H$_2$ or 3.0 km per kWh of H$_2$.

Based on our driving experience, a conventional ICE-powered car has an average driving distance of 25 miles per gallon of gasoline, or 10 km per liter of gasoline. Gasoline can be represented by C$_8$H$_{18}$, whose molar mass, density, and LHV are 114, 0.703 g cm^{-3}, and 5074.5 kJ mol^{-1}, respectively. Driving for 580 km will require 58 L of gasoline, corresponding to 1,814,980 kJ of energy (5074.5 kJ mol^{-1} × 58 L × 1000 cm^3 L^{-1} × 0.703 g cm^{-3}/114 g mol^{-1}). The energy within 6 kg of H$_2$ is 725,400 kJ (241.8 kJ mol^{-1} × 6 kg × 1000 g kg^{-1}/2 g mol^{-1}). Therefore, the efficiency of the fuel cell car is around 2.5 times that of a conventional ICE-powered car (1,814,980/725,400). The tank-to-wheel efficiency of the fuel cell vehicle is about 36% (73 kWh/201.5 kWh), and that of a conventional ICE-powered car is about 14% (36%/2.5).

If 6 kg of H$_2$ is stored at 5% wt by using a 700-bar lightweight reinforced cylinder, the storage device will weight 120 kg. If the stack itself is 100 kW, and the corresponding fuel cell system (stack plus balance of plant, or auxiliary, but excluding the battery mentioned below) has a specific power density of 0.5 kW kg^{-1} (the status in 2011 was 0.4 kW kg^{-1} and the 2017/2020 targets are 0.65/0.65 kW kg^{-1} based on US Department of Energy (DOE) reports; refer to Table 8.3 as well) (US DOE 2012), and an auxiliary battery energy capacity is 5 kWh and its specific energy density is 0.15 kWh kg^{-1}, then the entire fuel cell system will be around 353 kg.

As mentioned earlier, due to various losses, only around 73 kWh of energy out of the 201.5 kWh total energy in 6 kg of H$_2$ is sent out by the electrical motor to drive the vehicle. For an equivalent battery bank (without considering the 5 kWh auxiliary battery bank used in the fuel cell system mentioned above), when its discharging efficiency, the DC–DC (or DC–AC) converting efficiency, and the electrical motor efficiency are around 90%, respectively, its energy capacity should be around 100 kWh [73/(0.9 × 0.9 × 0.9)]. The total mass of the battery bank depends on its specific energy density, as shown in Figure 6.10. If the specific energy density is 0.15 kWh kg^{-1} based on the current Li-ion battery technology, the battery bank will weigh 700 kg. If we assume that the mass of the auxiliary components for the battery system is 100 kg, then the total battery system will be 800 kg, more than 2 times the fuel cell system. The volumetric energy density of the battery bank follows the same trend shown in Figure 6.10. At a volumetric energy density of 0.30 kWh L^{-1}, the volume of the battery system itself will be 330 L. The volumetric energy density of H$_2$ storage system has achieved 0.028 kg H$_2$ L^{-1} in 2010 (US DOE 2012), then the total volume of the H$_2$ storage system for 6 kg H$_2$ will be 214 L. Therefore, a fuel cell system is currently much lighter and possibly smaller than an equivalent battery system for a vehicle to drive the same distance.

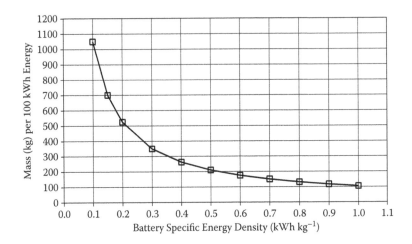

FIGURE 6.10
Total mass of a 100 kWh battery system versus its specific energy density.

6.4 Electrical Power Train

Since a passenger car only needs less than 20 kW of power during constant velocity cruising, and the other 60 kW of power is needed only for short-term acceleration and slope climbing, the power train system may be best composed of a fuel cell and an electrical energy storage device that can deliver a high-power output in a short time. Such a hybrid configuration can also prevent the stack from going extremely wide output voltage variations, which in turn can prolong the lifetime of the stack and increase the efficiency of the DC/DC converter or the DC/AC inverter. Supercapacitors can deliver extremely high-power output in seconds, and therefore may be the best candidates. Supercapacitors can also be charged in less than a minute and have a long life, being able to be charged and discharged in hundreds of thousands of cycles. The next candidate is a battery, such as Li-ion batteries and Ni-metal hydride batteries, which can offer several C discharge rates. Although their discharge rate (typically less than 5 C) and the number of charge–discharge cycles (around 1000) are much lower than those of supercapacitors, they have the advantages of (much) higher energy density, lower cost, and smaller volume. So, a fuel cell hybrid power train system could be composed of a fuel cell and a battery, a fuel cell and a supercapacitor, or a fuel cell, a battery, and a supercapacitor.

Figure 6.11 schematically shows a fuel cell–battery hybrid system configuration. During cruising, the power comes from the fuel cell, and during acceleration, the battery provides additional burst power. The power from both the fuel cell and the battery goes through a DC–DC converting or a

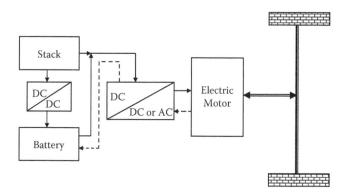

FIGURE 6.11
Schematic of a fuel cell/battery hybrid configuration.

DC–AC inverting process to supply the electrical motor with the right power (DC or AC, depending on whether it is a DC motor or an AC motor). During braking, the batteries will collect the regenerative braking power, which converts about 40% of the kinetic energy to electrical energy to enhance the total system efficiency. When the capacity of the battery becomes low during cruising, the stack will use some excess power to charge the battery in a controlled fashion. In Figure 6.11, the solid and dashed arrows represent the electrical flows, and the double-sided arrow represents the electrical and mechanical interchanges. The dashed arrows show the battery charging process through regenerative braking. It is important that the stack is never electrically charged; otherwise, it will be damaged in seconds. In such a configuration, the battery will automatically provide power to the DC–DC converter or DC–AC inverter when the output voltage of the stack drops to the battery's output voltage during acceleration. When acceleration is complete, the output voltage of the stack quickly becomes higher than the battery's voltage, and thus the battery will automatically stop sending power out.

In this configuration, batteries with a high discharging rate should be chosen in order to reduce the Ah, mass, and volume of the battery bank. For example, for a 60 kW power output at a voltage of 300 V, if the battery discharges at 1 C, the battery should be around 200 Ah; but if the battery can discharge at 5 C, the battery only needs to be around 40 Ah. This means that the 5 C battery will be around one fifth of the 1 C battery in Ah, mass, and volume.

A higher discharge rate will shorten the life of the battery. However, the high discharging process only affects the battery for a short time; its life should not be seriously affected.

If the battery cannot discharge at a high rate, its mass and volume will exceed those of a fuel cell system for generating the same level of power. In such a case, it is better to use the fuel cell system to provide all of the power for all driving situations and use the battery only to start up the fuel cell system and to collect the regenerative breaking power.

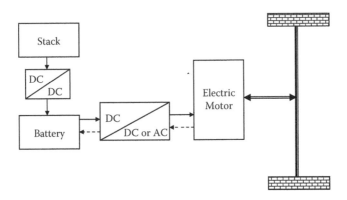

FIGURE 6.12
Schematic of another fuel cell/battery hybrid configuration.

In such a configuration, the DC outputs from the fuel cell stack and the battery are connected to a common electrical buss; therefore, only one DC–DC converter (or DC–AC inverter) is needed. Such a connection is possible because during the operation, the output voltages of the stack and the battery bank are in the voltage "plateau" regions, where the voltages do not change significantly and are mainly determined by the ohmic resistance.

Another way to hybrid a fuel cell and a battery bank is schematically shown in Figure 6.12. This configuration uses the battery itself to power the electrical motor completely; the battery is charged by the stack and collects the regenerative braking power. In this configuration, the battery is always in use, so its lifetime is a concern. Also, since the battery itself must be able to provide all the needed power, its capacity will be much larger than that shown in Figure 6.11. In addition, as the energy densities (both specific and volumetric) of batteries are much lower than those of a fuel cell based on current technologies, the mass and the volume of the battery bank will be significant, which in turn will reduce the travel distance and the cargo space of the vehicle. Therefore, this is not a good fuel cell–battery hybrid configuration.

Figure 6.13 schematically shows a fuel cell–supercapacitor hybrid configuration. It is similar to the fuel cell–battery hybrid configuration shown in Figure 6.11, but there is an additional DC–DC converter or DC–AC inverter between the supercapacitor and the electrical motor. The reason is that the voltage of a supercapacitor drops quickly during the discharge process according to Equation 6.5.

$$V = (2E/C)^{0.5} \qquad\qquad\qquad\qquad \text{(Eq. 6.5)}$$

where E and C are the energy and the capacitance of the supercapacitor, respectively. For example, when the energy drops by 50%, the voltage will drop by 30%. Without the additional DC–DC converter or DC–AC inverter, the fuel cell stack becomes the only energy provider because the voltage of

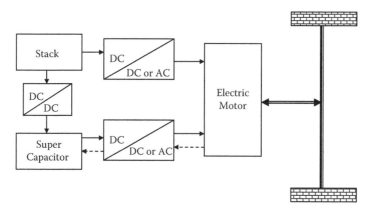

FIGURE 6.13
Schematic of a fuel cell/supercapacitor hybrid configuration.

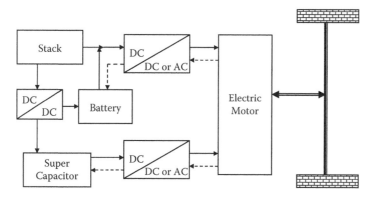

FIGURE 6.14
Schematic of a fuel cell/battery/supercapacitor hybrid configuration.

the supercapacitor quickly drops below that of the stack after it gives out a small amount of energy, which in turn could make the stack exceed its maximum power output, causing the entire system to collapse.

Figure 6.14 schematically shows a fuel cell–battery–supercapacitor hybrid configuration. This is a combination of the electrical connections shown in Figures 6.11 and 6.13. A single DC–DC converter is used for the stack to charge both the battery and the supercapacitor.

In configurations shown in Figures 6.13 and 6.14 the output voltages of the two DC/DC (or AC) converters (or inverters) must be set the same for evenly sharing the load. However, due to the potential intrinsic inaccuracy in the voltage set points of the two DC/DC (or AC) converters (or inverters) and the iR losses through the wire between the DC/DC (or AC) converters (or inverters) and the electric motor, the power sharing between the two DC/DC (or AC) converters (or inverters) may become tricky.

6.5 Fuel Cell System

In order to minimize the volume of the stack, the reactants are pressurized to 1–3 barg, and liquid-cooled stacks are used. The compression of incoming air is done through a compressor. Likely, the compressor is mechanically connected with an expender through a common shaft in a compressor–expender combination, as shown in Figure 6.15. As the hot and still-pressurized air from the cathode exhaust enters the expender, it expands and cools down through an isentropic process (i.e., constant entropy process). The energy lost from the air exhaust in the expender is transferred to the compressor through the shaft to help the compressor to compress the incoming air. Therefore, the additional energy input to the compressor can be reduced, lowering the system's parasitic energy loss. The liquid water collected from the cathode exhaust via the expender may be used to humidify air and/or H_2 for the stack operation.

In the early stage of fuel cell vehicle development, onboard reforming of hydrocarbon fuels, such as methanol, was considered in order for the vehicle to carry liquid fuels for convenience and to achieve longer driving distances. However, since onboard reforming increases system complexity, decreases system reliability, and takes quite a long time for the reformer to generate high-quality H_2, it was abandoned in around 2001. Since then, efforts have been focused on using pure H_2 for fuel cell vehicles.

Since H_2 has a very low volumetric energy density, onboard storing of H_2 becomes a challenge. High-pressure lightweight reinforced cylinders are the mainstream choice worldwide today. At 350 and 700 bar pressures, the H_2 mass percentage can reach more than 3% and nearly 5%, respectively. When storing 6 kg of H_2 onboard at 350 or 700 bars, the mass of the cylinder will be around 200 or 120 kg, respectively. The volume of each cylinder is

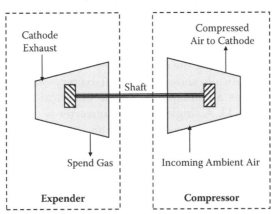

FIGURE 6.15
Illustration of a compressor–expender combination.

FIGURE 6.16
350-bar H_2 cylinder assembly for a fuel cell bus. Courtesy of Beijing Jonton Hydrogen Tech. Company.

typically around 70 and 150 L for cars and buses, respectively. For cars, the H_2 cylinder (1 or maximum 2) is typically placed beneath the rear passenger seats, while for buses, the cylinders (up to 10) are normally placed on the bus roof. Figure 6.16 shows an H_2 cylinder assembly that consists of eight 350-bar cylinders for fuel cell buses. It takes about 10 minutes to refuel all eight cylinders. There is a valve combination on each of the eight cylinders. The volume, length, and diameter of each cylinder are 150 liters, 1800 mm, and 385 mm, respectively. It was reported that the bus could travel 280 km per refueling.

Although H_2 is not more dangerous than other types of fuels, the H_2 storage cylinders have to pass various tests such as shooting, colliding, burning, permeation, and bursting. A high-pressure lightweight cylinder typically consists of an inner liner (aluminum or some special polymers) that is not permeable by H_2, wrapped with high-strength carbon fibers, followed by a fiberglass protective layer, and finally a smooth resin coating, as shown in Figure 6.17. Cuts on the protective coating cause more safety concerns than denting the cylinder. The strength and the quality of the carbon fibers are crucial. The wrapping of carbon fibers and fiberglass needs to be carefully designed in order for the resulting cylinder to endure high pressure evenly in different areas. Figure 6.18 shows a computer-simulated carbon fiber wrapping result.

The cylinder's bursting pressure is at least twice the H_2 storage pressure. Figure 6.19 shows a test result after such a cylinder burst under high pressure. Water was used as the testing medium, and the cylinder burst at a pressure of 1100 bars, 3.1 times the targeted H_2 storage pressure of 350 bars. When a regular steel cylinder bursts, it forms many metal pieces that could

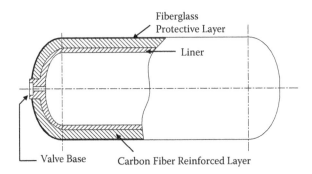

FIGURE 6.17
Schematic of a high-pressure lightweight H$_2$ cylinder. Courtesy of Tongji University.

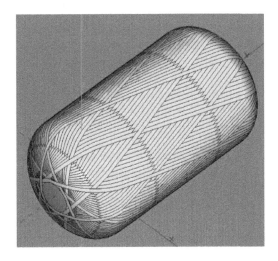

FIGURE 6.18
Computer-simulated carbon fiber wrapping result. Courtesy of Tongji University.

seriously injure a person, but when a lightweight high-strength cylinder bursts, it does not form flying pieces and is therefore safer than regular steel cylinders.

Figure 6.20 shows the fire burning test process of a lightweight high-pressure cylinder. Figure 6.21 shows the gunshot test result of a similar cylinder. After the gunshot, a bullet hole was created in the cylinder, but the cylinder did not burst or explode, resulting in neither fire nor explosion. Figure 6.22 shows a collision test process. The cylinders were filled with 350 bars of N$_2$ and the collision occurred at a speed of 50 km h^{-1}. At the moment of the collision, the gas outlet valve should be automatically shut off. This automatic shutoff function prevents H$_2$ from flowing out of the cylinder when a collision occurs, and thus protects the entire fuel cell vehicle system.

FIGURE 6.19
Burst testing result of a high-pressure lightweight H$_2$ cylinder. Courtesy of Tongji University.

FIGURE 6.20
Burning test process of a high-pressure lightweight H$_2$ cylinder. Courtesy of Beijing Jonton Hydrogen Tech. Company.

Refueling a fuel cell vehicle with high-pressure H$_2$ is as quick as refueling an ICE vehicle with gasoline. For example, refueling 6 kg of H$_2$ only takes a few minutes. The key for achieving such a short refueling time is two-fold. First, the refueling transfers the high-pressure H$_2$ from the fuel tanks in the refueling station to the fuel cylinders of the vehicle without a pressure boosting process that typically uses a compressor. As the H$_2$ pressure within the receiving tank increases, the temperature of the tank increases

FIGURE 6.21
Gunshot test result of a high-pressure lightweight H$_2$ cylinder. Courtesy of Beijing Jonton Hydrogen Tech. Company.

FIGURE 6.22
Colliding test process on H$_2$ cylinders. Courtesy of Beijing Jonton Hydrogen Tech. Company.

due to H$_2$ compression. The temperature increases faster in the beginning, and then it slows down. When the refueling is complete in a few minutes, the tank temperature is typically lower than about 80°C, which is below the safety temperature of the tank (regulated safety set point is 85°C). Second, the H$_2$ fueling coupler, which is a combination of the fueling nozzle that connects to the fuel source and the fueling receptacle that connects to a fuel

FIGURE 6.23
Refueling of a fuel cell bus. Courtesy of Beijing Jonton Hydrogen Tech. Company.

FIGURE 6.24
H_2 refueling coupler in use. Courtesy of Beijing Jonton Hydrogen Tech. Company.

receiver, is made reliable and intelligent and can safely withstand the fast flow of the high-pressure H_2. Figure 6.23 shows the refueling of a city bus, and Figure 6.24 shows the H_2 fueling coupler in use in refueling a minivan.

The rate of temperature increase is faster when H_2 is refueled at a higher rate. If the temperature reaches the regulated safety set point of 85°C the dispensing process will be halted automatically. Therefore, the maximum refueling rate is that before the temperature set point is triggered. The refueling rate is fastest in the beginning and slows down gradually as the refueling

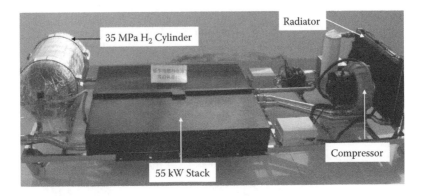

FIGURE 6.25
Major components of a fuel cell system developed by Shanghai Shenli High Tech. Company.
Courtesy of Shanghai Shenli High Tech. Company.

proceeds. The dispensing system meters and monitors the rate of refueling. The mass of H_2 within the cylinder can be controlled within ±0.5% accuracy.

Figure 6.25 shows the layout of the major components of a fuel cell subsystem developed by Shanghai Shenli High Tech. Company for passenger cars. The stack was 55 kW with a nominal electrical efficiency higher than 50%. The H_2 was stored in two cylinders at 35 MPa; one was 70 L and the other was 50 L. One-time H_2 refueling allowed the car to run for about 300 km. Currently, many car companies place the stack, the auxiliary power (battery or supercapacitor), and the H_2 storage cylinder(s) beneath the driver and passenger seats, while the other auxiliary components, such as the radiator and the compressor (and expender), are placed under the front hood. In such an arrangement, the fuel cell car will have similar passenger and cargo space as a regular car.

As a key module of the fuel cell system, the performance and reliability of the stack is crucial. As a world leader, Ballard Power Systems (Canada) has spent many years developing stacks. Table 6.7 lists the basic parameters of Ballard9 SSL™ stacks. At a rated power of 21.0 kW (current is 300 A), the specific and volumetric power densities of the stack are 1.24 kW kg^{-1} and 1.52 kW l^{-1}, respectively.

6.6 Examples

6.6.1 Tourist Cart

It is estimated that more than one thousand fuel cell vehicles have been developed worldwide so far. The number is small, but represents significant technology advancement achieved in the past one to two decades. Figure 6.26

TABLE 6.7

Basic Specifications of Ballard9 SSL™ Stacks

Parameter	Unit	Value					
Stack rated power	kW	3.8	4.8	10.5	14.3	17.2	21.0
Stack voltage at 300 A	V	12.8	16.0	35.0	48.0	57.4	70.2
Stack mass (without coolant)	kg	7.1	7.2	10.7	13	15	17
Stack core length	mm	92	104	174	220	255	302
Stack core width	mm	760					
Stack core height	mm	60					
Maximum current	A	300					
Startup temperature	°C	>2					
Environment temperature during operation	°C	−25–75					
Coolant inlet temperature	°C	2–68					
Storage temperature	°C	−40–60 (with dry-out procedure)					
		2–60 (without dry-out procedure)					
H_2 composition	%	>95 (prehumidification)					
H_2 inlet pressure	barg	1–2					
Air inlet pressure	barg	1–2					

Source: Adapted from Ballard Power Systems website, FCvelocity-9SSL Spec Sheet, http://www.ballard.com/files/PDF/Material_Handling/9SSL.pdf.

FIGURE 6.26
A tourist cart developed by Shanghai Shenli High Tech. Company. Courtesy of Shanghai Shenli High Tech. Company.

shows a tourist cart developed by Shanghai Shenli High Tech. Company. The electrical power was supplied by a PEMFC and a Pb-acid battery bank. The nominal power of the stack was 7 kW, and its electrical efficiency was higher than 50%. It used one 50 liter 25 MPa H_2 cylinder, and could run for more than 100 km with each H_2 refueling. The H_2 cylinder was placed at the back of the cart.

6.6.2 Car

Figure 6.27 shows a light car developed by South China University of Technology. The hybrid power was provided by a 5 kW fuel cell stack as the main power source and a 7 kWh Li-ion battery bank as the secondary power source, which provides extra energy when the car is accelerating or climbing. The light car could run for more than 300 km once the hydrogen cylinder was fully filled and the battery bank fully charged. The major parameters of the light car are listed in Table 6.8. The slope steepness is the percentage ratio of the height (H) to the horizontal length (L) of the slope. It relates to the slope angle in radians according to Equation 6.6.

$$\theta = \tan^{-1}(H/L) \qquad \text{(Eq. 6.6)}$$

Table 6.8 indicates that the largest slope steepness is 35%, which means that the slope angle is 19.2° [$(\tan^{-1}(0.35) \times 180/\pi$].

FIGURE 6.27
A light car developed by South China University of Technology (the inset highlights the stack). Courtesy of South China University of Technology.

TABLE 6.8

Major Parameters of a Light Car Developed by South China
University of Technology

Parameters	Unit	Value
Dimension of car (L × W × H)	mm	2880 × 1445 × 1570
Wheel base	mm	2190
Mass	kg	648
Highest velocity	km h^{-1}	80
Acceleration time from 0 to 50 km h^{-1}	s	7
Largest slope steepness to climb	%	35
Travel distance with full H$_2$ and battery bank	km	~300
Noise during acceleration	dB	48
Stack nominal power	kW	5
Stack peak power	kW	7
H$_2$ cylinder pressure	bar	100
H$_2$ cylinder volume	L	70
Li-ion battery bank capacity	Ah	200
Li-ion battery bank voltage	V	36

Source: Courtesy of South China University of Technology.

Based on the data in Table 6.8, we can estimate the maximum velocity that the light car could maintain when climbing a 19.2° slope. The stack peak power is 7 kW, assuming a 20% power loss due to parasitic power and DC–DC conversion; its output power is 5600 W (7000 × 0.8). If the 200 Ah Li-ion battery bank discharges at 2 C, its power output is 14400 W (36 V × 400 A). Then, the net power output from the power train system is 17800 W, (5600 W + 14400 W) × 0.9 − 200 W, where 200 W is the auxiliary power needed by the light car, and 0.9 is the efficiency of the electrical motor. Table 6.9 shows the power needs against the gravity and the total power need at different climbing velocities (without any acceleration). It is clear that about 85% of

TABLE 6.9

Power Needs versus Climbing Velocity at a Slope
of 19.2° for the Light Car Developed by South China
University of Technology

Velocity		Power Needs	Total Power Needs
km h^{-1}	m s^{-1}	against Gravity (W)	(W)
10	2.8	5801	6766
20	5.6	11602	13349
30	8.3	17404	19965
40	11.1	23205	26630
50	13.9	29006	33362
60	16.7	34807	40177

the power need is used to counteract gravity. This light car can climb a 19.2° slope at a 20–30 km h^{-1} speed with the Li-ion battery discharged between 2–3 C rates.

Table 6.10 lists the slope that a passenger car can climb at a constant velocity of 20, 40, and 60 km h^{-1}, respectively. The parameters in Table 6.1 are used for the calculation. It can be seen that if the power is 80 kW, the maximum slopes the car can climb are around 15°, 25°, and 60° at velocities of 60, 40, and 20 km h^{-1}, respectively. The results are plotted in Figure 6.28.

Table 6.11 lists the maximum velocity and the time to reach this velocity from rest for a car with 80 kW of power on different slopes. Again, the

TABLE 6.10

Power Needs of a Car at Different Climbing Speed versus Slope

	Power Needs (W) at Velocity of		
Slope (°)	60 km h^{-1}	40 km h^{-1}	20 km h^{-1}
5	27895	18211	9190
10	51440	33908	17038
15	74626	49365	24767
20	97275	64464	32316
25	119216	79091	39630
30	140281	93135	46652
40	179151	119048	59608
50	212704	141417	70793
60	239921	159561	79865

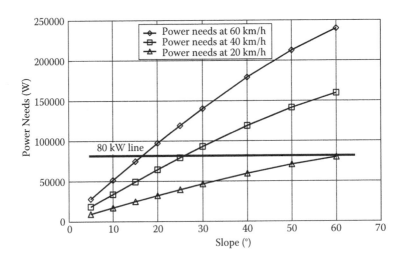

FIGURE 6.28
Power versus the slope that a car can climb at velocities of 60, 40, and 20 km h^{-1}, respectively.

TABLE 6.11

Achievable Maximum Velocity and Time Taken on Different Slopes for a Car with 80 kW Power

Velocity (km h⁻¹)	Time for Each 10 km h⁻¹ Velocity Increase at			
	20° Slope	15° Slope	10° Slope	5° Slope
10	0.20	0.19	0.18	0.17
20	0.54	0.47	0.41	0.36
30	1.22	0.90	0.71	0.58
40	3.31	1.68	1.12	0.83
50		3.55	1.72	1.13
60		14.36	2.70	1.48
70			4.63	1.92
80			10.25	2.48
90				3.25
100				4.37
Total time to maximum speed (s)	5.3	21.1	21.7	16.6

parameters listed in Table 6.1 are used for the calculation. At slopes of 10°, 15°, and 20°, the maximum velocities that a car can achieve from rest are 80, 60, and 40 km h⁻¹. At a slope of 5°, the car can achieve 100 km h⁻¹ or higher velocities. The total time to achieve speeds of 100, 80, 60, and 40 km h⁻¹ are 16.6, 21.7, 21.1, and 5.3 seconds on slopes of 5°, 10°, 15°, and 20°, respectively.

The maximum velocity and the total time to achieve it are plotted in Figure 6.29. The maximum velocity decreases linearly with the slope. Please

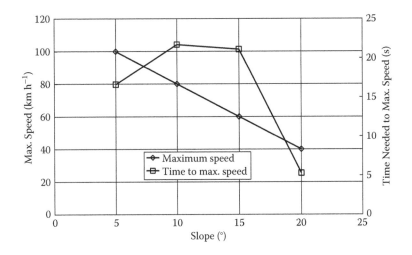

FIGURE 6.29

The maximum velocity and the total time to achieve it on 5°, 10°, 15°, and 20° slopes for a car with 80 kW output power.

FIGURE 6.30
Picture of an integrated 40 kW fuel cell system developed by Sunrise Power. Courtesy of Sunrise Power.

note that 100 km h^{-1} is not the maximum velocity that the car can achieve on a 5° slope.

Figure 6.30 shows a picture of an integrated 40 kW fuel cell system developed by Sunrise Power. Their goal is to integrate components of a fuel cell system in order to improve the reliability and reduce the size, manufacturing time, and cost of the system. The system used two stacks, each consisting of 140 unit cells. Figure 6.31 shows the mean cell voltage and the mean cell power density versus current density plots of Sunrise Power's 100-cell stacks. The mean cell voltage is 0.697 and 0.662 V at current densities of 800 and 1000 mA cm^{-2}, respectively. As shown in Figure 6.32, the unit cell voltage varies from 0.685 to 0.715 at a current density of 800 mA cm^{-2}.

6.6.3 Forklift

PEMFCs are also good candidates to use in powering forklifts. Currently, many forklifts operated in warehouses use Pb-acid batteries. As the capacity of the battery decreases with use, the lifting speed slows down, which in turn decreases productivity and increases the chance of operating errors. For each 8-hour shift, the battery bank needs to be replaced about 3 times, also resulting in a productivity decrease. The battery banks need to be charged for several hours, and therefore the warehouse has to provide space for storing batteries. When battery banks are replaced with fuel cells, those problems no longer exist. In contrast to vehicles for transportation, forklifts should use heavy steel cylinders for H$_2$ storage to keep the vehicles in balance during the lifting process.

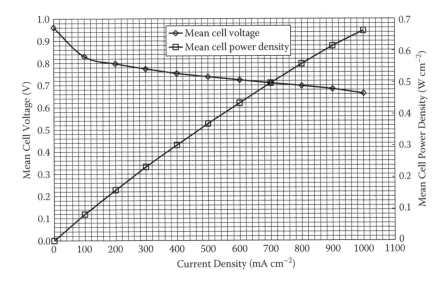

FIGURE 6.31
Mean cell voltage and power versus current density for Sunrise Power's 100-cell stack. Courtesy of Sunrise Power.

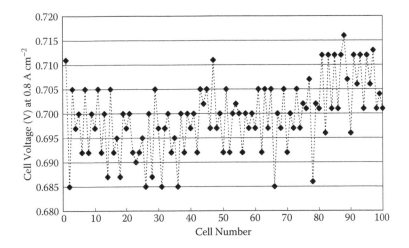

FIGURE 6.32
Unit cell voltage variation at a current density of 800 mA cm^{-2} for Sunrise Power's 100-cell stack. Courtesy of Sunrise Power.

6.6.3.1 Shanghai Shenli System

Figure 6.33 shows a forklift developed by Shanghai Shenli High Tech. Company. It used fuel cell–supercapacitor hybrid power with a targeted highest lifting weight of 2000 kg. The nominal power output of the fuel cell stack was 10 kW. The hydrogen cylinder was 50 L and the H_2 pressure was 350 bars.

The power needed by a forklift depends on the maximum weight it lifts, the lifting acceleration rate, and the lifting speed, according to Equation 6.7.

$$W = (mgv + mav)/\eta \qquad\qquad \text{(Eq. 6.7)}$$

where m, a, g, v, and η are the mass of the object (kg), the lifting acceleration rate, the acceleration rate of gravity, the lifting speed, and the motor efficiency (90%), respectively. Since a is about 100 times smaller than g, a is neglected in the following discussion. For lifting 2000 kg of weight, Table 6.12 shows the power needed versus the lifting speed. Obviously, the needed power is

FIGURE 6.33
A fuel cell forklift developed by Shanghai Shenli High Tech. Company. Courtesy of Shanghai Shenli High Tech. Company.

TABLE 6.12

Power versus Speed for Lifting
2000 kg Object

Lifting Velocity (m s⁻¹)	Power (W)
0.1	2178
0.2	4356
0.4	8711
0.5	10889
0.6	13067
0.8	17422
1.0	21778
1.5	32667
2.0	43556

TABLE 6.13

Lifting Speed versus Lifting Weight with
7 kW Fuel Cell System Power Output
and 90% Motor Efficiency

Lifting Weight (kg)	Lifting Velocity (m s⁻¹)
500	1.29
1000	0.64
1500	0.43
2000	0.32
3000	0.21
4000	0.16
5000	0.13

linearly related to the lifting speed. For example, to achieve a 0.5 and 1.0 m s⁻¹ lifting speed on a 2000 kg object, the power should be 10.9 and 21.8 kW.

For the 10 kW fuel cell stack shown in Figure 6.33, if we assume that the fuel cell system power output is 7 kW after taking into consideration the parasitic power and DC–DC converting (or DC–AC inverting) losses, and the efficiency of the motor is 90%, the maximum lifting speed versus lifting weight is listed in Table 6.13. It can be seen that to lift a 2000 kg object, the maximum lifting speed is 0.32 m s⁻¹. In a controllable lifting speed region, higher lifting speed means higher productivity, but it also means that more lifting power is needed.

The fuel cell forklift shown in Figure 6.33 (assuming the fuel cell system output power is 7 kW) carries 50 L of H_2 at 350 bars. It contains 579.2 moles of H_2 [(350 bar × 50 L)/(0.0831 bar L K⁻¹ mol⁻¹ × 298 K × 1.22)] at 298 K (1.22 is the compressibility factor of H_2 at 350 bars), which is equivalent to 38.9 kWh of energy (579.2 moles × 241.8 kJ mol⁻¹/3600 J Wh⁻¹). As discussed earlier, if 36% of the input H_2 energy is sent out by the motor (0.95 × 0.80 × 0.90 × 0.66/1.25,

TABLE 6.14

Plug Power GenDrive Systems (Courtesy of Plug Power)

Parameter	Unit	Model					
		1500	1600	1700	2300	2400	3300/3300-D
Nominal voltage	Vdc	36	36 or 48	36 or 48	36 or 48	36 or 48	24
Maximum continuous power	kW	8	8 or 10	8 or 10	8 or 10	8 or 10	1.8 or 3.2
Environment temperature	°C			−30–40			
H_2 mass	kg	1.5	1.6	1.8	1.2	1.2	0.72
H_2 pressure	bar			350			
Refueling time	min	<3	<3	<3	<2	<2	<1.5

Source: Adapted from Plug Power website, http://www.plugpower.com/AboutUs/ Documentation_Literature.aspx.

where 0.95, 0.80, 0.90, and 0.66 are the H_2 utilization, ratio of fuel cell system output power to the stack output power, motor efficiency, and average cell voltage, respectively) for lifting the object, then the total net lifting time is 2.0 h (38.9 kWh × 0.36/7 kW). If the lifting weight is 2000 kg, the total lifting distance is 2.3 km (2.0 h × 3600 s h^{-1} × 0.32 m s^{-1}/1000 m km^{-1}). If we assume that the forklift spends 1/5 of the time lifting objects and 4/5 of the time traveling between the original location and the final location of the object, then the forklift can work for about 10 hours per H_2 refueling. The super capacitator can be fully charged by the fuel cell stack during the non-lifting time period.

6.6.3.2 Plug Power System

Plug Power Inc. is a leader in developing fuel cell units used to power electric material-handling vehicles. The main product portfolio includes the GenDrive® Series 1000, 2000, and 3000, and their specifications are listed in Table 6.14. Basically, the nominal output voltage is at 24, 36, or 48 V, and the maximum continuous power at 1 hour rating ranges from 1.8, 3.2, 8, to 10 kW, respectively. Each forklift can work for an 8-hour shift. Compared to battery-powered forklifts, those fuel cell–powered forklifts increase productivity by up to 15% and lower the operational costs by up to 30%. Figures 6.34 and 6.35 show pictures of the GenDrive Series 1000 and 3000 fuel cell systems, respectively.

6.6.4 Data of Some Fuel Cell Vehicles

Table 6.15 presents the available data on 7 fuel cell vehicles. The information in the columns was provided by the following organizations: column 2, South China University of Technology; columns 3 and 4, Shanghai Shenli

FIGURE 6.34
Plug Power's GenDrive Series 1000 fuel cell. Courtesy of Plug Power.

FIGURE 6.35
Plug Power's GenDrive Series 3000 fuel cell. Courtesy of Plug Power.

TABLE 6.15

Parameters of Some Fuel Cell Vehicles

Parameter	Light Car by South China University of Technology	Car Used in 2010 Shanghai World Expo	Tourist Cart by Shanghai Shenli High Tech. Co.	Hyundai Tucson ix35 SUV	Toyota FCHV-adv Car	Nissan X-Trail FCV Car	Honda FCX-Clarity Car
H_2 volume (l)	70	120	50		Unknown		171
H_2 pressure (bar)	100	350	250	700	700	700/350	350
Stack power (kW)	5	55	7	100	90	90	100
Battery type	Li-ion	Li-ion	Pb-acid	Li-ion	$NiMH_x$	Unknown	
Battery capacity (kWh)	7						
Battery capacity (Ah)	200				Unknown		
Battery bank voltage (V)	36						
Supercapacitor	No	No	No	No		Unknown	
Travel distance (km)	~300	~300	~100	594[a]/635[b]	830[c]/760[d]/790[e]	500/370	620[c]

H₂ refueling time (min)	Unknown	3–5	3–5		Unknown		3–4
Dimension (L/W/H) (mm)	2880/1445/1570	Unknown		4410/1820/1655	4735/1815/1685	4485/1770/1745	4855/1845/1470
Total mass (kg)	650	Unknown		Unknown	1880	1860/1790	1630
Highest velocity (km h⁻¹)	80	Unknown		160	155	150	160
Motor power (kW)	Unknown			100 (AC)	90 (AC)	Unknown	
H₂ comp. factor	1.06	1.22	1.16	1.45	1.45	1.45/1.22	1.22
Moles of H₂ (mole)	267	1390	435	2800	Cannot estimate due to unknown H₂ cylinder volume		1981
Mass of H₂ (kg)	0.53	2.80	0.90	5.60			4.00
Estimated distance (km)ᶠ	290	272	87	545			389

a New European Driving Cycle.
b Federal Test Procedure.
c 10-15 Cycle Test Procedure.
d JC08 Test Procedure.
e LA4 Test Procedure.
f The estimated driving distance is based on the fuel cell itself except for the light car developed by South China University of Technology.

High Tech. Company; and column 5, Hyundai Motor Group. The information in the last three columns was made public by the corresponding companies during the 2011 Tokyo Fuel Cell Expo. "Unknown" means that the information is not known to the author. The last row shows the potential travel distance based on either the fuel cell itself when the battery/capacitor energy capacity is not known or the sum of the fuel cell and battery/capacitor when the battery/capacitor energy capacity is known. The parameters used in the calculation are based on those in Table 6.1. For example, the mass of a car is taken as 1500 kg and the mass of a tourist cart and the light car is taken as 500 kg. Also, a tourist cart, a light car, and a car are assumed to have a cruising velocity of 30, 30, and 60 km h^{-1} and stop every 50, 250, and 1000 m cruising distance, respectively; 40% regenerative power recovery is taken into account. The H$_2$ compressibility factor is also listed in the table for pressures of 100, 250, 350, and 700 bars, respectively. The reported travel distances are in quite good agreement with the estimated travel distances for the cases in columns 2 through 5. The travel distances of Hyundai, Toyota, Nissan, and Honda cars all pass 500 km per refueling. The Toyota FCHV-adv car reached 830 km per refueling using the 10-15 cycle test procedure. Although not listed here, Daimler-Benz displayed its F125 concept fuel cell cars during the 2011 Frankfurt Auto Show and believed that reaching 1000 km driving distance with one refueling was achievable. Figure 6.36 shows a picture of the Toyota FCHV-adv car and the "70 MPa" is located on the rear left side of the car.

FIGURE 6.36
Toyota FCHV-adv car.

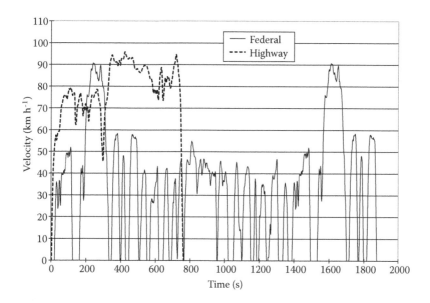

FIGURE 6.37
Federal and highway test procedures.

6.7 Vehicle Test Procedures

Figure 6.37 shows the Federal and the Highway test procedures, and Figure 6.38 shows the Japan 10-15 and the UN/ECE Elemental Urban test procedures. When reporting a driving distance the test procedure must be clearly stated, otherwise the results cannot be compared with each other.

6.8 Summary

In order to achieve a 10-second acceleration time from 0 to 100 km h^{-1} and a 5 km h^{-1} s^{-1} highway passing acceleration rate on flat roads, the power of a passenger car needs to be around 80 kW. When climbing hills of around 20°, about 85% of the power output is used to work against gravity.

With 6 kg of H$_2$ onboard, a car could be driven for around 1200 and 600 km at an uninterrupted cruising velocity of 30 and 110 km h^{-1}, respectively. The frequent stops and breaking that are characteristic of local driving dramatically reduce driving distances.

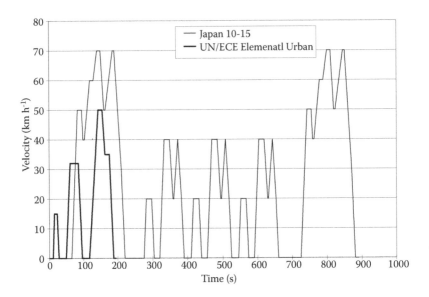

FIGURE 6.38
Japan 10-15 and UN/ECE elemental urban test procedures.

UTC Power announced on August 10, 2011 that one of its hybrid electrical transit buses powered by the PureMotion System Model 120 fuel cell power train surpassed 10,000 operating hours in a real-world driving situation with its original stacks and without cell replacement. This achievement indicates that fuel cell technology is probably able to meet the commercial needs. Nearly all major car manufacturers worldwide have announced that they plan to mass produce (larger than ca. 1000) fuel cell cars around 2015. Real-world driving has repeatedly shown that a passenger car can travel for more than 500 km per refueling, and refueling a car takes less than 5 minutes. Based on the data available, the US DOE estimated that by the end of 2011, the cost of an 80 kW fuel cell power train system would be $49 per kW if the annual production volume reached 500,000 vehicles. The Toyota FCHV-adv car can start at −30°C environmental temperature. All these facts indicate that after many years of development, fuel cell vehicles will be ready to enter the commercial stage soon.

A practical hurdle is not fuel cell technology, but the lack of widespread H_2 refueling stations. There are only around 300 such stations worldwide today, which is obviously far too few to support the commercial needs of fuel cell vehicles. Solving this issue requires close collaboration among governments (federal and local), car companies, power companies, and the public. It is believed that as the world crude oil reserve inches toward depletion, environmental issues are taken more seriously, and other renewable energy is more widely deployed, governments, companies, and the public will realize that fuel cells are the technology that must be used in vehicles.

7

Portable Power

7.1 Volumetric Energy Density of Various Fuels

The fuel cell is a strong contender as a power source for portable applications. Currently, the power source for portable applications is various batteries, such as Li-ion batteries, Ni-metal hydride batteries, Ni-Cd batteries, Zn-air batteries, and even Pb-acid batteries. Batteries are convenient to use, especially for low power loads (e.g., a few watts or less), but the time they can power the load is often quite short, and recharging of secondary batteries takes a few hours. A fuel cell has the potential to provide longer running time and with minimal refueling time.

The challenge is the fuel storage capacity of a fuel cell system. H_2 has the highest specific energy density, but the lowest volumetric energy density among all the fuels. The US Department of Energy (DOE) set the ultimate target for the volumetric H_2 storage system at 0.070 g cm^{-3} for light-duty vehicles. This can be used as a rough guide for portable applications. Table 7.1 lists the volumetric lower heating value (LHV) of H_2 with different storage means, or generated from hydrolysis of $NaBH_4$, or obtained by steam reforming methanol, and the volumetric LHV of liquid methanol, liquid ethanol, liquid octane (e.g., gasoline), gaseous CH_4, and C_3H_8. The results and the statements in Table 7.1 are only for the fuels themselves, without considering the volume of the containers and the auxiliary components. Clearly, even liquid H_2, whose density is 0.071 g cm^{-3}, can only barely meet the target. The target could only be met or exceeded by using hydrolysis of $NaBH_4$, steam reforming methanol, liquid methanol, liquid ethanol, and liquid gasoline. The H_2 volumetric LHV from MH_x, hydrolysis of $NaBH_4$, and steam reforming of methanol are estimated below.

If we assume MH_x itself stores 1.5% wt of releasable H_2 (many metal hydrides can store 2.0% wt of H_2, but about 0.5% cannot be released at about 1 barg) and its density is 5.0 g cm^{-3} (varying with types of metal hydride materials), then its H_2 storage capability is 0.075 g cm^{-3} (1.5% × 5.0 g cm^{-3}).

$NaBH_4$ hydrolysis is shown in Reaction 7.1.

$$NaBH_4 + 2H_2O = NaBO_2 + 4H_2 \qquad \text{(R. 7.1)}$$

TABLE 7.1

Volumetric Energy Density of Different Fuels

Fuel	LHV (kJ/mol)	Density (kg/L)	Molar Density (mole/L)	Volumetric LHV (kJ/L)	Ratio of LHV to DOE Target
DOE target – H_2	–241.8	7.00E-02	35.00	–8463.0	1.000
H_2 @ 1 bar (g)[a]	–241.8	8.08E-05	0.04	–9.8	0.001
H_2 @ 125 bar (g)[a]	–241.8	9.35E-03	4.68	–1130.4	0.134
H_2 @ 350 bar (g)[a]	–241.8	2.32E-02	11.60	–2804.9	0.331
H_2 @ 700 bar (g)[a]	–241.8	3.90E-02	19.50	–4715.1	0.557
LH_2 (l) – H_2	–241.8	7.10E-02	35.50	–8583.9	1.014
MH_x (s) – H_2[b]	–241.8	7.50E-02	37.50	–9067.5	1.071
$NaBH_4$ hydrolysis (aq.) – H_2[c]	–241.8	9.92E-02	49.60	–11993.3	1.417
MeOH steam reforming (l) – H_2[d]	–241.8	1.02E-01	51.00	–12331.8	1.457
CH_3OH (l)	–638.1	7.90E-01	24.69	–15753.1	1.861
C_2H_5OH (l)	–1234.0	7.87E-01	17.11	–21112.1	2.495
CH_4 @ 1 bar (g)	–802.8	6.56E-04	0.04	–29.3	0.003
C_3H_8 @ 1 bar (g)	–2043.2	1.91E-03	0.04	–88.7	0.010
C_8H_{18} (l)	–5074.5	7.03E-01	6.17	–31292.8	3.698

[a] H_2 compressibility factors are 1, 1.08, 1.22, and 1.45 at 1, 125, 350, and 700 bars, respectively.
[b] Assuming that MH_x itself stores 1.5% wt releasable H_2 and its density is 5.0 g cm^{-3}.
[c] Assuming that the $NaBH_4$ solution contains about 30% wt $NaBH_4$, the density of its aqueous solution is 1.43 g cm^{-3}.
[d] Assuming that MeOH steam reforming is carried out with equal moles of MeOH and H_2O.

Even though it is a solid, the density of $NaBH_4$ is only 1.07 g cm^{-3}. If we assume that the density of an aqueous solution typically containing 30% wt of $NaBH_4$ is 1.43 g cm^{-3} without considering the volume change of water when $NaBH_4$ is added, 1 cm^3 of such a solution contains about 0.43 g $NaBH_4$ and 1.0 g H_2O; 0.41 g (0.43g × 36$_{H_2O}$/37.8$_{NaBH_4}$) H_2O will react with $NaBH_4$ at a 2:1 stoichiometric ratio to produce H_2, generating 0.046 g (0.43 × 4/37.8) and 0.046 g (0.41 × 2/18) of H_2 from $NaBH_4$ and H_2O, respectively. Therefore, the H_2 volumetric density is 0.092 g cm^{-3}. At 25°C, the solubility of $NaBH_4$ in water is 0.55 g cm^{-3} (corresponding to about 35% wt of $NaBH_4$). So, the maximum amount of H_2 generated from such a solution could reach around 0.116 g cm^{-3} [(0.55 × 4/37.8) + (0.55 × 36/37.8) × 2/18].

Reaction 7.2 shows the steam reforming process of methanol.

$$CH_3OH + H_2O = CO_2 + 3H_2 \qquad \text{(R. 7.2)}$$

The reaction requires equal moles of methanol and water. One liter of methanol contains 24.69 moles of methanol (1 L × 1000 cm^3 L^{-1} × 0.79 g cm^{-3}/32 g mol^{-1}). The volume of water with corresponding moles will be 0.444 L [(24.69 mole × 18 g mol^{-1})/(1 g cm^{-3} × 1000 cm^3 L^{-1})]. So, the total volume of methanol plus water is 1.444 L (assuming the volumes of methanol and water are additive). The amount of H_2 produced is 148.1 g (24.69 mole × 3 moles H_2 per mole of CH_3OH × 2 g mol^{-1}). Therefore, the volumetric H_2 density is 0.102 g cm^{-3} (148.1 g/1444 cm^3).

Although liquid methanol, ethanol, and gasoline all possess volumetric energy densities that are 1.9, 2.5, and 3.7 times of that of the DOE target based on the LHV, only methanol can be directly oxidized in a fuel cell. In conjunction with other advantages of methanol, such as being biodegradable and easy to produce, carry, and transport, a direct methanol fuel cell (DMFC) becomes the mainstream choice for portable applications.

Table 7.2 lists the major parameters of methanol. For liquid methanol, when the product water is in the gaseous form, its LHV_{liq} is −638.1 kJ mol^{-1}, or 4.35 Wh cm^{-3}, or 5.53 Wh g^{-1}. The volumetric energy density is much higher than that of H_2 at ambient pressures. This LHV value is derived from the higher heating value (HHV) when both the methanol and water are in the liquid form. The HHV when both the methanol and water are in the liquid form is −726.1 kJ mol^{-1}, and the enthalpy of evaporation of water is 44 kJ mol^{-1};

TABLE 7.2

Properties of Methanol

Parameter	Unit	Value
Chemical formula	—	CH_3OH
Density at 25°C	g cm^{-3}	0.7872
Molar mass	g mol^{-1}	32.04
Molarity	mol l^{-1}	24.7
Specific heat capacity at 25°C	J g^{-1} K^{-1}	2.53
Thermal conductivity at 25°C	W m^{-1} K^{-1}	0.20
Enthalpy of combustion at 25°C (both MeOH and product water in liquid form)	kJ mol^{-1}	−726.1
Higher heating value (both MeOH and product water in liquid form)	kJ mol^{-1}	−726.1
	J g^{-1}	−22,662
	Wh cm^{-3}	−4.96
	Wh g^{-1}	−6.29
Lower heating value (MeOH in liquid form and product water in vapor form)	kJ mol^{-1}	−638.1
	J g^{-1}	−19,916
	Wh cm^{-3}	−4.35
	Wh g^{-1}	−5.53
Vapor pressure at 25°C	kPa	17
Autoignition temperature	°C	385
Flash point temperature	°C	11
Flammable range in air by volume	%	6–37
Boiling point	°C	65
Enthalpy of vaporization at STP	kJ mol^{-1}	37.43
Melting point	°C	−98

therefore, the LHV with liquid methanol but gaseous water is −638.1 kJ mol^{-1} (−726.1 + 44x2) (combustion of 1 mole of methanol produces 2 moles of water). If methanol is in the gaseous form, its LHV$_{gas}$ will be around −675.5 kJ mol^{-1} (−638.1 kJ mol^{-1} minus the enthalpy of methanol evaporation, that is, 37.43 kJ mol^{-1}).

7.2 Direct Methanol Fuel Cells

In a DMFC, methanol is directly oxidized at the anode, as shown by Reaction 7.3. Each methanol molecule requires one water molecule for the reaction to proceed. Without water, the reaction cannot proceed, and this must be remembered when designing a DMFC system. The product is gaseous CO_2, and it must be vented out so it does not impede the diffusion of methanol to the anode catalyst layer. At the cathode, oxygen combines with protons and electrons to form water, as shown by Reaction 7.4. The overall reaction is shown by Reaction 7.5, where each methanol molecule reacts with 1.5 O_2 molecules to produce 1 CO_2 molecule and 2 H_2O molecules.

$$\text{Anode:} \quad CH_3OH + H_2O = CO_2 + 6H^+ + 6e^- \qquad \text{(R. 7.3)}$$

$$\text{Cathode:} \quad 1.5O_2 + 6H^+ + 6e^- = 3H_2O \qquad \text{(R. 7.4)}$$

$$\text{Overall:} \quad CH_3OH + 1.5O_2 = CO_2 + 2H_2O \qquad \text{(R. 7.5)}$$

When methanol and water are both in the gaseous form, $\Delta H°$ and $\Delta G°$ of Reaction 7.5 are −675.6 and −689 kJ per mole of methanol, respectively. Therefore, the thermodynamic voltage of the overall reaction under the standard condition (but both methanol and water are in the gaseous form) is

$$E° = -\Delta G°/(6F) = 689000/(6 \times 96485) = 1.19 \text{ V}$$

The voltage based on $\Delta H°$ is 1.17 V [(675600/(6/96485)], and this voltage is used to calculate fuel cell electrical efficiency.

A DMFC is quite similar to a proton exchange membrane fuel (PEMFC) in stack structure and components. They both use a PEM for transporting the protons and Pt-based catalysts at the cathode. The anode catalyst for a DMFC is typically a Pt-Ru alloy that has higher CO tolerance than Pt alone, and this is similar to the PEMFC when H_2 contains trace amounts of CO. In the intermediate steps during methanol oxidation, some CO-like species will form, which can seriously poison the anode catalyst. The presence of Ru helps the removal of the CO-like species from the Pt surface through Reaction 7.6.

$$Pt-CO_{ad} + Ru-OH_{ad} = Pt + Ru + CO_2 + H^+ + e^- \qquad \text{(R. 7.6)}$$

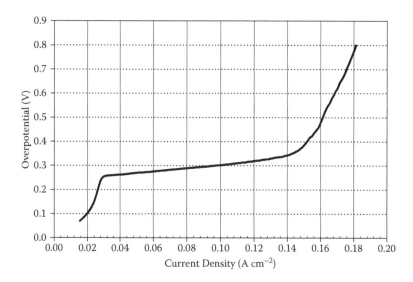

FIGURE 7.1
Methanol oxidation overpotential.

It is believed that Ru more easily (i.e., at lower potentials) forms an –OH group on its surface, which in turn oxidizes the –CO group adsorbed on the neighboring Pt surface.

Due to the slow oxidation kinetics, methanol will hardly be oxidized with a reasonable reaction rate at potentials lower than about 0.25 V. This can be seen in Figure 7.1, which shows the methanol oxidation process. Once the voltage was higher than 0.25 V, the oxidation current increased significantly. When the current density went beyond about 0.14 A cm⁻², a steep overpotential increase started due to the mass transport resistance for this particular experiment. For current densities between 0.03 and 0.14 A cm⁻², the overpotential versus the current density was linear, primarily due to the iR loss.

Because of the significant activation losses from both the methanol oxidation reaction (around 0.25 V loss) and the oxygen reduction reaction (around 0.25 V loss), the cell voltage drops to around 0.65 V when a DMFC generates a very small current density. For a working DMFC operated under 100°C, its voltage is normally kept around 0.40 V, which means that the electrical efficiency is about 33% with the remaining 67% as heat.

Due to methanol crossover the OCV of a DMFC can hardly be higher than 0.70 V. The methanol crossover current density is 100~200 mA cm⁻² depending on the condition, and these experimental results are in good agreement those estimated according to methanol diffusion (see below). At such a crossover current density the cathode voltage losses due to both methanol oxidation reaction (MOR) and oxygen reduction reaction (ORR) are significant. Those losses can be estimated by using Equation 2.22 in Chapter 2 on R. 7.3 and R. 7.4 for MOR and ORR respectively, and the summation will be the total losses at OCV for a DMFC. Please refer to Table 2.5 in Chapter 2 that shows the ORR losses at different H_2 crossover current densities with low Pt loading electrode (0.5 mg cm^{-2}). The Pt loading at the cathode of a DMFC is about 4 to 8 times higher than 0.5 mg cm^{-2} and thus the ORR voltage loss will be smaller than that shown in Table 2.5 (please refer to Table 2.4). In order to reduce the impact of methanol crossover to the cathode (i.e., making the cathode catalyst inactive to methanol oxidation), methanol tolerant catalysts that are still possess high activity toward ORR are widely investigated. These catalysts are normally Pt-based alloys such as PtPd.

Fick's first law of diffusion (Equation 2.28) can again be used to estimate the limiting current densities arising from methanol crossover at open circuit. We choose four bulk methanol concentrations in water, 0.5, 1.0, 1.5, and 2.0 M. We know that a fully hydrated Nafion PEM with equivalent 1100 contains 22 water molecules per $-SO_3H$ group and the concentration of $-SO_3H$ is 1.8 M, which means that the PEM has a water concentration of 39.6 M (22×1.8 M) when fully saturated. Assuming methanol and water will be taken into the PEM by the $-SO_3H$ groups based on their relative bulk concentrations, the methanol concentrations within the PEM will be ca. 0.36, 0.73, 1.11 and 1.49 M from the solutions with bulk methanol concentrations of 0.5, 1.0, 1.5, and 2.0 M, respectively (the molar mass is 32 and 18, and the density is 0.79 and 1.0 for neat methanol and water, respectively). The diffusion coefficient of methanol in Nafion is 4.9×10^{-6} cm^2 s^{-1} at 60°C (Kauranen 1996). Within those numbers in hand and assuming all the methanol arriving at the cathode side is oxidized (e.g., its concentration at the cathode/PEM interface is 0), the limiting methanol crossover flux and thus the current density can be calculated using Equations 2.28 and 2.30. The results are shown in Table 7.3. It can be seen that even with the thickest Nafion membrane, the methanol limiting crossover current densities range from 58 to 242 mA cm^{-2} when the bulk methanol concentration changes from 0.5 to 2.0 M. Those crossover current

TABLE 7.3

Methanol Limiting Crossover Flux and Current Density at Open Circuit through PEM with Different Thicknesses

PEM Thickness (μm)	MeOH Crossover Flux (mmol cm^{-2} s^{-1})				MeOH Crossover Current Density (mA cm^{-2})			
Bulk MeOH Conc. (M)	0.5M	1.0M	1.5M	2.0M	0.5M	1.0M	1.5M	2.0M
MeOH Conc. in PEM (M)	0.36	0.73	1.11	1.49	0.36	0.73	1.11	1.49
25	0.00071	0.00143	0.00217	0.00293	409	827	1256	1694
50	0.00035	0.00071	0.00108	0.00146	204	414	628	847
87	0.00020	0.00041	0.00062	0.00084	118	238	361	487
125	0.00014	0.00029	0.00043	0.00059	82	165	251	339
175	0.00010	0.00020	0.00031	0.00042	58	118	179	242

densities are significant considering the fact that a DMFC working current density is typically lower than 200 mA cm^{-2}. Besides PEM thickness, the temperature also significantly affects the crossover current density, higher temperatures lead to larger crossover current densities. During the operation of a DMFC, since the methanol concentrations at the anode/PEM interface are much lower than those shown in Table 7.3 due to the consumption of methanol within the anode, the actual methanol crossover current densities will be much lower than those shown in Table 7.3, but they are still significant with at least 20% methanol being lost. The actual methanol loss can be gauged by measuring the amount of CO_2 in the gas phase and the amount of methanol in the liquid phase (after condensation) coming out of the cathode with the exhaust. It is possible that not all the methanol is oxidized to CO_2 at the cathode, especially when the methanol crossover rate is high.

7.3 Conventional DMFC System Architecture

Figure 7.2 shows a functional block diagram of a conventional DFMC system. The regular arrows and the bold arrows represent the mechanical (or fluid) connections and the electrical connections, respectively. A pump sends methanol aqueous solution, normally 0.5 to 1.5 M (1 M methanol solution contains about 3% wt of methanol, and the remaining 97% is water), to the stack from a methanol mixing tank; the anode exhaust, a mixture of methanol, CO_2, and water, is pumped back to the methanol mixing tank where the

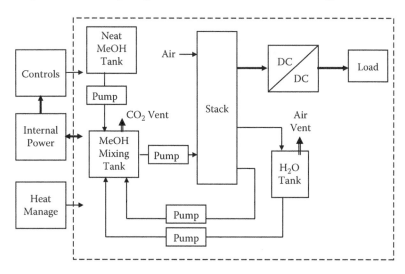

FIGURE 7.2
Functional block diagram of a conventional DMFC system.

gaseous product CO_2 is vented out as much as possible. Without venting the CO_2 out, its pressure (or concentration) will increase with time, impeding methanol diffusion to the catalyst layer, which then results in lower fuel cell performance. The cathode exhaust, which contains water and unreacted air, goes through a water–gas separator, and the resulting liquid water is sent to the methanol mixing tank. If enough water is collected, the fuel cell system will achieve a water balance to eliminate the otherwise periodic addition of water. Reaction 7.3 shows that each methanol molecule consumes one water molecule, and Reaction 7.4 shows that three water molecules are produced at the cathode side, so if one third of the water produced at the cathode is recovered, the system will be able to achieve a water balance. There is a methanol concentration sensor in the methanol mixing tank, and when the methanol concentration is lower than a preset value, neat methanol will be sent to the mixing tank. Air can get into the cathode of the stack either through natural breathing or via a device such as a pump or a blower. The stack output is regulated by a DC–DC converter to power the load. There is an internal power supply source, such as a battery or a supercapacitor; it starts the DMFC system and provides additional power to the load when there is a power need surge. The double-sided arrow means that it can provide the power needs for the fuel cell system and that it may be charged by the fuel cell system (e.g., from the DC–DC converter). The temperature control of the stack and other devices is achieved through a heat management module. The operation of the entire system is coordinated by a control module.

The major drawbacks of such a conventional DMFC system include lower energy density, higher system complexity, and cathode flooding. The mixing tank and the water recovery tank increase the system volume. In order to minimize their impact on the system volumetric energy density, they should be made as small as possible. The system needs about 4–5 pumps: one pump sends neat methanol to the methanol mixing tank, one pump sends methanol solution to the stack, one pump recirculates the anode exhaust to the methanol mixing tank, one pump sends the recovered water from the cathode exhaust to the methanol mixing tank, and potentially one more pump sends air into the stack. Every pump needs to be properly controlled, increasing the controls, the system complexities, and the system cost. For such a DMFC system configuration, a large amount of water will diffuse through the PEM to the cathode side because of the extremely high water concentration in the anode fuel mixture, leading to serious flooding of the cathode. Sending air via natural breathing to the cathode cannot handle such an amount of water at the cathode; therefore, forced air is needed via a pump, blower, or a compressor. Also, the air stoichiometric ratio must typically be higher than 3 to alleviate the seriousness of the cathode flooding, which further increases parasitic power losses. Of course, air should not be humidified before entering the stack. Further, a significant amount of methanol transports through the PEM to the cathode via diffusion and electroosmotic drag by protons, which is called the methanol crossover. The crossover causes significant fuel loss and lowers the cathode voltage by forming a mixed potential.

7.4 DMFC System Using Neat Methanol by MTI Micro Fuel Cells

7.4.1 How to Make It Work

In order to overcome those drawbacks, MTI Micro Fuel Cells Inc. developed a much simpler system whose functional block diagram is shown in Figure 7.3. Neat methanol is directly sent to the stack without using a methanol mixing tank, and the anode and the cathode exhaust recirculation loops are eliminated. Such a system needs a maximum of two pumps, one for sending neat methanol to the stack and the other for sending air to the stack. Since neat methanol has a detrimental impact on the PEM, a methanol diffusion film (MDF) is used to help control the rate of neat methanol reaching the catalyst layer. Methanol is nearly completely reacted (i.e., oxidized) within the anode catalyst layer when the reaction rate is as fast as the methanol diffusion rate, and therefore, little methanol would touch the PEM. Experimental results have shown that the methanol crossover rate through the PEM could be controlled at around 5%, far less than that when a dilute methanol solution is used as in the conventional system setup shown in Figure 7.2. Due to the absence of water in the neat methanol fuel, cathode flooding is not a problem. Therefore, the MTI system not only lowers the system cost and complexity, and increases the fuel utilization, cell efficiency, and system energy density, but also enhances the system's durability and lifetime.

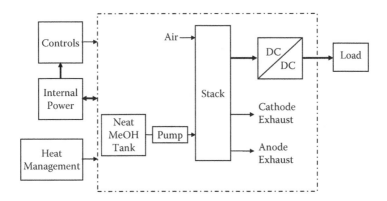

FIGURE 7.3
Functional block diagram of an MTI DMFC system.

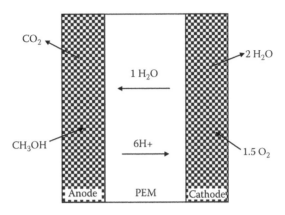

FIGURE 7.4
Schematic representation of ideal water distribution in a neat methanol DMFC.

Air is not externally humidified for the MTI system either. In other words, unhumidified ambient air is sent to the fuel cell stack. We know that without adequate humidification the cathode reaction will be impaired, the resistance of the PEM will be higher, and the anode reaction cannot proceed without water, as shown in Reaction 7.3. All those water needs are met by effectively using the water produced at the cathode, as shown in Figure 7.4. Reaction 7.3 shows that each methanol molecule needs one water molecule for the reaction to occur, while Reaction 7.4 shows that three water molecules are produced at the cathode for each methanol molecule to react at the anode. So, if one water molecule diffuses through the PEM to the anode side, it will be able to support the anode reaction, leaving two excess water molecules, as shown by Reaction 7.5, to be taken out by the air exhaust. If the incoming unhumidified ambient air needs to be humidified, it can be achieved by using the hot and humid exhaust air through a membrane exchange process.

Clearly, it is important for enough water to diffuse through the PEM from the cathode to the anode for a DMFC using neat methanol to work. Since a thinner PEM will facilitate water diffusion, we can estimate what the PEM thickness should be for the anode to get enough water through diffusion. Table 7.4 shows the water production rate at the cathode and the required water flux for the anode; the latter is one third of the former according to Reactions 7.3 and 7.4. The water diffusion rate through a Nafion-type PEM with different thicknesses is shown in Table 2.10 of Chapter 2. The water flux through 175 μm Nafion is 0.0059 mmol cm^{-2} s^{-1}. Compared with the anode water needs in Table 7.4, such a PEM can meet the needs of a neat methanol DMFC at current density of 3.4 A cm^{-2}. The water flux through a thinner membrane is even higher. A DMFC is rarely operated at current densities higher than 0.5 A cm^{-2} due to the high overpotentials, so all the commercial PEMs with thicknesses ranging from 25 to 175 μm can easily meet the anode water needs of a neat methanol DMFC.

TABLE 7.4

Water and Air Situations of a DMFC at Different Current Densities (assuming that one third of the water produced at the cathode diffuses to the anode to support the anode reaction)

Current Density ($A\ cm^{-2}$)	H_2O Produced (mmol $cm^{-2}\ s^{-1}$)	H_2O Needed by Anode (mmol $cm^{-2}\ s^{-1}$)	H_2O Needed Taken Out (mmol $cm^{-2}\ s^{-1}$)	O_2 Needed at 1 Stoich. Ratio (mmol $cm^{-2}\ s^{-1}$)	Air Needed at 1 Stoich. Ratio (mmol $cm^{-2}\ s^{-1}$)	Air Remaining at 1 Stoich. (mmol $cm^{-2}\ s^{-1}$)	Vapor% in Cathode Exhaust
0.05	0.00026	0.00009	0.00017	0.00013	0.00062	0.00049	26.2
0.1	0.00052	0.00017	0.00035	0.00026	0.00123	0.00097	26.2
0.15	0.00078	0.00026	0.00052	0.00039	0.00185	0.00146	26.2
0.2	0.00104	0.00035	0.00069	0.00052	0.00247	0.00195	26.2
0.25	0.00130	0.00043	0.00086	0.00065	0.00308	0.00244	26.2
0.3	0.00155	0.00052	0.00104	0.00078	0.00370	0.00292	26.2
0.4	0.00207	0.00069	0.00138	0.00104	0.00494	0.00390	26.2
0.5	0.00259	0.00086	0.00173	0.00130	0.00617	0.00487	26.2
0.6	0.00311	0.00104	0.00207	0.00155	0.00740	0.00585	26.2
0.7	0.00363	0.00121	0.00242	0.00181	0.00864	0.00682	26.2
0.8	0.00415	0.00138	0.00276	0.00207	0.00987	0.00780	26.2
0.9	0.00466	0.00155	0.00311	0.00233	0.01110	0.00877	26.2
1	0.00518	0.00173	0.00345	0.00259	0.01234	0.00975	26.2
1.5	0.00777	0.00259	0.00518	0.00389	0.01851	0.01462	26.2
2	0.01036	0.00345	0.00691	0.00518	0.02468	0.01949	26.2
2.5	0.01296	0.00432	0.00864	0.00648	0.03085	0.02437	26.2
3	0.01555	0.00518	0.01036	0.00777	0.03702	0.02924	26.2
3.4	0.01762	0.00587	0.01175	0.00881	0.04195	0.03314	26.2

Table 7.4 also shows the air situation at 1 times the stoichiometric ratio. Referring to Table 2.2 in Chapter 2 for the water % (vapor volume in air) at saturated vapor pressure at different temperatures, the two net water molecules from the oxidation of each methanol molecule can be taken out by the remaining air, even at 1 times the stoichiometric ratio if a neat methanol DMFC works at 67°C or higher temperatures. Then, the power consumption by the air supply device can be very low, in distinct contrast with conventional DMFCs that need more than 3 times the air stoichiometric ratios. For a working fuel cell, the air stoichiometric ratio is always kept higher than 1 in order to supply excess O_2 to the cathode, so for a neat methanol DMFC operated at 67°C or higher temperatures, cathode flooding should not happen, and the air exiting the cathode may not be fully saturated. At such temperatures, methanol will be in the vapor form because the boiling point of methanol is 65°C. If the DMFC operates at 57°C and 47°C, the air stoichiometric ratio should be around 2 and 3, respectively. From this simple analysis, a neat

methanol DMFC should operate at a temperature that is higher than 65°C due to considerations of both faster reaction kinetics and easy air supply.

7.4.2 DMFC Charger

Figure 7.5 shows a schematic of an MTI DMFC system as a charger for portable electronics. The cartridge is hot-swappable, that is, the system can continue supplying power while the cartridge is being replaced. The replacement only takes a few seconds. During the replacement process, the internal battery will supply power to the load. The system operates at a temperature higher than 70°C, and thus it is the methanol vapor that gets into the anode catalyst layer.

Table 7.5 lists the major parameters of such a DMFC system. The total volume of the system is around 200 cm^3, like a small pocket booklet, for ease of carrying. It uses neat methanol, and the hot-swappable methanol cartridge can provide about 30 Wh net power. It can provide power instantly, with its output voltage, current, and power up to 5 V, 1 A, and 5 W, respectively. It can operate at environment temperatures from 0°C to 40°C using ambient air.

Experiments have shown that Nafion membranes with a thickness ranging from 25 to 175 μm can all be used in the MTI DMFCs. This is in consistent with the water flux analysis mentioned earlier. Figure 7.6 shows the V-I and W-I curves of a single cell using a 50 μm Nafion membrane during a 6001-hour test. During the first 4998 hours of the test, the methanol flow rate was set at 0.38 cm^3 h^{-1}, corresponding to a limiting current of 1.51 A if all the methanol participated in the reaction. The limiting currents during the test

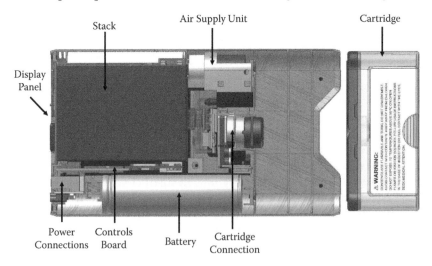

FIGURE 7.5

Schematic of an MTI DMFC system. Courtesy of MTI Micro Fuel Cells.

TABLE 7.5

Major Parameters of the MTI DMFC System

Parameter	Unit	Value
System volume	cm³	~200
Startup time	ms	Instant
Fuel	—	Neat methanol
Methanol concentration	M	24.7
Output power	W	Up to 5
Output voltage	V	Up to 5
Output current	A	Up to 1
Cartridge replacement	—	Hot swappable
Energy provided by 1 cartridge	Wh	~30
Operating temperature range	°C	0–40
Oxidant	—	Ambient air

Source: Courtesy of MTI Micro Fuel Cells.

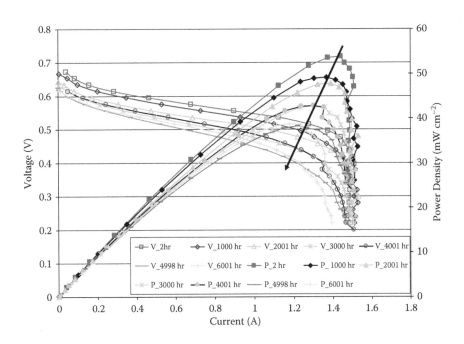

FIGURE 7.6

Performance change during 6001 hours of single cell test using a 50 μm PFSA membrane. Courtesy of MTI Micro Fuel Cells.

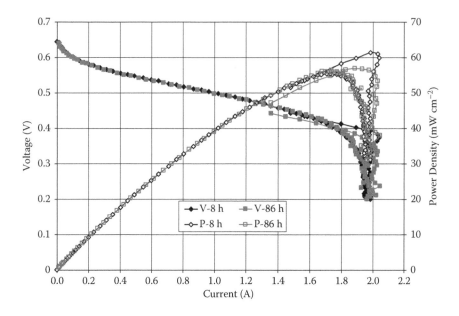

FIGURE 7.7
Performance change during 86 hours of single cell test using a 175 μm PFSA membrane. Courtesy of MTI Micro Fuel Cells.

did appear at about 1.51 A. Afterward, the methanol flow rate was decreased to 0.35 cm³ h⁻¹, corresponding to a theoretical limiting current of 1.39 A. The limiting current from the test result at the 6001-hour mark did show similar current. The fuel cell performance decayed with time, with the V-I curves shifting downward in a nearly parallel fashion in the iR loss region, indicating that the decay was dominated by the slowdown of the reaction kinetics, most likely caused by the decay of both the anode and the cathode catalyst layers. The peak power output decreased from about 55 to 38 mW cm⁻² in 6001 hours, with the peak power position shifting toward lower current with time.

Figure 7.7 shows the V-I and W-I curves of a single cell using a 175 μm Nafion membrane during 86 hours of testing. The methanol flow rate was set at 0.50 cm³ h⁻¹, corresponding to a theoretical limiting current of 1.99 A. The limiting current obtained during the test was very close to this current, indicating again that all the methanol could be oxidized at the limiting current with the test setup. The cell showed no decay during the 86 hours of testing.

The limiting current density and the peak power density are determined by the neat methanol feed rate, as shown in Figure 7.8. When the neat methanol flow rate was doubled, the peak power density nearly doubled, from 47 to 85 mW cm⁻². In the lower current density region (e.g., far away from the limiting current density region), the cell performance was lower with a higher methanol feed rate because higher unreacted methanol concentrations led to higher methanol crossover rates, which in turn lowered the cathode potential.

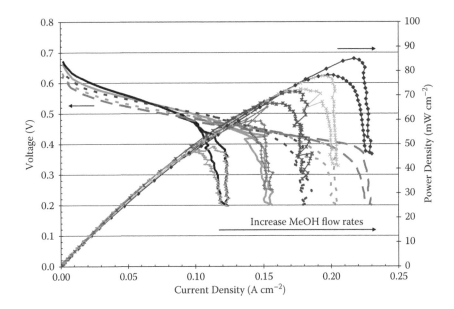

FIGURE 7.8
Performance versus neat methanol feed rate. Courtesy of MTI Micro Fuel Cells.

In practical operation, the current cannot be set at the limiting current. Instead, it should be set at a current slightly lower than the limiting current, and as the cell decays with time, the current set point should be decreased accordingly. The ratio of the working current to the limiting current represents the methanol utilization. Part of the unused methanol diffuses to the cathode side and the other part is taken out by the CO_2 exhaust at the anode. For automatic operation, it is better to automatically trace the peak power and allow the fuel cell to work near the peak power.

MTI DMFCs also showed good durability in stacks. Figure 7.9 shows the average unit cell output power change in about 2700 hours. In the first 1800 hours, the average cell voltage was set at 0.45 V, and afterward it was set at 0.425 V to accommodate the downward shift of the V-I curves. The average cell decay rate was 4.3% per 1000 hours.

7.5 DMFC Systems Developed by SFC Energy

SFC Energy AG (formerly Smart Fuel Cell) is a world leader in developing DMFC systems as battery chargers for various applications. It has achieved sales of more than 25,000 units, which is one of the world's highest fuel cell sales figures. Its product portfolio includes the EFOY 600, 900, 1200, 1600,

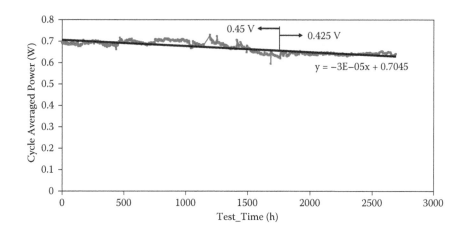

FIGURE 7.9
Unit cell performance change in a stack during 2700 hours of testing (50 μm PFSA membrane). Courtesy of MTI Micro Fuel Cells.

and 2200 with output power of 25, 38, 50, 65, and 90 W (between 2006 and 2011), respectively; the EFOY Comfort 80, 140, and 210 with output power of 40, 72, and 105 W (since May 2011), respectively; and the EFOY Pro 600, 1600, 2200, and 2200XT with output power of 25, 60, 90, and 90 W, respectively. Figure 7.10 shows the EFOY 1600 DMFC system.

The DMFC systems developed by SFC Energy bear great similarities. The specifications of the EFOY Pro DMFC systems are listed in Table 7.6. These systems are used to charge 12 or 24 V Pb-acid batteries. The nominal

FIGURE 7.10
Picture of EFOY 1600 DMFC system. Courtesy of SFC Energy.

TABLE 7.6

Specifications of SFC Energy EFOY Pro DMFC Systems

Parameter		EFOY Pro 600/1600/2200/2200XT
Nominal power output (W)		25/65/90/90
Nominal voltage output (V)		12 or 24
Nominal methanol consumption (L kWh^{-1})		0.9
Weight (kg)		6.8/8.3/8.6/9.0
Dimension L × W × D (mm)		433 × 188 × 278
Noise level at 1 m/7 m distance (dB)		39/23
Operating temperature (°C)		−20–45
Startup temperature (°C)		5–45
Storage temperature (°C)		1–45
Altitude (m)		Up to 1500
Automatic charging-on threshold (V)	12 V Battery	<12.3
	24 V Battery	<24.6
Automatic charging-off threshold (V)	12 V Battery	>14.2
	24 V Battery	>28.4
Nominal charging current (A)	12 V Battery	2.1/5.4/7.5/7.5
	24 V Battery	1.05/2.7/3.75/3.75
Recommended battery capacity (Ah)	12 V Battery	10–100/40–250/60–350/60–350
	24 V Battery	5–50/20–120/30–175/30–175
User interface		With text display at the unit or via remote control
Data interface		RJ-45 plug for accessories
Warranty	Month	24/24/24/24
	Hour	3000/3000/3000/4500
Fuel cartridge (M5/M10/M28)	Volume (L)	5/10/28
	Weight (kg)	4.3/8.4/22
	Dimension L × W × D (mm)	190 × 145 × 283/230 × 193 × 318/420 × 280 × 360
	Capacity (kWh)	5.5/11.1/31.1

Source: Adapted from SFC Energy website, Data Sheet: EFOY Pro Series Fuel Cells USA/CAN, http://www.efoy-pro.com/sites/default/files/110512_data_sheet_industry_efoy_pro_us.pdf.

methanol consumption rate is about 0.9 L kWh^{-1}, which means that the overall electrical efficiency of the system is 25.5% [(1/0.9 L kWh^{-1})/4.35 kWh L^{-1}]. Since the average cell voltage of a DMFC can hardly be higher than 0.45 V during operation, if we assume that the fuel utilization is 90%, the DC–DC converting loss is 10%, and the system's parasitic power loss is 10%, then the overall efficiency of a DMFC system will reach 28% [(0.45/1.17) × 0.90 × 0.90 × 0.90]. Clearly, the DMFC systems developed by SFC Energy have among the highest electrical efficiency based on the current technology. Also, the 3000 to 4500 operational hours of warranty offered by SFC Energy is very impressive.

The EFOY Pro can be operated onsite, or locally with a remote controller, or remotely through an Internet connection. Through an interface adaptor, the data interface on the fuel cell system can be connected to a computer. Adding a modem, a remote computer can also be connected, making remote controlling, diagnostics, programming, software updating, monitoring, and operation available. After an EFOY system is connected to both the battery to be charged and the fuel cartridge, turning on the On/Off switch on the fuel cell system will put the fuel cell system into an automatic operation mode. When the voltage of the battery decreases below 12.3 or 24.6 V for 12 or 24 V batteries, the fuel cell system will automatically start charging the batteries until their voltages reach 14.2 or 28.4 V, respectively. Afterward, the fuel cell system goes into standby mode until the next charging need is detected. It takes about 20 minutes for the system to reach its nominal output power from a cold start. Once charging is started, the system is programmed to run for at least 30 minutes to protect the fuel cell components. If an operator tries to turn off the system before it runs for 30 minutes, the system will keep running until 30 minutes is reached. When the ambient temperature drops to near 0°C while the system is in standby mode, the system will automatically start the antifreeze mode, provided both the batteries and the fuel cartridge are connected to the fuel cell system. In this mode, the fuel cell will run to generate heat and electricity, and the heat will keep the fuel cell above freezing temperatures. SFC Energy estimates that continuously running in the antifreeze mode for 5 months will consume about 10 L of methanol under central European winter weather conditions, which implies that (without considering the battery) the average output power of the stack in the antifreeze mode is nearly 12 W [(10 L × 4.35 kWh L^{-1} × 1000 W kW^{-1})/(5 months × 30.5 day $month^{-1}$ × 24 h day^{-1})].

When selecting a fuel cell system, whether or not its power output should be compatible with the final load depends on the situation. If the fuel cell is only used to charge a battery, and the total operation time depends on the capacity of the battery, the fuel cell power output can be lower than the load. In contrast, if the fuel cell is not only for charging the battery, but also for supporting the load when the battery runs out, then the fuel cell output power cannot be lower than the load. The reason is that for long-time continuous operation, the battery will finally become an energy relay after its energy runs up, and the power it sends to the load cannot be more than that it receives from the fuel cell. For example, for an EFOY Pro 600 system, the nominal power output of the system is 25 W. SFC Energy recommends that the capacity of the 12 V Pb-acid battery be 10 to 100 Ah. If the battery discharges at 0.1 C, the discharging current will be 1 to 10 A, and the battery's output power is around 12 to 120 W, respectively. In other words, at 0.1 C discharge rate, the 100 Ah battery bank can power a load of 120 W. However, since the fuel cell can only supply 25 W output power to the battery, a 120 W load can only be continuously powered by the combination of a 25 W fuel cell and a 12 V/100 Ah battery for about 12–13 hours (nearly 10 hours from

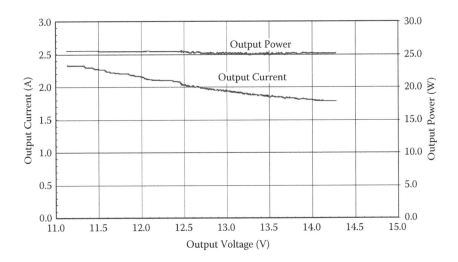

FIGURE 7.11
Voltage versus power and current for the SFC Energy EFOY 600 DMFC system. Courtesy of SFC Energy.

the battery and 2~3 hours from the fuel cell). In such a situation, the size of the methanol tank basically becomes irrelevant. However, if the load is 12 W, a 12V/10Ah battery bank itself can power it for 10 hours, and the 25 W fuel cell system carrying 5 liter neat methanol with 25% electrical efficiency can power it for 453 hours (25% x 4.35 kWh L^{-1} x 5 L x 1000 W kW^{-1}/12 W). Therefore, in order to utilize the long running time advantage of a fuel cell, its rated output power should meet or exceed the power need of the load.

Figure 7.11 shows the voltage versus power and current of the 25 W EFOY 600 DMFC system. Its output current declines with an increase in the output voltage, and the output power stays above 25 W with little change. Based on this figure, it appears that the SFC EFOY system operates in a constant power mode. The battery voltage was lower in the beginning of the charging process, and thus the fuel cell system generated higher output currents. The charging current decreased as the battery voltage increased when the charging proceeded. Such a charging strategy can facilitate the charging process without causing damage to the battery to be charged.

7.6 Summary

The consumer market offers the largest opportunity for portable power sources. Both the functionalities and the power requirements of portable electronics are increasing as more and more people use and depend on portable electronics, and the mobile workforce intends to work on the go, but batteries have not been able to keep up with this trend. There is a good opportunity for fuel cells to fill in the power gap because they enable quick refueling without plugging into a power outlet and enable electronics to run much longer than batteries per refueling. Unfortunately, technical challenges associated with H_2 storage, such as low H_2 volumetric storage capacity, make H_2 fuel cells a difficult choice for portable power sources.

It is widely believed that the DMFC will be among the fuel cells to be first commercialized. Liquid methanol is easy to make, transport, store, and refuel; it has a high volumetric energy density (4.35 Wh cm^{-3} based on LHV) and is biodegradable; it has also achieved regulatory acceptance in many countries.

Compared to conventional DMFCs, which feed the anode with a dilute aqueous methanol solution, feeding neat methanol directly to the anode offers additional advantages. The latter system does not require a water storage tank, a methanol solution mixing tank, a methanol concentration sensor, a methanol recirculation loop, and a cathode exhaust water condensation loop. It can use ambient air with a low air stoichiometric ratio slightly higher than 1 without incurring cathode flooding issues, it can achieve a methanol utilization rate higher than 80%, and it has demonstrated an acceptable performance decay rate of around 4% per 1000 operating hours.

However, DMFCs do suffer some drawbacks such as lower electrical efficiency and higher catalyst loadings as compared to H_2 fuel cells. Efforts should be continued on developing anode catalysts with improved methanol oxidation kinetics, cathode catalysts with a high tolerance to methanol, membranes with lower methanol permeation rates, and strategies to reduce the methanol crossover rate. Attention should also be given to other direct-feed fuel cells using other liquid fuels (such as formic acid).

8

Perspectives

8.1 Status and Targets

8.1.1 DOE Funding

Due to the strategic importance of fuel cells, some countries such as the USA and Japan have set up specific development plans. Although the plans need periodical modifications based on the actual developmental status, they provide clear guidance and targets, and make the R&D efforts systematic and focused. Table 8.1 and Figures 8.1 and 8.2 show the funding situation by the Fuel Cell Technologies Office of the US Department of Energy (DOE) from 2006 to 2013 (2013 is a request). The major funded areas include fuel cell systems R&D, hydrogen fuel R&D, technology validation, safety, codes & standards, system analysis, manufacturing R&D, market transformation, and education. The first three areas account for more than ~80% of the total funding. The funding increased significantly from 2006 to 2007, peaked in 2008, and dramatically decreased from 2010 to 2011. Figure 8.3 shows funding distribution to sub-areas under the fuel cell systems R&D and the hydrogen fuel R&D in 2008 and 2009. Within the fuel cell systems R&D, most of the funding has been used for the development of stacks (and their key components). Within H_2 R&D, more funding has been directed to the H_2 storage over H_2 production & delivery. The reasons may be that these areas are most challenging and they are the major bottlenecks for commercializing fuel cells.

There is a lot of valuable information on the DOE websites (http://www1.eere.energy.gov/hydrogenandfuelcells/ and http://www.hydrogen.energy.gov/), and it is strongly suggested that interested readers check these websites from time to time. In the past few years, information updating appeared to be not as frequent as in the previous years, but it is still the best place to learn about the overall situation of fuel cells.

About 10 years ago the DOE set up technical targets for various fuel cell applications such as transportation, distributed power, portable power, and auxiliary power. Some of the targets have been changed since then based on the developmental status. The most recent updates in various areas can be found at this DOE website: http://www1.eere.energy.gov/hydrogenandfuelcells/mypp/pdfs/fuel_cells.pdf.

TABLE 8.1

Yearly Total Funding from the Fuel Cell Technologies Office of the US DOE
($ in thousands)

R&D Areas	2006	2007	2008	2009	2010	2011	2012	2013 (Request)
FC System R&D	33336	55633	60419	80068	75609	41916	43556	38000
Hydrogen Fuel R&D	34431	67430	80978	67823	45750	32122	33785	27000
Technology Validation	33301	39413	29612	14789	13005	8988	8987	5000
Market Transformation	0	0	0	4747	15005	0	3000	0
Safety, Codes & Standards	4595	1349	15442	12238	8653	6901	6893	5000
Education	495	1978	3865	4200	2000	0	0	0
System Analysis	4787	9637	11099	7520	5408	3000	2925	3000
Manufacturing R&D	01928	4826	4480	4867	2920	1941	2000	
Total Funding	**110945**	**189511**	**206241**	**195865**	**170297**	**95847**	**101087**	**80000**

Source: US DOE, http://www1.eere.energy.gov/hydrogenandfuelcells/budget.html

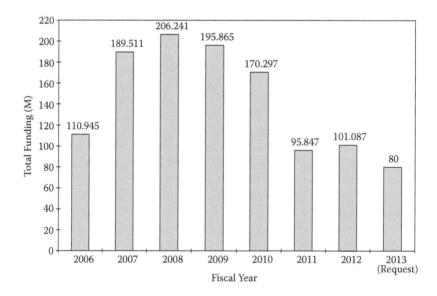

FIGURE 8.1

Yearly Total Funding from the Fuel Cell Technologies Office of the US DOE from 2006 to 2013.

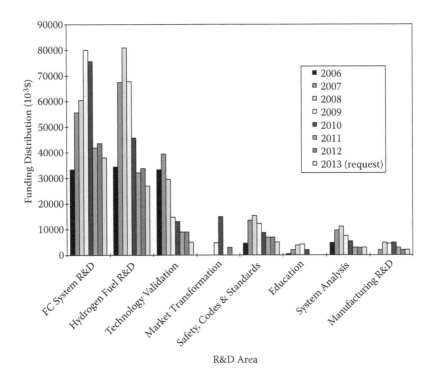

FIGURE 8.2
Funding Distribution in Major R&D Areas from 2006 to 2013.

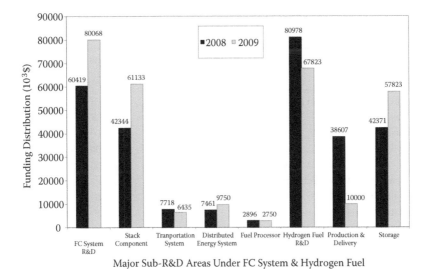

FIGURE 8.3
Funding Distribution under the Fuel Cell System R&D and the H_2 R&D for 2008 and 2009.

8.1.2 CHP System

Table 8.2 lists the technical perspectives on 1–10 kW_e residential combined heat and power (CHP) proton exchange membrane fuel cell (PEMFC) systems operated on natural gas. It is assumed that the standard utility natural gas is delivered to the PEMFC system at the typical residential distribution line pressures, and regulated AC power based on the lower heating value (LHV) of the fuel is produced by the fuel cell system.

The cost estimate is based on manufacturing 50,000 units per year. Compared with higher temperature fuel cells such as the solid oxide fuel cells (SOFCs), a regular PEMFC can hardly be operated at temperatures higher than 90°C, making the electrical efficiency relatively lower and the heat energy recovery more difficult. Reaching a combined electrical and heat energy efficiency to 87.5% by 2015 is not a trivial task. With a targeted system electrical efficiency of 40.5% in 2015, the efficiency of the stack will be around 47.7% because the sum of the parasitic power loss and the DC-AC inverting power loss can easily reach 15% of the stack power output. Then, the maximum heat efficiency is 52.3% (100%–47.7%). Meanwhile, ca. 5% of the fuel will be wasted in the fuel processing steps (the energy loss during fuel processing was reported to be larger than 20% (Lee 2007)) and another 5% is wasted during the purging. Therefore, the overall CHP efficiency will be 82.8% (40.5% + 52.3% – 5% – 5%), which is about 5 percentage lower than the 87.5% target. In other words, the 2015 target of 87.5% CHP efficiency may not be achievable at all, needless to say the 2020 target of 90% efficiency. Hwang

TABLE 8.2

Technical Perspectives on 1–10 kW_e Residential Combined Heat and Power Systems Operated on Natural Gas

Parameter		Unit	Year		
			2011 Status	2015 Targets	2020 Targets
Electrical efficiency at rated power		%	34–40	40.5	>45
CHP energy efficiency		%	80–90	87.5	90
Cost (based on 50,000 units yearly)	2 kW_{avg} system	$ kW⁻¹	NA	1200	1000
	5 kW_{avg} system		2300–4000	1700	1500
	10 kW_{avg} system		NA	1900	1700
Time from 10% to 90% rated power		min	5	3	3
Startup time from 20°C		min	~30	30	20
Degradation with cycling per 1000 hours		%	2	0.5	0.3
Operation lifetime		h	12,000	40,000	60,000
System availability		%	97	98	99

Source: Adapted from US DOE, *Fuel Cell Technologies Program: Technical Plan—Fuel Cells*, updated in March 2012, http://www1.eere.energy.gov/hydrogenandfuelcells/mypp/pdfs/fuel_cells.pdf.

and Zou (Hwang 2010) reported a CHP system with maximum combined efficiency of 81% (maximum electrical and heat efficiencies are 40% and 48%, respectively, but not at the same time). In order to increase the total energy efficiency, some of the heat energy from electronic and electrical components and the DC-AC inverter should be utilized. In the winter time, this kind of heat might be able to be collected for space heating, but in the summer time, it is nearly impossible to use this kind of heat.

The cold startup time will be determined by the reforming technology used and by the quality of the reformate that is required by the fuel cell stack. A steam reforming process is typically slower than a partial oxidation process. If the stack can handle higher CO concentrations, the startup time can be shorter using the same reforming technology. It might be necessary to install an H_2 storage cylinder onsite to supply H_2 during the startup of the fuel cell system, or to use a high-capacity battery bank. Fortunately, startup time is not crucial for CHP applications because once the system is started, it will be kept running in most cases.

Regarding system costs, the balance of plant (BOP) cost per kW is expected to be higher for a 2 kW system than for a 5 or 10 kW system, and therefore it is not clear why the targeted system cost per kW is set higher with an increase in the system power output. The targeted degradation rate of 0.5% per 1000 in 2015 is very challenging. If we assume that the BOL average cell voltage is 0.70 V, it is only allowed to decrease by 3.5 mV after 1000 operational hours.

8.1.3 Cars and Buses

The targets for PEMFC cars are listed in Table 8.3. One of the difficult targets to achieve appears to be the low-temperature unassisted startup at −30°C. Obviously, graphite plates cannot be used due to their high thermal capacity and mass. The plates have to be made with thin metallic materials with sufficient corrosion protection. For assisted startup at −40°C, an easy and effective way to keep the stack temperature higher should be determined.

The target for the system energy efficiency (60% at 25% rated power) appears to be in conflict with the target for the stack efficiency (65% at 25% rated power), because the loss between the two is only allowed to be around 8% (5%/65%). In other words, the combined losses from parasitic power need, anode purging, and DC-AC (or DC-DC) converter are only 8%, while about 20% loss is the nom currently (~10% from parasitic power need, >3% from anode purging, >5% from DC-AC(or DC) converting). Setting the system energy efficiency at 55% at 25% rated power may be more meaningful

Over years of R&D, the system cost has been reduced significantly, as shown in Figure 8.4. The cost reduced from $275 to $108 and to $49 per kW from 2002 to 2006 and to 2011, about a 50% reduction every 4 years. The target of $30 per kW^{-1} by 2017 does not appear to be overly optimistic. At this cost, a 100 kW fuel cell system costs only $3,000, similar to a normal internal combustion engine (ICE) system.

TABLE 8.3

Technical Targets for 80 kW$_e$ Fuel Cell Systems for Automotive Applications

System Parameter		Unit	2011 Status	2017/2020 Targets
Energy efficiency at 25% rated power		%	59	60/60
Volumetric power density		W L^{-1}	400	650/850
Specific power density		W kg^{-1}	400	650/650
Cost (based on 500,000 systems produced per year)		$ kW$_e$$^{-1}$	49	30/30
Cold startup time to 50% of rated power from 20°C ambient temperature		s	<10	5/5
Cold startup time to 50% of rated power from −20°C ambient temperature		s	—	30/30
Startup energy from 20°C ambient temperature		MJ	—	1/1
Startup energy from −20°C ambient temperature		MJ	7.5	5/5
Durability in automotive drive cycle		h	2500	5000/5000
Low temperature enabling assisted start		°C	—	−40/−40
Low temperature enabling unassisted start		°C	−20	−30/−30
Stack	Volumetric power density	W L^{-1}	2200	2250/2500
	Specific power density	W kg^{-1}	1200	2000/2000
	Efficiency @ 25% rated power	%	65	65/65
	Cost	$ kW$_e$$^{-1}$	22	15/15
	Durability with cycle	h	2500	5000/5000

Source: Adapted from US DOE, *Fuel Cell Technologies Program: Technical Plan—Fuel Cells,* updated in March 2011,: http://www1.eere.energy.gov/hydrogenandfuelcells/mypp/pdfs/fuel_cells.pdf.

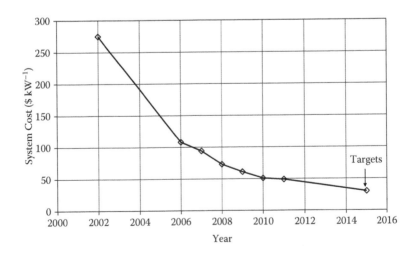

FIGURE 8.4

80 kW fuel cell system cost for cars based on 500,000 production unit per year. (S. Satyapal, US DOE, "DOE Fuel Cell R&D Progress," IPHE Meeting, Shanghai, China, September 21, 2010.)

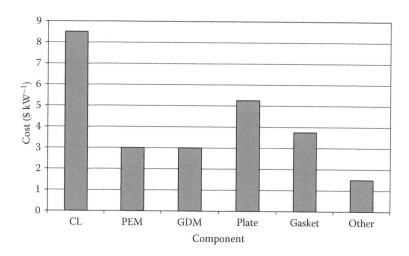

FIGURE 8.5

Cost of Stack Components per kW. (From: Dimitrios Papageorgopoulos, US DOE, "Status of DOE EERE's Fuel Cell Research and Development Efforts," Fuel Cell Seminar & Expo, San Antonio, TX. October 19, 2010.)

Within the system, about half of the cost arises from the stack, and the other half from the BOP. The cost of the major components of the $25 per kW stack (2010 status) is plotted in Figure 8.5. The largest cost comes from the catalyst layers due to the use of precious metals, followed by the plates because of protective coatings against corrosion. The cost of the PEM, GDM, and gaskets are quite similar.

Table 8.4 shows the technical targets for PEMFC buses. The target year is not specified. It is not clear how difficult it will be to achieve the targets in the lifetime of the power train (25,000 hours) and the road call frequency (10,000 miles per road call) based on the 2011 status. For a midsized bus carrying about 30 people, a power output of around 160 kW is basically enough (please refer to Chapter 6), and therefore, if the cost of the fuel cell system for a 80 kW car can achieve $30 per kW according to the 2017 target (Table 8.3), the power train cost for the bus should only be around $4800; if we double the power need to 320 kW for a larger bus, the power train cost would be at about $9600, still far less than the targeted $200,000. Even though the yearly production volume of buses is much lower than that of cars that leads to a higher cost per kW for buses, the $200,000 target is still too high. It is expected that this target may be revised to a significantly lower number (<$50,000?).

8.1.4 H₂ Storage

For transportation applications, the volumetric and gravimetric energy density of an H_2 storage system is important. A fuel cell car should be able to run for about 500 km per refueling in order to compete with current ICE-powered

TABLE 8.4

Technical Targets for Fuel Cell Buses

Parameter	Unit	2011 Status	Target
Bus lifetime	y/h	To be determined (TBD)/10,000	12/50,000
Power train lifetime	y/h	6/10,000	6/25,000
Bus availability	%	70	90
Number of refuelings per day	—	1	1
Refueling time	min	<10	<5
Bus cost	$	2,000,000	600,000
Power train cost	$	1,000,000	200,000
Road call frequency (all/power train)	mile	1,900/2,400	4,000/10,000
Operation time (hours per day/days per week)	—	19/7	20/7
Range (miles)	mile	>300	300
Fuel economy (miles/gallon of diesel equivalent)	—	6.5	8

Source: Adapted from Dimitrios Papageorgopoulos, US DOE, "Discussion of Fuel Cell Bus Technical and Cost Target RFI Submission," 2011 National Fuel Cell Workshop, New Orleans, LA, October 4, 2011.

cars. The technical targets for light-duty vehicles are listed in Table 8.5. The most challenging target appears to be the cost of $2–$4 per gasoline gallon equivalent (gge) at the pump. Based on the LHV, 1 kg of H_2 (LHV = 500 moles × 241.8 kJ mol^{-1} = 120,900 kJ) is basically equivalent to 1 gallon of gasoline (LHV = 4 L × 31292.8 kJ L^{-1} = 125,171 kJ). As described in Chapter 3, the C Series electrolyzer with an efficiency of 59% developed by Proton OnSite requires about 6 kWh of electricity to generate 1 Nm^3 H_2 (= 40 moles), which is about 0.08 gge (40/500). The cost of 6 kWh grid AC electricity is about $0.6 in the United States, and thus the electricity cost equivalent to 1 gge will be around $7.50 ($0.6/0.08 gge). This is about 2 times the price of 1 gallon of gasoline in the United States. Therefore, to achieve the targeted $2–$4 gge^{-1} cost, the cost of the grid electricity should be around $0.025–$0.050 per kWh.

Fortunately, since the H_2 fuel cell car will be about 2.5 times as efficient as the gasoline-based ICE cars, the H_2 price can be 2.5 times that of gasoline on similar amounts of the LHV without affecting end users. For example, if the gasoline price is $4 per gallon, the price of H_2 can be set at $10 per kg. In Europe, the gasoline price is about $9 per gallon, and therefore the H_2 can be sold for $22.5 per kg, which is already significantly higher than the cost of H_2 production. Based on those analyses, the H_2 cost target at the pump can be set about 3 times higher than the numbers in Table 8.5.

8.1.5 Portable System

Table 8.6 shows the DOE technical targets for portable fuel cell systems. The cost is based on production volumes of 50,000, 25,000, and 10,000 units per

TABLE 8.5

H_2 Storage System Technical Targets for Light-Duty Vehicles

Parameter	Unit	2010 Status	2017 Target	Ultimate
Gravimetric energy capacity	kWh kg^{-1}	1.5	1.8	2.5
Gravimetric H_2 capacity	kgH$_2$ kg^{-1}	0.045	0.055	0.075
Volumetric energy capacity	kWh L^{-1}	0.9	1.3	2.3
Volumetric H_2 capacity	kgH$_2$ L^{-1}	0.028	0.040	0.070
H_2 purity (dry basis)	%	—	99.97	99.97
H_2 cost at pump	$ gge^{-1}	—	2–4	2–4
Min./max. delivery temperature	°C	—	–40/85	–40/85
Operational cycle life	Cycles	—	1500	1500
"Well"-to-vehicle efficiency	%	—	—	60
5 kg H_2 fueling time	min	—	3.3	2.5
Min. H_2 full fueling rate	kgH$_2$ min^{-1}	—	1.5	2.0
Discharging rate from 10% to 90% load	s	—	15	15
Response time from 90% to 0% load	s	—	0.75	0.75

Sources: Adapted from Ned Stetson, US DOE, "An Overview of U.S. DOE's Activities for Hydrogen Fuel Cell Technologies," Materials Challenges in Alternative & Renewable Energy, Clearwater, FL, February 27, 2012; and US DOE, *Fuel Cell Technologies Program: Technical Plan—Storage*, 2011 Interim Update, http://www1.eere.energy.gov/hydrogenandfuelcells/mypp/pdfs/storage.pdf.

TABLE 8.6

Technical Targets for Portable Fuel Cell Systems

Parameter	Units	<2 W/10–50 W/100–250 W Systems		
		2011 Status	2013 Targets	2015 Targets
Specific power	W kg^{-1}	5/15/25	8/30/40	10/45/50
Power density	W L^{-1}	7/20/30	10/35/50	13/55/70
Specific energy	Wh kg^{-1}	110/150/250	200/430/440	230/650/640
Energy density	Wh L^{-1}	150/200/300	250/500/550	300/800/900
Cost[a]	$ W^{-1}	150[b]/15/15	130[b]/10/10	70[b]/7/5
Durability	h	1500/1500/2000	3000/3000/3000	5000/5000/5000

[a] The cost is based on production volumes of 50,000, 25,000, and 10,000 units per year for <2 W, 10–50 W, and 100–250 W systems, respectively.

[b] Cost of each <2 W system.

Source: Adapted from US DOE, *Fuel Cell Technologies Program: Technical Plan—Fuel Cells*, updated in March 2012, http://www1.eere.energy.gov/hydrogenandfuelcells/mypp/pdfs/fuel_cells.pdf.

year for <2 W, 10–50 W, and 100–250 W systems, respectively. The specific and volumetric energy densities of liquid methanol are 5.53 Wh g^{-1} and 4.35 Wh cm^{-3}, respectively, based on the LHV of 638.1 kJ mol^{-1}. If the electrical efficiency of a DMFC is 30%, then the effective specific and volumetric energy densities of methanol itself will be 1600 Wh kg^{-1} and 1300 Wh L^{-1}, respectively, which are 2.5 (1600/650) and 1.4 (1300/900) times those of the 2015 targets. Obviously, a system carrying more methanol will more easily meet those targets. For example, the energy density target of 900 Wh L^{-1} can be easily met when the volume of liquid methanol is around 70% of the total system volume (900/1300). Based on this fact, targets for the specific energy and energy density are met easily by carrying more liquid fuels. The specific power and the power density, however, are more crucial.

8.2 Challenges

Commercialization of fuel cells still faces several challenges. One challenge arises from the reliability, lifetime, and cost of the fuel cell itself. Cost is related to production volume. At the currently extremely low production volume, the cost at large production volumes can only be estimated through projection, which may carry significant errors. The reliability and lifetime are related to the cost because a fuel cell system can be made more reliable and have a longer lifetime when more costly components are used. Achieving the reliability, lifetime, and cost targets together is quite challenging. Compromise should be made in the early commercialization process. It is quite acceptable for many end users to spend slightly more money on a cleaner and equally reliable technology. If the reliability is poor, causing a lot of problems for the end users, the product has little attractiveness even if it is cheaper.

Another challenge comes from the lack of widespread H_2 infrastructure. Large-scale production of H_2 is not that costly, and neither are building H_2 refueling stations or setting up H_2 pipelines, but they require coordinated efforts nationwide. The federal government must provide the vision and set up the goals to change or modify the existing energy supply infrastructure. Producing H_2 through electrolysis of water can use excess or unusable wind power, solar power, hydro power, and even nuclear power, which will use those renewable energies more effectively.

A further challenge arises from the erroneous belief by many people that H_2 is more dangerous than other fuels. As discussed in Chapter 3, from the physical to the chemical properties and to the real-world test results, H_2 is not more dangerous than other fuels. Quite the contrary, H_2 is often safer than most of the common fuels routinely used by us. As public education grows and more and more H_2-powered products are demonstrated and used, we should be able to overcome this challenge eventually.

8.3 Perspectives

Fuel cell technology has made great progress in the past few decades. Many technical results taken for granted today were unimaginable in the past. The cost has been significantly reduced, the reliability has been drastically increased, and the lifetime is becoming competitive with traditional incumbent technologies. When more people realize the potential detrimental impact of global warming, air pollution, and depletion of fossil fuels on human beings, more people will be energized to join the race to develop, advance, and use fuel cells and related H_2 technologies. Optimistically, this may take fewer than 10 years based on the current status of fuel cell technology itself. Pessimistically, this may take up to 40 years when the world oil reserve has depleted and many coastal cities have been submerged in oceans. The choice is ours!

Appendix A: Terminology

Some of the terms listed below do not appear in this book but will be encountered during broader reading. Since the International Electrotechnical Commission (IEC), http.//www.iec.ch, is working on fuel cell terminologies under Technical Committee 105, readers are strongly recommended to refer to the IEC's thorough explanations.

absolute temperature (K): Temperature used in thermodynamics; equals 273.15 plus the degrees Celsius (°C).

acceptance test: Test typically performed by manufacturers to show that the products meet their specifications.

AC–DC converter: A device that converts AC power to DC power.

activation loss or activation overpotential: Deviation of voltage from the thermodynamic equilibrium value when there is a current passing through due to the presence of an energy barrier to the reaction to proceed.

activation of electrode: Steps or processes that make an electrode achieve its highest potential performance.

activation polarization: Phenomenon of voltage deviation from the thermodynamic equilibrium value as a result of a current passing through that pushes the reaction pairs away from their thermodynamic equilibrium states.

air bleed: Addition of several percentage points of air into the fuel stream that gets into the stack's anode in order to mitigate anode catalyst poisoning (by species such as carbon monoxide) via catalytic oxidation of the poison within the anode.

air supply module (ASM): All parts and components, starting with those receiving air from the environment to those sending cathode exhaust out to the environment.

air-cooled stack: Stack whose cooling is achieved by movement of air.

alkaline fuel cell (AFC): A fuel cell using an alkaline solution or a membrane as the electrolyte with hydroxide (OH^-) as the charge transport species.

ambient condition: Condition at normal temperatures and pressures.

anode: Electrode where an oxidation reaction takes place.

anodic potential scan: Potential scans in the direction from lower to higher potentials.

backup power: Power used when a regular power system, such as the grid power, fails.

balance of plant (BOP): All the supporting/auxiliary components except for the stack within a fuel cell system.

battery: A chemical energy storing and electrical power generating device that converts the chemical energy stored inside to electrical energy; it consists of an anode, a cathode, an electrolyte between the two electrodes, a reducing chemical at the anode, and an oxidizing chemical at the cathode.

battery discharging: The process of a battery providing electrical power by consuming internally stored reactants.

biomass: Biological materials from living or dead organisms; plants, grass, and municipal waste are all biomass.

bipolar plate: Thermally and electrically conductive plate that has the fuel flow fields on one side and the oxidant flow fields on the other side; it separates individual cells in a stack, acts as a current collector, and provides mechanical support for the electrodes or the membrane electrode assemblies.

blower: A gas-moving and -compressing device that can compress air up to about 25 kPa; pressure boost ability is higher than a fan but lower than a compressor.

buss or buss bar: Strips or bars of highly electrically conductive metals, such as copper, used to distribute current from the sources to loads and back to the sources with minimal iR loss; there are two strips or bars for a complete circuit, one for sending current from one electrode of the source to the load and the other for returning the currents from the load to the other electrode of the source.

buss voltage: The voltage difference between a pair of buss bars.

carbon cloth: Porous cloth woven with flexible carbon filaments. When used in fuel cells as the GDM, it is typically filled with a mixture of carbon powder and PTFE.

carbon paper: Porous paper made of highly graphitized carbon fibers typically bound together with resin.

Carnot limit: The theoretical efficiency limit of a heat engine to convert heat energy to mechanical energy. It is the ratio of the temperature difference between the combustion chamber and the exhaust from the chamber to the temperature of the combustion chamber (unit of temperature in Kelvin).

casting: A manufacturing process that first applies liquid materials into a mold and is then followed by solidifying the materials.

catalyst: Substance that accelerates a reaction by lowering the activation energy barrier of the reaction without itself being consumed.

catalyst-coated membrane (CCM): Membrane whose surfaces are coated with an anode and a cathode catalyst layer on either side.

catalyst-coated substrate (CCS) or catalyst-coated backing (CCB): Carbon paper or carbon cloth with one surface coated with a catalyst layer.

catalyst layer: A thin layer composed of catalysts and other necessary components; it is normally coated on the surface of either a membrane or a substrate.

catalyst loading: Amount of catalyst incorporated into an electrode per its geometric unit area.

catalyst poisoning: Inhibition of the catalytic ability of a catalyst due to the adsorption of a substance on the surface of the catalyst.

catalyst sintering: Growth of catalyst particles in size.

catalyst specific activity: Current generated per unit mass of catalyst at a given voltage (commonly agreed voltage being 0.90 V for the oxygen reduction reaction).

catalyst support: Substance on top of which the catalyst is deposited.

cathode: Electrode where a reduction reaction takes place.

cathodic potential scan: Potential scans in the direction from higher to lower potentials.

cell voltage monitoring (CVM): Simultaneously measurement of the voltage difference between all neighboring cathode and anode plates within a stack.

charge transfer coefficient: A parameter signifying the fraction of overpotential that is utilized in activating reaction species to overcome the energy barrier of an electrochemical reaction.

charging of metal hydride: Process of letting hydrogen be stored in metal hydride at a suitable pressure and temperature.

check valve: A valve that allows fluid to flow in one direction.

chemical energy: Enthalpy change of formation, equivalent to thermal or heat energy.

chloralkali process: Electrolysis of $NaCl$ aqueous solution to form $NaOH$, Cl_2, and H_2; Cl^- is oxidized to form Cl_2 at the positive electrode. $2Cl^- = Cl_2 + 2e^-$; H_2O is reduced to form OH^- and H_2 at the negative electrode: $2H_2O + 2e^- = 2OH^- + H_2$; overall reaction is $2NaCl + 2H_2O = 2NaOH + H_2$; an OH^- selective membrane separates the two half reaction chambers and allows OH^- to transport from the negative electrode side to the positive electrode side.

co-flow: Fluids flow in the same direction.

coking: Formation of a carbon solid, such as filaments from a raw fuel such as CH_4, on top of catalysts in a reformer.

cold start: Start stack or fuel cell system at ambient temperature.

cold state: State without power input to or output from the system; the temperatures of the system and the environment are the same.

combined heat and power (CHP) system: System supplying both heat and electrical power.

combined heat and power efficiency: Ratio of total useable electrical and heat energy outflow from a fuel cell system to the total enthalpy inflow to the fuel cell system.

communications module (COM): All parts and components responsible for communicating among parts or components within a fuel cell system and for communicating between the fuel cell system and the external modules.

compressibility factor of hydrogen: A factor that is larger than 1 to correct for the amount of pressurized H_2 calculated according to the ideal gas law $PV = nRT$.

compressor: A mechanical device that increases significantly the pressure of a gas by reducing its volume.

concentration polarization: Polarization caused by lower reactant concentration at the reaction sites than its bulk concentration.

condenser: Equipment that condenses vapor to liquid water and thus separates the latter from other gaseous components.

conditioning: Steps or processes that make an electrode achieve its highest potential performance.

constant current operation: Operation at a constant output current.

constant power operation: Operation at a constant output power.

constant voltage operation: Operation at a constant output voltage.

control module (CM): Combination of software and hardware responsible for controlling the operation of a fuel cell system including modules that are not part of the fuel cell system but affect the operation of the fuel cell system.

conversion efficiency of a raw fuel: The percentage of the raw fuel that is converted to H_2 and CO during fuel reforming.

coolant: A fluid that takes heat from one item and transfers it to another item so that the temperature of the first item is lowered.

coolant plate: Thermally and electrically conductive plate that has flow fields on one side for the flow of a coolant.

coolant tank: Container for storing coolant.

coolant temperature at the stack inlet: Temperature of coolant at the point of entering a stack.

coolant temperature at the stack outlet: Temperature of coolant at the point of leaving a stack.

CO oxidative stripping: Removing the CO monolayer from the electrode surface by scanning the potential from a value that is not high enough to cause CO oxidation to a value that is high enough to completely oxidize CO; the oxidation reaction is: $CO + H_2O = CO_2 + 2H^+ + 2e^-$.

core/shell particle: A composite particle with a core made of one material and a shell made of another material with catalytic activity; the shell can be as thin as less than one monolayer.

counter-flow: Fluids flow in opposite directions.

cracking of ammonia: The process of converting ammonia (NH_3) to N_2 and H_2 at a suitable temperature with the aid of a catalyst.

cross-flow: Fluids flow at an angle typically perpendicular to each other.

crossover: Reactant's passing through the electrolyte separating the anode from the cathode.

current collector: Conductive material at the two ends of a stack, one collects electrons generated by the stack and sends them to the load, and the other receives electrons returned by the load.

current density: Current per geometric electrode surface area.

cyclic voltammetry: A popular electrochemical technique with the potential of the working electrode ramped linearly versus time and switched the ramping direction at the preset high and low potential limits.

DC–AC inverter: A device that converts DC power to AC power of desired voltage and frequency.

DC–DC converter: A device that converts DC power from one voltage to another voltage.

dead-ended flow: A fluid flow configuration where the exit for the exhaust is closed for most of the time and is opened periodically for a very short time (a few hundred milliseconds) to let the exhaust out.

degradation rate: Rate at which the performance deteriorates over time, typically expressed in μV loss per hour at a constant current or current density.

desulfurization: Process to remove sulfur compounds from the raw fuel.

desulfurizer: Device to remove sulfur compounds from the raw fuel.

diffusion: Movement or transport of a species from higher concentration to lower concentration driven by the concentration gradient.

direct formic acid fuel cell (DFAFC): A fuel cell with formic acid oxidized directly at the anode.

direct liquid fuel cell (DLFC): A fuel cell with a liquid fuel oxidized directly at the anode.

direct methanol fuel cell (DMFC): A fuel cell with methanol oxidized directly at the anode.

discharging capacity of a battery: The total amount of charge released by a battery at a specified discharging current and condition before reaching its lower cutoff voltage; it is the discharging current times the discharging duration.

discharging of metal hydride: Process of metal hydride releasing H_2 for use.

discharging rate of a battery: Ratio of discharging current to that at 1 C discharging current.

electrical efficiency: Ratio of net electrical power produced by a fuel cell system to the total enthalpy inflow to the fuel cell system.

electrical power train: All components and parts that generate electrical power and deliver it to the wheels of a vehicle, including the engine, electrical motor, transmission, drive shafts, differentials, and the final drive.

electrocatalyst: Catalyst that accelerates the rate of an electrochemical reaction without itself being consumed.

electrochemical reaction: Chemical reaction occurring at the interface of electrode and electrolyte involving separation or combination of electrons and ions; the overall reaction includes one half reaction at anode/electrolyte interface and the other half reaction at cathode/electrolyte interface; ions transport through the electrolyte between the two electrodes, and electrons move via an external circuit between the two electrodes.

electrochemically active surface area (ECSA): Total surface area of the electrocatalyst within the unit geometric electrode surface area that will catalyze an electrochemical reaction; typically measured using cyclic voltammetry.

electrode: An electronic conductor (or semiconductor) through which an electric current enters or leaves the electrochemical cell as a result of an electrochemical reaction occurring at the interface between it and an electrolyte.

electrode area: Geometric surface area of the active portion of an electrode.

electrode flooding: Phenomenon of the reactant diffusion to the catalyst surface or product diffusion away from the catalyst surface being impeded by the existence of excess liquid water in a fuel cell.

electrolysis of water: Process of water being split into H_2 and O_2 under the input of external electrical energy. H_2O is oxidized to form H^+ and O_2 at the positive electrode. $2H_2O + 4e^- = 4H^+ + O_2$; H^+ transports through the electrolyte to the negative electrode side, where it combines with electrons to form H_2: $2H^+ + 2e^- = H_2$; overall reaction is $2H_2O = 2H_2 + O_2$.

electrolyte: Substance containing ions that are moveable under an electrical field.

electrolyzer: Device to convert water to H_2 and O_2 with input of external electrical energy.

emergency shutdown: Shutdown action or process carried out manually by activating the emergency switch on the fuel cell system, in case of emergency such as component or system malfunction, H_2 leakage, fire, and explosion, to avoid equipment damage and/or personnel injury.

end plate: Outmost insulating plate at either end of a stack to transmit the compression force to the stack components and to allow fluids to get into and out of the stack via internal ports, ducts, or manifolds.

energy conversion: Process of converting energy from one form to another form.

enthalpy of vaporization: Thermal energy needed to evaporate liquid to gas.

equivalent weight (EW): Mass (g) of an ionomer such as PFSA that contains one mole of $-SO_3H$.

exchange current density: Current density of the forward or the backward reaction when an electrode is at thermodynamic equilibrium with 0 net current flowing through.

exhaust management module (EMM): All parts and components responsible for cleaning unreacted H_2 in the anode exhaust before sending it out to the environment.

exothermic process: Reaction process releasing heat to the environment.

expanded PTFE (ePTFE): Thin PTFE film with fine pores after being stretched; porosity and density are about 95% and 0.1 g cm^{-3}, respectively.

external leak: Leak of a fluid to outside the stack in a fuel cell.

external reforming: Reforming process performed outside of a stack.

extrusion: A manufacturing process involving pushing or drawing a material through a die of a desired gap.

Faraday constant: The charge of 1 mole of electrons, which is 96485 coulombs.

Fick's first law: Relates diffusive flux to concentration gradient. $J = -D \, dc/dx$, where D is the diffusion coefficient and dc/dx is the concentration gradient of diffusive species.

flow field: Overall feature created on the surface of a plate for distribution of fluids.

flow-field channel: Channels created on the surface of plates for the flow of fluids such as H_2, air, and coolant.

flow-through operation: Operation to keep the fluid inlet and outlet ports open so that the fluid can flow through continuously.

forced ventilation: Moving and replacing air by mechanical means.

formic acid: Substance with the chemical formula CHOOH.

fossil fuels: Fuels such as coal, petroleum, and natural gas formed from buried dead organisms through millions of years of natural processes such as anaerobic decomposition.

freeze–thaw test: Testing by alternately subjecting an object to below and above the water freezing temperature.

fuel: Substance releasing energy when combusted or reacted.

fuel cell: Electrochemical converting device that converts chemical energy directly into electrical energy with the reactants supplied from outside of the reactor called stack.

fuel cell system electrical efficiency: Ratio of the electrical power sent out to the load by a fuel cell system to the chemical energy (i.e., the enthalpy) of the fuel entering the fuel cell system.

fuel cell vehicle: Electric vehicle using a fuel cell as the primary power source for propulsion.

fuel conversion efficiency: Percentage of fuel being converted to desired products during fuel processing.

fuel starvation: Amount of fuel is less than that needed for stoichiometric reaction.

fuel supply module (FSM): All parts and components starting from the fuel storage tank to the fuel inlet port on the end plate of a stack.

fuel utilization: Ratio of the amount of fuel that generates current to the total amount of fuel entering the stack.

fueling coupler: Combination of the fueling nozzle that connects to the fuel source and the fueling receptacle that connects to a fuel receiver.

full load operation: Operation at the rated output current or power.

gas diffusion electrode (GDE): Porous electrode that allows fluids to get in or out during the electrochemical reaction process.

gas diffusion medium (GDM): Thermally and electrically conductive porous substrate, such as carbon paper or carbon cloth, that allows fluids to get in and out, and mechanically supports the membrane or the CCM.

gasket: A deformable material placed between two mating parts to mechanically prevent leakage of fluids from or into the two parts under compression.

geothermal energy: Heat energy stored inside the Earth that can be harnessed for use.

Gibbs free energy: Energy available to do external work, excluding any work associated with changes in volume and pressure.

graphitization: Process of converting amorphous carbon to highly graphitized carbon through high-temperature treatments.

grid-dependent operation: Operation of a fuel cell system that is connected to the grid.

grid-independent operation: Operation of a fuel cell system that is not connected to the grid.

gross power: Output power of a stack at its terminals.

guiding pins: Solid rods vertically positioned on one compression plate that guide the assembling of stack components by allowing the pins go through the corresponding holes on those components; the guiding pins can be removed after the stack is built.

H_2 crossover: H_2 passing through the electrolyte from the anode side to the cathode side.

H_2 infrastructure: Large enough (such as city, provincial, national, and international wide) physical and organizational structures, including H_2 production, transportation, storage, and utilization.

heat efficiency: Ratio of net usable heat collected from a fuel cell system to the total enthalpy inflow to the fuel cell system.

heat engine: An engine that converts heat energy to mechanical energy through combustion of a fuel.

heat exchanger: A device to effectively transfer heat from one medium to another.

heat management module (HMM): All parts and components responsible for maintaining all components and parts in workable temperature ranges within a fuel cell system.

higher heating value (HHV): Enthalpy change of a reaction with the product water in liquid form.

hot bonding: Bonding two parts together in the presence of a heat source that softens or even melts certain components in the parts to be bonded.

hot or warm start: Start shortly after a stack or a fuel cell system shuts down when the temperature of the stack or the fuel cell system components (such as the stack, the reformer, the WGS reactor, and the coolant) is still hotter or warmer than the environmental temperature.

hot swappable: Replacement of a part without affecting system operation.

humidification: Process of increasing the water content of a gaseous reactant stream.

humidifier: Apparatus capable of increasing the water content of a gaseous reactant stream.

hybrid fuel cell system: A power system with a fuel cell as the primary power and another energy device, such as a battery or a supercapacitor, as the auxiliary or supporting power source.

hydration level of membrane: Amount of water taken by membrane with respect to the total amount of water that can be taken by the membrane.

hydrazine: A chemical with the formula N_2H_4; extremely toxic and dangerously unstable.

hydrocarbon: Organic compounds primarily consisting of elemental hydrogen and carbon.

hydrodesulfurization (HDS): Process of allowing a sulfur compound to first react chemically with H_2 to be converted to H_2S, which is in turn converted to metal sulfide and H_2 by reacting with a metal such as Zn.

hydrogen storage: Storing H_2 for subsequent use.

hydrophilic: Attracting water with a surface contact angle lower than $90°$.

hydrophobic: Repelling water with a surface contact angle higher than $90°$.

hydroxide exchange membrane: Polymer membrane with OH^- as the mobile ions under an electrical field.

idling state: State of no power output but prepared to generate power when a need is detected.

indirect methanol fuel cell (IDMFC): Methanol is reformed and processed to an H_2-rich gas mixture that is in turn sent to a fuel cell anode.

inlet hydrogen pressure: Pressure of H_2 gas at the point of entering a stack.

internal combustion engine (ICE): Combustion of a fuel with an oxidant (typically air) within a chamber to cause expansion of gases that in turn applies forces to a moving part such as a piston to do mechanical work. Cars using gasoline as the fuel rely on a spark to ignite the fuel–oxidant mixture, while a diesel engine using diesel as the fuel depends on the heat of highly compressed oxidant to ignite the fuel–oxidant mixture.

internal current: Electrical (e.g., electronic) current between anode and cathode that passes through the electrolyte separating them.

internal leak: Leak of a fluid from one chamber to another chamber within a stack.

internal power supply module (IPM): All parts, components, and power sources responsible for providing power for needs of either the fuel cell itself or the load, or both before the stack generates power or generates enough power.

internal reforming: Reforming process performed within a stack.

iR loss: Voltage loss caused by ohmic resistance.

leak rate: Rate of a fluid getting out of its designated enclosure.

lifetime: Time duration that an object, such as a membrane, an MEA, a stack, or a fuel cell system, can perform the desired function.

limiting current: Maximum current when concentration of a reactant drops to zero at the electrode surface due to its complete consumption at the surface of the electrode.

liquefaction of H_2: Process of converting H_2 from gas form to liquid form under a suitable pressure and an extremely low temperature ($< -240°C$); theoretically about 1/10 of the LHV of H_2 is needed, but in reality about 1/3 of the LHV of H_2 is needed.

liquid-cooled stack: Stack with its cooling achieved by movement of liquid.

load: Electrical current or power needed by a device to function.

load cycling: Changing load (power or current) among different values during the operation of a fuel cell.

load-following operation: Fuel cell system automatically adjusts its output (electrical power or heat power) according to the demand.

lower heating value (LHV): Enthalpy change of a reaction with water produced in the vapor form.

manifold: Conduit within a stack that receives the incoming fluid and evenly distributes it among all the cells, or collects the fluid from all the cells and then leads it out.

mass activity: Current at a specified potential per unit mass of a catalyst.

mass transport resistance loss or concentration overpotential: Deviation of voltage from the thermodynamic equilibrium value due to lower than bulk reactant concentration on the surface of an electrode.

membrane electrode assembly (MEA): Assembly consisting of a membrane and a gas diffusion electrode on either side.

metal hydride: Certain metals, such as Pd, Ti, Mn, Ni, Mg, and La-series metals or their alloys, that can store quite an impressive amount of H_2 through physiochemical processes.

methane: Substance with the chemical formula CH_4, the simplest organic compound.

methanol: Substance with the chemical formula CH_3OH; the simplest alcohol.

methanol diffusion film (MDF): A thin film that controls the diffusion of methanol from the source to the anode catalyst layer in a DMFC.

microporous layer (MPL): A thin layer made from a mixture of carbon powder and PTFE that is typically applied to the surface of a GDM; it is located between the GDM and the catalyst layer.

molten carbonate fuel cell (MCFC): Fuel cell using a molten salt, such as a mixture of Li_2CO_3 and Na_2CO_3 or K_2CO_3 and Li_2CO_3, as the electrolyte with CO_3^{2-} as the charge transport species.

monolayer: A single atomic or molecular thick layer on top of another substance.

monopolar plate: Thermally and electrically conductive plate that has fuel or oxidant flow fields on one side and the other side is either flat or consists of flow fields for coolant distribution. It separates individual cells in a stack, acts as a current collector, and provides mechanical support for the electrodes or the membrane–electrode assemblies.

Nafion: A perfluorosulfonic acid copolymer manufactured by DuPont, typically with an equivalent weight of 1100 g equiv^{-1}.

natural ventilation: Air moves naturally without using any mechanical means.

Nernst equation: An equation that relates the equilibrium potential to the activities of reaction species: $E = E^\circ + (RT/nF)\ln(a_{ox}/a_{red})$.

ohmic polarization or ohmic resistance loss: Deviation of voltage from the thermodynamic equilibrium value caused by iR losses.

onboard reforming: Process of converting a raw fuel such as CH_4, methanol and gasoline to an H_2-rich gas mixture with low CO concentrations on a vehicle that carries the raw fuel and uses the resulting H_2–gas mixture in a fuel cell to generate electrical energy.

open-circuit voltage (OCV): Voltage between the two terminals of a power source without external current flow.

operational state: State with electrical power output (for fuel cell) or input (for electrolyzer).

osmotic drag coefficient: Number of water molecules taken by each proton when it moves from anode to cathode during the operation of a PEMFC.

osmotic drag of water: Phenomenon of water molecules taken by protons when the latter move from anode to cathode during the operation of a PEMFC.

outlet hydrogen pressure: Pressure of H_2 gas at the point of leaving the stack.

output voltage: Voltage between the two terminals of a power source when it supplies power to a load.

oxidant: Substance capable of receiving electrons during a reaction.

oxidant utilization: Ratio of the amount of oxidant that generates current to the total amount of oxidant entering a stack.

parasitic load or power: Power consumed by the fuel cell system itself. It arises from components and devices such as sensors, control boards, pumps, fans, blowers (or compressors), solenoid valves, contactors, switches, and the power losses when currents pass through certain components such as diodes and wires.

partial oxidation reforming (POX): A hydrocarbon fuel such as CH_4 reacts with O_2 to form CO and H_2 stoichiometrically: $CH_4 + 0.5O_2 = CO + 2H_2$.

perfluorinated: All H atoms covalently bonded to carbon atoms are replaced by F atoms.

perfluorosulfonic acid (PFSA): Perfluorinated copolymer with a pendent branch ending in a sulfonic acid group.

phosphoric acid fuel cell (PAFC): Fuel cell that uses liquid phosphoric acid as the electrolyte with protons as the charge transport species; the acid is typically hosted by a porous matrix such as SiC and PBI.

polarization: Deviation of voltage from the thermodynamic equilibrium value of an electrode when there is a current passing through due to activation, ohmic resistance, or mass transport resistance.

polarization curve: Voltage versus current (or current density) plot.

polybenzimidazole (PBI): A polymer with a basic repeating unit of $-C_6H_3N_2HC-$, possessing very high thermal and chemical stabilities.

polytetrafluoroethylene: A polymer with a repeating unit of $-CF_2CF_2-$ formed from tetrafluoroethylene monomer $CF_2 = CF_2$.

porosity: Ratio of the total volume occupied by pores to the total volume of the entire object.

portable fuel cell system: Fuel cell system intended to be carried to various locations for use.

power conditioning module (PCM): All parts and components responsible for converting the stack DC power output to the power desired by the load.

preferential oxidizer (Prox): A reactor that preferentially oxidizes CO over H_2 with an oxidant such as air in the presence of a suitable catalyst.

pressure regulator: A device that lowers the pressure of an incoming gas.

primary battery: Battery that is not rechargeable and is discarded after the stored reactants are consumed.

proton exchange membrane (PEM): A solid polymer membrane that conducts protons when hydrated.

proton exchange membrane fuel cell (PEMFC): A fuel cell using a PEM as the electrolyte.

pump: A device to move fluids by mechanical action.

purging: Action or process of flushing out unwanted fluids such as the anode exhaust from the stack.

purging duration: Time duration when purging occurs, typically on the order of several hundred milliseconds, during the operation of a fuel cell.

purging frequency: Time interval between two adjacent purgings; or purging times per unit time.

rated current: Highest current specified by manufacturers at which a device such as a stack or fuel cell system is designed to operate continuously.

rated power: Designed or recommended highest power output under intended operation continuously.

raw fuel: Fuel from a supplier without any treatment by the user.

reactants: Chemicals that can undergo chemical or electrochemical reactions under the test condition.

reactant recirculation: Process to send the unreacted reactant into its inflow stream.

reductant: Substance capable of losing electrons during a reaction.

reformate: H_2-rich gas mixture after a non-H_2 raw fuel undergoes fuel processing steps.

reformer: Reactor to convert a non-H_2 raw fuel, such as methane, to a gas mixture of H_2 and CO.

reforming: Chemical reaction process of converting a hydrocarbon fuel such as CH_4 to an H_2-rich gas mixture.

refueling: Process of transferring fuel such as H_2 from a storage tank to a receiving tank.

regenerative fuel cell: Electrochemical cell able to function as both a fuel cell and an electrolyzer.

reliability: Ability of a device to perform its required functions under stated conditions either for a specified period of time or for a specified number of times.

renewable energy: Naturally replenishable energy such as solar energy, wind energy, hydro energy, tidal wave energy, and geothermal energy.

resistance overpotential: Potential deviation from the thermodynamic equilibrium value due to iR losses.

response time: Period of time from receiving a demand to meeting the demand.

roughness factor: Ratio of actual surface area to geometric surface area due to the surface roughness of the object.

routine test: Regular test based on a routine procedure.

saturated vapor pressure: Maximum vapor pressure at a specific temperature and pressure.

secondary battery: Battery that is rechargeable to convert products back to reactants through the input of electrical energy.

sensor: A device that responds to a physical quantity and often converts it to a signal that is conveniently read by an observer or an instrument.

series connection: One cell's anode connects to the neighboring cell's cathode.

short stack: Stack with fewer unit cells than a full-scale stack, but possesses the major characteristics of the latter.

shutdown: Action or process that turns a system from the power-generating state to the no-power-generating state.

shutdown time: Period of time from receiving a shutdown demand to complete shutdown.

single cell: A cell that contains only one anode and one cathode separated by electrolyte.

single cell test: Test carried out on a single cell.

solar energy: Energy from sunlight that is harnessed into useful power forms such as heat and electricity.

solenoid valve: A valve that opens or closes mechanically using electrical or magnetic energy.

solid oxide fuel cell (SOFC): Fuel cell that uses a solid oxide conductor such as yttria-stabilized zirconia (YSZ) as the electrolyte with O^{2-} as the charge transport species.

specific energy density: Energy output per unit mass under intended operation.

specific power: Power output per unit mass under intended operation.

stack: Assembly of two or more unit cells.

stack terminals: Metallic plates connected to stack current collectors for leading the stack current and voltage out.

stack test: Test carried out on a stack.

stacking: Process of assembling a stack by placing individual components on top of each other.

standard condition: Condition at temperature 25°C and pressure 1 bar.

standby state: State of no power output but prepared to generate power when a need is detected.

startup time: Period of time from receiving a startup demand to achieving certain operational state. The operational state can either mean the moment of having power output, or the moment of having certain percentage (e.g., 80%) power output, or the moment of having all the required power output; so in practice it should be clearly stated to which the start-up time refers.

stationary fuel cell system: Fuel cell system intended to be fixed in place for use.

steady state: State with parameters that are constant over time.

steam reforming (SR): Process of a hydrocarbon fuel such as CH_4 reacting with steam to form CO and H_2 stoichiometrically: $CH_4 + H_2O = CO + 3H_2$.

stoichiometric ratio: Ratio of the total amount of a reactant to the amount that participates in the reaction.

sulfonic acid group: Acid group with the chemical formula $-SO_3H$.

supercapacitor: Electrochemical energy storage and converting device consisting of two porous electrodes separated by electrolyte. Under an external electrical power supply, negative ions adsorb (concentrate) on the surface of the positive electrode to form an electrochemical double layer, and positive ions adsorb on the surface of the negative electrode to form another electrochemical double layer. During discharge, the negative ions on the surface of the positive electrode and the positive ions on the surface of the negative electrode move into the electrolyte as electrons move from the negative electrode to the positive electrode via an external circuit. Since the electrochemical double layers are in three dimensions, the capacitance of a supercapacitor is much higher than that of a conventional capacitor. A supercapacitor has higher power density but lower energy density than a battery. A supercapacitor can be charged and discharged hundreds of thousands of times because it only involves adsorption and desorption of ions on the electrode surfaces. If a capacitor also involves oxidation and reduction of electrode materials, it is called a pseudo-capacitor.

supported catalyst: Catalyst particles anchored on a support.

Tafel equation: A mathematical equation, $\Delta E_A = a + b\log j$, formulated by Swiss chemist Julius Tafel in 1905 to relate the overpotential of an electrochemical reaction to the current density when the overpotential is larger than about 116 mV/n_r, where n_r is the number of electrons involved in the rate-determining step of the reaction.

telecommunications station: All components that are connected to a wireless medium in a network for receiving and transmitting radio frequencies for wireless-enabled devices such as cell phones to communicate.

temperature cycling: Process of changing the temperature of a cell among preset values to evaluate the impact of the temperature change on the performance and the durability of the cell.

theoretical efficiency of a fuel cell: Ratio of the Gibbs free energy change to the enthalpy change of a reaction.

theoretical efficiency of an electrolyzer: Ratio of the enthalpy change to the Gibbs free energy change of a reaction.

thermal stability: Stability against temperatures (typically high temperatures).

thermodynamic voltage: The voltage under thermodynamic equilibrium conditions, which equals the voltage based on the Gibbs free energy change.

three-phase region or boundary: Region within a catalyst layer that are accessible to electrons, protons, and reactants so that an electrochemical reaction can take place.

tubular humidifier: Apparatus consisting of many thin-walled tubules, made of materials such as Nafion, that allow water to transport through. When inlet drier and cooler air passes through the inside (or outside) of the tubules and wetter and hotter air exhaust passes through the outside (or inside) of the tubules, some heat and moisture are transferred from the latter to the former.

underpotential deposition: Electroreduction of a species at a potential less negative than its thermodynamic reduction potential (typically due to favorable interaction between the depositing species and the electrode) to form up to a monolayer deposit on the electrode.

unit cell: Each individual single cell in a stack.

water balance: Maintaining water within a fuel cell at suitable levels for proper operation of the fuel cell.

water–gas shift (WGS) reactor: Reactor where steam reacts with CO to form CO_2 and H_2: $CO + H_2O = CO_2 + H_2$.

wind energy: Energy generated by natural wind that is harnessed into useful power forms such as electrical and mechanical.

Water Management Module (WMM): All parts and components responsible for supplying the water need of a fuel cell system.

Appendix B: Brief Introduction to Fuel Cell Developers

The following is only a brief introduction to the companies, universities, and institutes that generously provided some data and pictures for this book. Readers can check their websites for detailed information.

Ballard Power Systems: A world leader in developing and manufacturing proton-exchange membrane fuel cells, particularly fuel cell stacks; Ballard Power Systems is located in Burnaby, British Columbia, Canada, and was founded by Dr. Geoffrey Ballard in 1979. The company went public in 1993 on the Toronto Exchange, and was listed on NASDAQ (symbol BLDP) in 1995. Company website: http://www.ballard.com.

Beihan University: A Chinese key state university established in 1952 and located in Beijing, China. Beihan University had a total student enrollment of around 24,000 in 2011. University website: http://www.buaa.edu.cn/.

Beijing General Research Institute of Nonferrous Metals: The largest Chinese institute of research into nonferrous metals with around 3300 employees. The institute is a leading institute in developing metal hydride materials, was established in 1952, and is located in Beijing, China. Institute website: http://www.grinm.com.

Beijing Jonton Hydrogen Tech. Company: A Chinese company aiming at developing and manufacturing equipment for production, pressurization, storage, and utilization of gases such as H_2. It was established in 2010 and is located in Beijing, China. Company website: http://www.jonton.net.

Dalian Institute of Chemical Physics, Chinese Academy of Sciences: A research institute focusing on fundamental and applied research and technology transfers in broad areas such as H_2, fuel cells, energy storage, renewable energy, catalysis, advanced materials, and energy strategy. The institute was established in 1949 and is located in Dalian, Liaoning, China. Institute website: http://www.dicp.ac.cn.

East China University of Science and Technology: A Chinese key state university with a total student enrollment of more than 27,700. It was established in 1952 and is located in Shanghai, China. University website: http://www.ecust.edu.cn.

IdaTech: A world leader in developing and manufacturing PEM fuel cell systems for emergency and primary applications, particularly using reformed methanol as the fuel. IdaTech was founded in 1996 with headquarters located in Bend, Oregon, USA, and was acquired by Ballard Power Systems in 2012. Company website: http://www.ballard.com.

Jiansu Tianniao High Tech. Company: A private Chinese company specializing in the fabrication of high-strength carbon fibers, Kevlar fibers, glass fibers, carbon cloth, carbon paper, carbon felt, airplane braking pads, and Kevlar cloth for various applications. Jiansu Tianniao High Tech. Company is located in Yixing, Jiangsu, China. Company website: http://www.jstianniao.com.

MTI Micro Fuel Cells Inc.: A world leader in developing and manufacturing portable micro direct methanol fuel cells (DMFCs) up to a few watts using neat methanol as the fuel. It is located in Albany, New York, USA. Company website: http://www. mtimicrofuelcells.com.

Plug Power Inc.: A leader in developing and manufacturing proton exchange membrane fuel cell (PEMFC) systems, currently used specifically for motive power applications, including electric lift trucks. Plug Power was established in 1997 with company headquarters located in Latham, New York, USA, and is listed on NASDAQ with the symbol PLUG. Company website: http://www.plugpower.com.

Proton OnSite (formerly Proton Energy): A world leader in developing and manufacturing PEM-based electrolyzers. Proton OnSite is located in Wallingford, Connecticut, USA, and was founded in 1996. Company website: http://www.protononsite.com.

SFC Energy AG (formerly Smart Fuel Cells): A world leader in developing and manufacturing DMFC systems as battery chargers for various applications. SFC Energy AG was founded by Dr. Manfred Stefener in 2000 with headquarters located in Germany. Company website: http://www.sfc.com.

Shanghai Jiao Tong University: A Chinese key state university established in 1896 as one of the earliest Chinese universities. The university is located in Shanghai, China, with a total student enrollment of around 41,300 in 2011. University website: http://www.sjtu.edu.cn.

Shanghai Shenli High Tech. Company: A leading Chinese PEMFC and vanadium fluid cell developer and manufacturer. It is located in Shanghai, China, and was founded by Dr. Liqing Hu in 1998. Company website: http://www. sl-power.com.

South China University of Technology: A Chinese key state university with a total student enrollment of over 80,000 in 2011. South China University of Technology was established in 1952 and located in Guangzhou, Guangdong, China. University website: http://www.scut.edu.cn.

Sunrise Power: A leading Chinese company in developing and manufacturing PEMFC components such as catalysts, membranes, MEAs, and stacks, and fuel cell systems for stationary and motive applications. Sunrise Power was established in 2001 with headquarters located in Dalian, Liaoning, China. Company website: http://www.fuelcell.com.cn.

Tongji University: A Chinese key state university with a total student enrollment of more than 35,000. Tongji University is a leading Chinese university in developing lightweight high-pressure H_2 cylinders. It was established in 1907 as one of the earliest Chinese universities and is located in Shanghai, China. University website: http://www.tongji.edu.cn.

Wuhan Intepower Fuel Cells Co., Ltd.: A Chinese company aimed at developing, manufacturing, and marketing PEMFCs for telecommunications backup power applications. It was founded in 2008 with headquarters located in Wuhan, Hubei, China. Company website: http://www.intepower.com.

Wuhan University of Technology: A Chinese key state university with a total student enrollment of more than 52,700. It was established in 2000 by combining three local universities. Wuhan University of Technology is a leading Chinese university in developing catalyst-coated membranes (CCMs) for PEMFCs. It is located in Wuhan, Hubei, China. University website: http://www.whut.edu.cn.

References

Alberti, G., R. Narducci, and M. Sganappa. 2008. Effects of hydrothermal/thermal treatments on the water-uptake of Nafion membranes and relations with changes of conformation, counter-elastic force and tensile modulus of the matrix. *J. Power Sources* 178: 575–583.

Bard, A. J. and Larry R. Faulkner, *Electrochemical Methods: Fundamentals and Applications*, 1980, New York: John Wiley & Sons.

Beaver, M. G., H. S. Caram, and S. Sircar. 2010. Sorption enhanced reaction process for direct production of fuel-cell grade hydrogen by low temperature catalytic steam-methane reforming. *J. Power Sources* 195: 1998–2002.

Cheekatamarla, P. K. and C. M. Finnerty. 2006. Reforming catalysts for hydrogen generation in fuel cell applications. *J. Power Sources* 160: 490–499.

Costamagna, P., S. Grosso, and R. D. Felice. 2008. Percolative model of proton conductivity of Nafion membranes. *J. Power Sources* 178: 537–546.

Curtin,D. E., R. D. Lousenberg, T. J. Henry, P. C. Tangeman, and M. E. Tisack. 2004. Advanced materials for improved PEMFC performance and life. *J. Power Sources* 131: 41–8.

Debe, M. K., A. K. Schmoeckel, G. D. Vernstrom, and R. Atanasoski. 2006. High voltage stability of nanostructured thin film catalysts for PEM fuel cells. *J. Power Sources* 161: 1002–1011.

Gevantman, L. H. 1994, Solubility of selected gases in water. In *CRC Handbook of Chemistry and Physics (1913–1995), 75th Edition (Special Student Edition)*, editor-in-chief. D. R. Lide, **6**-3–**6**-6. Boca Raton: CRC Press.

Gottesfeld, S. and J Pafford. 1988. A New Approach to the Problem of Carbon Monoxide Poisoning in Fuel Cells Operating at Low Temperatures. *J. Electrochem. Soc.* 135(10): 2651–2652.

Gurvich, L. V., V. S. Iorish, V. S. Yungman, and O. V. Dorofeeva. 1994. Thermodynamic properties as a function of temperature. In *CRC Handbook of Chemistry and Physics (1913–1995), 75th Edition (Special Student Edition)*, editor-in-chief, David R. Lide, **5**-48–**5**-71. Boca Raton: CRC Press.

Haug, A. T. and R. E. White, 2000. Oxygen diffusion coefficient and solubility in a new proton exchange membrane. *J. Electrochem. Soc.* 147(3): 980–3.

Hayden, B. E. and J. Suchsland. 2009. Support and Particle Size Effects in Electrocatalysis, in *Fuel Cell Catalysis—A Surface Science Approach*, editor, M. T. M. Koper, 567–592. Hoboken, New Jersey: John Wiley & Sons.

Hiramitsu, Y., N. Mitsuzawa, K. Okada, and M. Hori. 2010. Effects of ionomer content and oxygen permeation of the catalyst layer on proton exchange membrane fuel cell cold start-up. *J. Power Sources* 195: 1038–1045.

Hou, J., H. Yu, L. Wang, D. Xing, Z. Hou, P. Ming, Z. Shao, and B. Yi. 2008. Conductivity of aromatic-based proton exchange membranes at subzero temperatures. *J. Power Sources* 180: 232–237.

Hwang, J. J. and M. L. Zou. 2010. Development of a proton exchange membrane fuel cell cogeneration system. *J. Power Sources* 195: 2579–2585.

Ishikawa, Y., T. Morita, K. Nakata, K. Yoshida, and M. Shiozawa. 2007. Behavior of water below the freezing point in PEFCs. *J. Power Sources* 163: 708–712.

Jiang, J. and A. Kucernak. 2005. Investigation of fuel cell reactions at the composite microelectrode/solid polymer electrolyte interface. I. Hydrogen oxidation at the nanostructured Pt/Nafion membrane interface. *J. Electroanal. Chem.* 567: 123–137.

Jung, H. Y., S. Park, and B. N. Popov. 2009. Electrochemical studies of an unsupported PtIr electrocatalyst as a bifunctional oxygen electrode in a unitized regenerative fuel cell. *J. Power Sources* 191: 357–361.

Kauranen, P. S. and E. Skou. 1996. Methanol permeability in perfluorosulfonate proton exchange membranes at elevated temperatures. *J. Appl. Electrochem.* 26: 909–917.

Kerr, J. A. 1994. Strength of chemical bonds. In *CRC Handbook of Chemistry and Physics (1913–1995), 75th Edition (Special Student Edition)*, editor-in-chief, David R. Lide, **9**-51–**9**-73. Boca Raton: CRC Press.

Larminie, J. and A. Dicks. 2003. Fuel cell systems explained (2nd Edition). Chichester: John Wiley & Sons.

Lee, D., H. C. Lee, K. H. Lee, and S. Kim. 2007. A compact and highly efficient natural gas fuel processor for 1-kW residential polymer electrolyte membrane fuel cells. *J. Power Sources* 165: 337–341.

Li, X., X. Qiu, H. Yuan, L. Chen, and W. Zhu. 2008. Size-effect on the activity of anodic catalyst in alcohol and CO electrooxidation. *J. Power Sources* 184: 353–360.

Lide, D. R. (Editor-in-Chief), 1994, CRC Handbook of Chemistry and Physics (1913–1995), 75th Edition (Special Student Edition), CRC Press.

Maillard, F., S. Pronkit, and E. R. Savinova. 2009. Size Effects in Electrocatalysis of Fuel Cell Reactions on Supported Metal nanoparticles, in *Fuel Cell Catalysis—A Surface Science Approach*, editor, M. T. M. Koper, 507–566. Hoboken, New Jersey: John Wiley & Sons.

Miller, T. M. 1994. Electron affinities. In *CRC Handbook of Chemistry and Physics (1913–1995), 75th Edition (Special Student Edition)*, editor-in-chief, David R. Lide, **10**-180–**10**-191. Boca Raton: CRC Press.

Mock, P. and S. A. Schmid. 2009. Fuel cells for automotive powertrains—A techno-economic assessment. *J. Power Sources* 190: 133–140.

Motupally, S., A. J. Becker, and J. W. Weidner, 2000. Diffusion of water in Nafion 115 membranes. *J. Electrochem. Soc.* 147(9): 3171–7.

Önsan, Z. I. and A.K. Avci. 2011. Reactor Design for Fuel Processing, in *Fuel Cells: Technologies for Fuel Processing*, edited by D. Shekhawat, J. J. Spivey, and D. A. Berry, 451–516. Amsterdam, The Netherlands: Elsevier.

Parthasarathy, A., S. Srinivasan, A. J. Appleby, and C. R. Martin, 1992. Temperature dependence of the electrode kinetics of oxygen reduction at the platinum/Nafion interface—a microelectrode investigation, *J. Electrochem. Soc.* 139(9):1992: 2530–7.

Pineri, M., G. Gebel, R. J. Davies, and O. Diat. 2007. Water sorption-desorption in Nafion membranes at low temperature, probed by micro X-ray diffraction. *J. Power Sources* 172: 587–596.

Rajalakshmi, N., T.T.Jayanth, and K.S.Dhathathreyan. 2004. Effect of Carbon Dioxide and Ammonia on Polymer Electrolyte Membrane Fuel Cell Stack Performance. *Fuel Cells* 3(4):177–180.

Reiser, C. A., L. Bregoli, T. W. Patterson, J. S. Yi, J. D. Yang, M. L. Perry, and T. D. Jarvi. 2005. Reverse-Current Decay Mechanism for Fuel Cells. *Electrochem. Solid-State Lett.* 2005 8(6): A273–A276

Shi, W., B. Yi, M. Hou, F. Jing, and P. Ming. 2007. Hydrogen sulfide poisoning and recovery of PEMFC Pt-anodes. *J. Power Sources* 165: 814–818.

Tabe, Y., M. Saito, K. Kukui, and T. Chikahisa. 2012. Cold start characteristics and freezing mechanism dependence on start-up temperature in a polymer electrolyte membrane fuel cell. *J. Power Sources* 208: 366–373.

Tang, H., Z. Qi, M. Ramani, and J. F. Elter. 2006. PEM fuel cell cathode carbon corrosion due to the formation of air/fuel boundary at the anode. *J. Power Sources* 158: 1306–12.

US DOE. 2012. Fuel Cell Technologies Program: Technical Plan—Fuel Cells, http://www1.eere.energy.gov/hydrogenandfuelcells/mypp/pdfs/fuel_cells.pdf.

Vanysek, P. 1994. Electrochemical series. In *CRC Handbook of Chemistry and Physics (1913–1995), 75th Edition (Special Student Edition)*, editor-in-chief, David R. Lide, 8-21–8-31. Boca Raton: CRC Press.

Wang, Q., C.-S. Cha, J. Lu, and L. Zhuang. 2012. Ionic conductivity of pure water in charged porous matrix. *ChemPhysChem* 13: 514–519.

Wikipedia. Hydrogen safety. http://en.wikipedia.org/wiki/Hydrogen_safety.

Wild, P. J., R. G. Nyqvist, F. A. Bruijn, and E. R. Stobbe. 2006. Removal of sulphur-containing odorants from fuel gases for fuel cell-based combined heat and power applications. *J. Power Sources* 159: 995–1004.

Index